Study Guide for Ruric E. Wheeler's MODERN MATHEMATICS

Eighth Edition

Study Guide for
Ruric E. Wheeler's
MODERN MATHEMATICS

Eighth Edition

Terry Goodman
Central Missouri State University

Brooks/Cole Publishing Company
Pacific Grove, California

Brooks/Cole Publishing Company
A Division of Wadsworth, Inc.

© 1992 by Wadsworth, Inc., Belmont, California 94002. All rights reserved. No part of this book may be reproduced, stored in a retrieval system, or transcribed, in any form or by any means—electronic, mechanical, photocopying, recording, or otherwise—without the prior written permission of the publisher, Brooks/Cole Publishing Company, Pacific Grove, California 93950, a division of Wadsworth, Inc.

Printed in the United States of America

10 9 8 7 6 5 4 3 2 1

ISBN 0-534-16607-5

Sponsoring Editors: *Jeremy Hayhurst, Faith B. Stoddard*
Editorial Assistants: *Nancy Champlin, Sarah Wilson*
Cover Design: *Vernon T. Boes*
Cover Illustration: *John Edeen*
Typesetting: *George Bergeron, Supplinc*
Printing and Binding: *Malloy Lithographing, Inc.*

To the Student

This Study Guide is designed to be used as review and additional explanation for the concepts discussed in Ruric E. Wheeler's *Modern Mathematics*, Eighth edition. You should refer to the relevant sections in this Study Guide once you've completed the corresponding section or chapter in your text. Where you are not able to see how the Study Guide solution is obtained you should refer to your text for explanation and worked examples similar to those contained here.

The Study Guide is designed for use with a pencil or pen. Always make a serious effort to answer the question in the space provided on the left side of the page before you refer to the answer or solution on the right side. You should cover the right side as you go to allow yourself to think the problem through before you refer to the answer. If you work through the text thoughtfully in this way you will strengthen your mathematical problem-solving skills and be better prepared for examinations and your eventual role as a mathematics teacher.

Good luck in your course and your career.

Acknowledgments

Thanks for encouragement and help in producing this Study Guide are due to Marc Edwards, Faith Stoddard, and Jeremy Hayhurst at Brooks/Cole Publishing Company. George Bergeron at Supplinc deserves credit for his careful design, electronic art, and formatting work.

This text was authored using Mathwriter™ 2.0, a remarkable technical wordprocessor for the Macintosh™, which is published and distributed by Brooks/Cole Publishing Company.

Terry Goodman

Contents

Chapter 1 Critical Thinking, Problem Solving, and Logic

Section 1 Critical Thinking and Inductive Reasoning 1
Section 2 An Introduction to Problem Solving 7
Section 3 Critical Thinking and an Introduction to Logic 11
Section 4 Critical Thinking and Conditionals 16
Section 5 Quantifiers 22
Section 6 Making Use of Deductive Logic 27

Chapter 2 Tools for Problem Solving: Sets and Numbers

Section 1 An Introduction to Sets 32
Section 2 Cartesian Products and Relations 37
Section 3 The Number of Elements in a Set 42
Section 4 Whole Number Addition, Subtraction, and Order 48
Section 5 Whole Number Multiplication and Division 53

Chapter 3 Numeration Systems

Section 1 History of Numeration Systems 58
Section 2 Using Exponentials in Addition and Subtraction Algorithms 63
Section 3 Multiplication and Division Algorithms 69
Section 4 Patterns for Nondecimal Bases 74
Section 5 Computations in Different Bases 79

Chapter 4 The System of Integers and Elementary Number Theory

Section 1 The System of Integers 85
Section 2 Integer Multiplication and Division 91
Section 3 Divisibility 96
Section 4 Primes, Composites, and Factorization 99
Section 5 Greatest Common Divisor and Least Common Multiple 104
Section 6 Modular Arithmetic 107

Chapter 5 Introduction to the Rational Numbers

Section 1 The Set of Rational Numbers 111
Section 2 Addition and Subtraction of Rational Numbers 117
Section 3 Multiplication and Division of Rational Numbers 124
Section 4 A Return to Problem Solving 131
Section 5 An Introduction to Decimals 136
Section 6 The Arithmetic of Decimals 141

Chapter 6 From Rational Numbers to Real Numbers

Section 1 Ratio and Proportion 145
Section 2 The Language of Percent 150
Section 3 The System of Rational Numbers 155
Section 4 The Real Number System 160

Chapter 7 Consumer Mathematics

Section 1 Some Comparisons of Interest Rates 166
Section 2 A Calculator Approach to Annuities 172
Section 3 Present Value of an Annuity 176
Section 4 Annual Percentage Rate 180

Chapter 8 Introduction to Probability Theory

Section 1 The Language of Probability 183
Section 2 Empirical Probability and the Fundamental Principle of Counting 188
Section 3 Counting Techniques Using Permutations and Combinations 193
Section 4 Properties of Probability 198
Section 5 Expected Probability and the Probability of Compound Events .. 203

Chapter 9 The Uses and Misuses of Statistics

Section 1 Frequency Distributions and Graphical Representations 211
Section 2 What is Average? 220
Section 3 How to Measure Scattering 227
Section 4 The Normal Distribution 235

Chapter 10 Informal Geometry

Section 1 Some Basic Ideas of Geometry 240
Section 2 Lines, Planes, and Angles 245
Section 3 Simple Closed Curves 251
Section 4 Geometric Patterns in Nature and Art 256
Section 5 Simple Closed Surfaces 263

Chapter 11 Measurement and the Metric System

Section 1 Measurement and the International Metric System 269
Section 2 Congruence and Measure 273
Section 3 Formulas for Perimeter and Area 278
Section 4 Surface Area and Volume 284
Section 5 Other SI Units of Measure 289

Chapter 12 Additional Topics of Geometry

Section 1 Congruence and Triangles 293
Section 2 Congruence and Right Triangles 299
Section 3 Justifications of Constructions 306
Section 4 Similar Geometric Figures 313
Section 5 Topological Equivalence 319
Section 6 Traversing a Network 323

Chapter 13 Coordinate Geometry and Transformations

Section 1	Coordinate Geometry	328
Section 2	Introduction to Reflections or Flips	335
Section 3	Slides or Translations	343
Section 4	Rotations and Successive Motions	351
Section 5	Transformations, Congruence, and Similarity	359

Chapter 14 Algebra and Geometry

Section 1	Functions	364
Section 2	Expressing Ideas with Linear Functions	370
Section 3	Slopes and Linear Equations	376
Section 4	Systems of Linear Equations	383
Section 5	Quadratic Functions and the Parabola	389

Chapter 15 Introduction to Computers

Section 1	Introduction to the BASIC Language	397
Section 2	Some BASIC Statements	400
Section 3	Solving Problems Using the Computer	403
Section 4	More BASIC Tools for Problem Solving	406
Section 5	Introduction to Logo	409
Section 6	Looping in Logo - The REPEAT Command	413

Chapter 1
Critical Thinking, Problem Solving, and Logic

Section 1 Critical Thinking and Inductive Reasoning

Cover the right side of the page and work on the left, then check your work

1. Using estimation, determine whether the sum of the first six terms of the geometric sequence will be greater than 300,000.

    2, 20, 200, 2000, . . .

    The sum of the first six terms of the sequence will be less than 300,000.

    (a) Method 1:

    The first six terms are

    2, 20, 200, 2000, 20000, 200000

    By using estimation, we can see that the sum of the first five terms will be less than 100,000. This added to 200,000 will be less than 300,000.

    (b) Method 2:

    Recall that
    $$S_n = \frac{a(r^n - 1)}{r - 1}$$
    can be used to find the sum of the first n terms of a geometric sequence. In this example, $a = 2$, $r = 10$, and $n = 6$. Then we have $S_6 = \frac{2(10^6 - 1)}{10 - 1} = \frac{2}{9} \cdot$ (about 1,000,000) < 300,000

    Note that $\frac{3}{9}$, or $\frac{1}{3}$, of 1,000,000 would be about 333,333.

2. Find a pattern and complete the blanks.

    1, 3, 7, 15, 31, ___ , ___ , ___

    | 1, 3, 7, 15, 31, **63**, **127**, **255**

3. Find a pattern and complete the blanks.

   2, 4, 8, 14, 22, 32, ___ , ___ , ___

   | 2, 4, 8, 14, 22, 32, **44, 58, 74**

4. The following is an arithmetic sequence. Fill in the blanks.

   4, 13, 22, 31, 40, ___ , ___ , ___

   | 4, 13, 22, 31, 40, **49, 58, 67**
   | The arithmetic sequence has a constant difference of **9**. The term following 40 will be 40 + **9**, the next term 49 + **9**, and so on.

5. The following is a geometric sequence. Fill in the blanks.

   2, 10, 50, 250, 1250, ___ , ___ , ___

   | 2, 10, 50, 250, 1250, **6250, 31250, 156250**
   |
   | The geometric sequence has a constant ratio of **5**. The term following 1250 will be 1250 × **5**, the next term 6250 × **5**, and so on.

6. What is the sum of the first seven terms of the following sequence?

   3, 6, 12, 24, 48, ___ , ___ , ___

   | $S_7 = \dfrac{3(2^7 - 1)}{2 - 1} = \dfrac{3(128 - 1)}{1} = \dfrac{3(127)}{1} = 381.$

7. Construct an arithmetic sequence whose common difference is 6 and first term is 4.

   | 4, 10, 16, 22, 28, . . .

8. Construct a geometric sequence whose first term is 3 and common ratio is 7.

   | 3, 21, 147, 1029, 7203, . . .

Each term is generated by multiplying the preceding term by **7** (the common ratio).

9. Find the first five terms of the sequence whose nth term is given by $n^2 + 2$.

3, 6, 11, 18, 27
When $n = 1$, then $n^2 + 2 = 1^2 + 2 = 3$. To find the other terms, we can use:

n	$n^2 + 2$
2	6
3	11
4	18
5	27

10. Ms. McKee is paid $1000 at the end of the first month on the job. Each month after that, she is paid $30 more than in the preceding month.

(a) What is Ms. McKee's monthly salary at the end of the first year on the job?
(b) How much will she have earned after the first year on the job?
(c) How long will she have to work before her monthly salary will be $1300?

(a) $1,330
(b) $13,980
(c) 11 months

We found earlier that the nth term in an arithmetic sequence can be given by $a + (n - 1)d$. In this example, $a = 1000$ and $d = 30$. Thus, we find:

Part (a): $1000 + (12 - 1)30 = 1000 + (11)30 = 1000 + 330 = \$1,330$

Part (b): We must find the sum of the first 12 terms of the sequence

1000, 1030, 1060, 1090, 1120, . . .

Part (c): We want to know what value of n (number of months) will produce a monthly salary of $1,300. We must solve the following equation,
$$1000 + (n - 1)\,30 = 1300.$$
Solving, we find:
$$1000 + 30n - 30 = 1300$$
$$970 + 30n = 1300$$
$$30n = 330$$
$$n = 11 \text{ months}$$

11. Find an expression for the nth term of this sequence.

7, 13, 19, 25, 31, ___ , ___ , ___

nth term given by:

$$7 + (n - 1)\,6 \quad \text{or}$$
$$6n + 1$$

12. What are the next three triangular numbers?

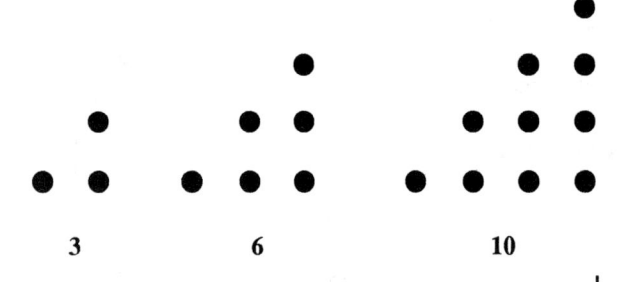

3   6   10

The next three triangular numbers are 15, 21, 28.

13. Ten people are in a room. If each person shakes hands with every other person in the room, how many different handshakes will there be?

There will be 45 different handshakes. We can solve this problem by using a pattern. We can make a table such as the one below and obtain more information.

Number People	Number Handshakes
2	1
3	3
4	6
5	10

Looking at the table, we see that the number of handshakes increases by 2, then by 3, then by 4, and so on. If this pattern continues, then when the number of people is equal to 10, the number of handshakes will be equal to 45. You could simply continue the table.

We will investigate other ways to solve this and similar problems. Compare the numbers for the handshakes column with the triangular numbers of problem 12.

14. Find the sum of the first 100 consecutive odd counting numbers.

The sum of the first 100 consecutive odd counting numbers is 10,000. Again, a table will help us find a pattern.

Numbers to be Added	Sum
1	1
1 + 3	4
1 + 3 + 5	9
1 + 3 + 5 + 7	16
1 + 3 + 5 + 7 + 9	25

Note that the sum of the first two odd counting numbers is 2^2, the sum of the first three counting numbers is 3^2, and so on. Thus the sum of the first 100 counting numbers is 100^2 or 10,000.

15. Consider the following computations.

$2^3 - 1^3 = 8 - 1 = 7$
$3^3 - 2^3 = 27 - 8 = 19$
$4^3 - 3^3 = 64 - 27 = 37$
$5^3 - 4^3 = 125 - 64 = 61$

Without cubing the numbers, decide what $8^3 - 7^3$ will equal.

$8^3 - 7^3 = 169$
Consider the following:
$2^3 - 1^3 = 8 - 1 = 7$
    difference is 12
$3^3 - 2^3 = 27 - 8 = 19$
    difference is 18
$4^3 - 3^3 = 64 - 27 = 37$
    difference is 24
$5^3 - 4^3 = 125 - 64 = 61$
For the pattern to continue, $6^3 - 5^3$ will be 91 (difference of 30 from $5^3 - 4^3$) and finally, $8^3 - 7^3$ will be 169.

Section 2 An Introduction to Problem Solving

Cover the right side of the page and work on the left, then check your work

1. Translate the following verbal phrases into mathematical expressions or sentences.

   (a) The sum of x and y multiplied by 4.
   (b) The product of 3 times x and 5 is equal to 10.
   (c) The difference of 4 times a number and 9 times another number is 0.
   (d) If 5 folders cost $3, what is the price per folder?

   (a) $4(x + y)$
   (b) $(3x)5 = 10$
   (c) $4x - 9y = 0$
   (d) $\dfrac{3}{5} = x$

2. Express as a mathematical sentence and solve on the domain of counting numbers.

   (a) What number multiplied by 7 equals 91?
   (b) The sum of two numbers is 19. One number is 7 more than the other. What are the numbers?
   (c) Nine more than a number divided by 9 is 18.
   (d) The product of two numbers is 36, the smaller is one-fourth of the other.

   (a) $7x = 91$
   $x = 13$
   (b) $x + (x + 7) = 19$
   $2x + 7 = 19$
   $2x = 12$
   $x = 6$
   (c) $\dfrac{x + 9}{9} = 18$
   $x + 9 = 162$
   $x = 153$
   (d) $x(4x) = 36$
   $4x^2 = 36$
   $x^2 = 9$
   $x = 3 \qquad 4x = 12$

3. How many squares are there on a regular 8×8 checkerboard?

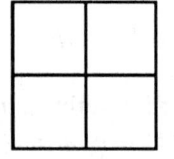

8　　CHAPTER 1　Critical Thinking, Problem Solving and Logic

1　large square
4　1 × 1 squares

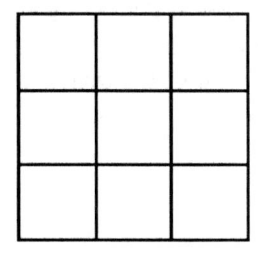

1　large square
4　2 × 2 squares
9　1 × 1 squares

8 × 8 checkerboard:

1 + 4 + 9 + 16 + 25 + 36 + 49 + 64
= 204

4. How many four digit numbers have the same digits as 1991?

6　1199
　　1919
　　9119
　　9911
　　9191
　　1991

5. A pen and a ruler together cost $3. The pen costs 80¢ more than the ruler. How much does the pen cost?

$1.90　　　$x + (x + .80) = 3.00$
　　　　　　　　$2x = 2.20$
　　　　　　　　$x = \$1.10$
　　　　　　　$x + 80 = \$1.90$

6. How can you cook an egg for exactly 15 minutes if all you have is a 7-minute and an 11-minute timer?

Start both timers together. When the 7-minute timer goes off, start cooking the egg. There are four minutes left on the 11-minute timer. When the 11-

Section 2 Introduction to Problem Solving   **9**

      minute timer goes off, start it again: 4 minutes plus 11 minutes equals 15 minutes.

7. A rancher has a rectangular pasture that he wants to fence. He wants the fenced area to be 80 feet longer than the width. If he has 1080 feet of fencing material, what should the length and width of the pasture be?

      width = 230
      length = 310
      $2x + 2(x + 80) = 1080$
      $2x + 2x + 160 = 1080$
      $4x = 920$
      $x = 230$
      $x + 80 = 310$

8. A mystery number is between 150 and 270. The product of the digits of the number is 81. What is the number?

      199

9. How many different choices of three odd numbers will yield seven as the sum?

      $1 + 1 + 5 = 7$
      $1 + 3 + 3 = 7$

10. You have only one 5 L container and one 3 L container. How can you measure exactly 4 L of water if neither container is marked for measuring?

      Fill the 3 L container and pour into the 5 L container. Then fill the 3 L container again and pour into the 5 L container. Empty the 5 L container and pour what is left in the 3 L container (1 L) into the 5 L container. Fill the 3 L container again and pour into the 5 L container. You now have 1 L + 3 L = 4 L.

11. Eight consecutive odd numbers are added to obtain a sum of 24,064. What are the numbers?

      $x + (x + 2) + (x + 4) + (x + 6) + (x + 8) + (x + 10) + (x + 12) + (x + 14) = 24{,}064$
      3001
      3003
      3005

10   CHAPTER 1   Critical Thinking, Problem Solving and Logic

3007	$8x + 56 = 24{,}064$
3009	$8x = 24{,}008$
3011	$x = 3001$
3013	
3015	

12. A baseball and football together weigh 1.25 pounds. A baseball and basketball together weigh 2.5 pounds. A football and basketball together weigh 3.25 pounds. What is the weight of each ball?

> Baseball: 0.25 pounds
> Basketball: 2.25 pounds
> Football: 1.00 pounds

13. If it takes a grandfather clock 15 seconds to chime "six" how long does it take the clock to chime "twelve"?

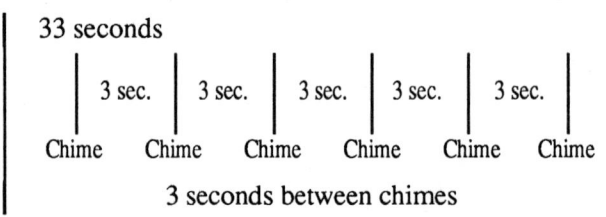

> 33 seconds
>
> | 3 sec. | 3 sec. | 3 sec. | 3 sec. | 3 sec. |
> |---|---|---|---|---|
> | Chime | Chime | Chime | Chime | Chime | Chime |
>
> 3 seconds between chimes

14. Twelve out of every 15 students ride a bus to school. Twelve out of every 18 students live in Warrensburg. 240 students ride a bus to school. How many students do not live in Warrensburg.

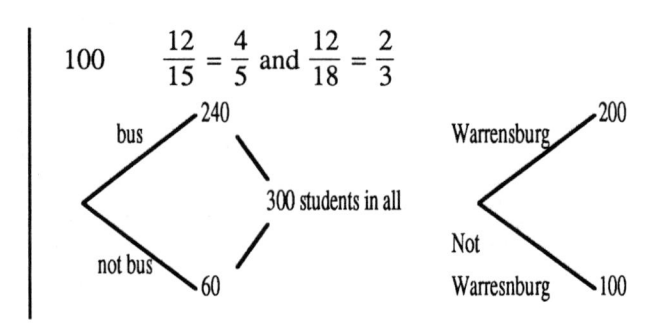

> 100 $\dfrac{12}{15} = \dfrac{4}{5}$ and $\dfrac{12}{18} = \dfrac{2}{3}$
>
> bus — 240
> not bus — 60
> 300 students in all
> Warrensburg — 200
> Not Warresnburg — 100

15. How many triangles of perimeter 10 are there if the length of each side must be a counting number?

> Eight.
> 1, 1, 8   2, 3, 5   3, 4, 3
> 1, 2, 7   2, 4, 4
> 1, 3, 6   2, 6, 2
> 1, 4, 5

Section 3 Critical Thinking and an Introduction to Logic

Cover the right side of the page and work on the left, then check your work

1. Determine which of the following are statements and classify each statement as either true or false.

   (a) No smoking.
   (b) $4 + 1 > 7$.
   (c) $x - 7 = 14$.
   (d) Texas is a country.
   (e) $4 + 8 = 12$.

   (a) Not a statement
   (b) Statement - False
   (c) Not a statement
   (d) Statement - False
   (e) Statement - True

2. If p is F and q is T, find the truth value for each of the following.

   (a) $\sim p \vee q$
   (b) $\sim(p \vee q)$
   (c) $\sim p \vee \sim q$
   (d) $\sim(p \wedge q)$

   (a) T
   (b) F
   (c) T
   (d) T

3. Find the truth values of the statements in Exercise 2 if p is T and q is F.

   (a) F
   (b) F
   (c) T
   (d) T

4. Determine the truth value for each of the following statements.

   (a) $4 + 7 = 11$ and $9 - 3 = 4$
   (b) $4 + 7 = 11$ or $9 - 3 = 4$
   (c) $4 \cdot 3 > 10$ and $3 + 9 \neq 11$
   (d) $20 \div 4 = 5$ or $20 \div 5 = 4$

(a) F
(b) T
(c) T
(d) T

5. Translate the following statements into symbolic form, using R, S, T, U, ∧, ∨, and ~, where:

   R: Students like mathematics.
   S: Students study for long hours.
   T: Students like pizza.
   U: Students go to parties.

   (a) Students like mathematics and students do not like pizza.
   (b) Students study for long hours or students go to parties.
   (c) Students do not go to parties and students like mathematics.
   (d) Students do not like pizza or students do not like mathematics.

(a) R ∧ ~T
(b) S ∨ U
(c) ~U ∧ R
(d) ~T ∨ ~R

6. Using the statements of Exercise 5, translate the symbolic statements into English sentences.

   (a) R ∧ ~U
   (b) ~ (S ∨ ~T)
   (c) R ∨ (S ∧ T)
   (d) ~T ∧ U

(a) Students like mathematics and students do not go to parties.
(b) It is false that students study for long hours or students do not like pizza.
(c) Students like mathematics, or students study for long hours and students like pizza. (Note the use of the comma.)
(d) Students do not like pizza and students go to parties.

7. Assume that for the statements in Exercise 5 we have the following:

Statement R is true
Statement S is false
Statement T is true
Statement V is false

Determine the truth value of each of the following.

(a) Students like mathematics or students study for long hours.
(b) ~V ∧ S
(c) Students do not like pizza and students do not go to parties.
(d) ~ (R ∧ ~T)

(a) T
(b) F
(c) F
(d) T

8. Complete the following truth table.

p	q	$\sim p$	$\sim q$	$\sim p \vee \sim q$	$p \vee \sim q$
T	T				
T	F				
F	T				
F	F				

p	q	$\sim p$	$\sim q$	$\sim p \vee \sim q$	$p \vee \sim q$
T	T	F	F	F	T
T	F	F	T	T	T
F	T	T	F	T	F
F	F	T	T	T	T

9. Let p represent the statement "Mathematics is fun" and q the statement "successful students study". Translate each of the following sentences into symbols.

(a) Mathematics is fun and successful students do not study.

(b) Mathematics is not fun or successful students study.
(c) It is false that mathematics is fun and successful students do not study.

(a) $p \wedge \sim q$
(b) $\sim p \vee q$
(c) $\sim (p \wedge \sim q)$

10. Construct a truth table for $\sim (p \vee \sim q)$.

p	q	~q	p ∨ q	~(p ∨ ~q)
T	T	F	T	F
T	F	T	T	F
F	T	F	F	T
F	F	T	T	F

11. If p is T, q is F, and r is T, determine the truth value for each of the following.

(a) $(\sim q \wedge p) \vee r$
(b) $(r \wedge \sim p) \vee (r \wedge \sim q)$
(c) $\sim [(p \vee q) \wedge r]$
(d) $p \wedge (\sim q \vee r)$

(a) T
(b) T
(c) F
(d) T

12. Answer the following true or false.

(a) A conjunction is true when the simple statements have the same truth value.
(b) The disjunction $p \vee q$ is true when p is true and q is false.
(c) The negation of a statement is always false.

(a) False. When the simple statements are both true.
(b) True.
(c) False. Negation of a false statement is true.

13. Given the statements p: 6 is an even number, q: 6 is a multiple of 3, and r: 6 is a factor of 14, write each of the following in symbolic form.

(a) 6 is an even number and 6 is not a factor of 14.
(b) 6 is not a multiple of 3 or 6 is not a factor of 14.
(c) 6 is not a factor of 14, and 6 is an even number or 6 is a multiple of 3.
(d) 6 is not a factor of 14 and 6 is not an even number.

(a) $p \wedge \sim r$
(b) $\sim q \vee r$
(c) $\sim r \wedge (p \vee q)$
(d) $\sim r \wedge \sim p$

14. Classify each of the compound statements from Exercise 13 as true or false.

(a) T
(b) F
(c) T
(d) F

15. Using the statements from Exercise 13, write the following in English sentences.

(a) $\sim p \vee \sim q$
(b) $p \vee (\sim q \wedge r)$
(c) $(p \vee q) \wedge (\sim p \vee r)$

(a) 6 is not an even number or 6 is not a multiple of 3.
(b) 6 is an even number, or 6 is not a multiple of 3 and 6 is a factor of 14.
(c) 6 is an even number or 6 is a multiple of 3, and 6 is not an even number or 6 is a factor of 14.

Section 4 Critical Thinking and Conditionals

Cover the right side of the page and work on the left, then check your work

1. Let p be the statement "It is sunny", q be the statement "It is raining", and r be the statement "It is summer." Write the following in symbolic notation.

   (a) If it is sunny, then it is summer.
   (b) If it is not raining, then it is summer.
   (c) If it is not summer, then it is not raining.
   (d) If it is not raining, then it is not sunny.

   (a) $p \to r$
   (b) $\sim q \to r$
   (c) $\sim r \to \sim q$
   (d) $\sim q \to \sim p$

2. In Exercise 1, assume p and r are true and q is false. Classify each of the conditionals in Exercise 1 as true or false.

   (a) T
   (b) T
   (c) T
   (d) F

3. Write each of the following statements in the form "If … then".

   (a) Squares are rectangles.
   (b) A wet bird does not fly at night.
   (c) Prime numbers are odd.
   (d) People who use Sparkle toothpaste have fewer cavities.

   (a) If a figure is a square, then it is a rectangle.
   (b) If a bird is wet, then it does not fly at night.
   (c) If a number is prime, then it is odd.
   (d) If a person uses Sparkle toothpaste, then he/she has fewer cavities.

4. State the converse, inverse, and the contrapositive of the following.

(a) If you like mathematics, then you like this book.
(b) If a quadrilateral has opposite sides parallel, then it is a parallelogram.

(a) Converse: If you like this book, then you like mathematics.
Inverse: If you do not like mathematics, then you do not like this book.
Contrapositive: If you do not like this book, then you do not like mathematics.

(b) Converse: If a figure is a parallelogram, then it is a quadrilateral with opposite sides parallel.
Inverse: If a quadrilateral does not have opposite sides parallel, then it is not a parallelogram.
Contrapositive: If a figure is not a parallelogram, then it is not a quadrilateral with opposite sides parallel.

5. Using the notation of Exercise 1, translate the following into sentences.

(a) $\sim p \rightarrow p$
(b) $\sim r \rightarrow \sim q$
(c) $(\sim p \land q) \rightarrow r$

(a) If it is not raining, then it is sunny.
(b) If it is not summer, then it is not raining.
(c) If it is not sunny and it is raining, then it is summer.

6. Using the truth values of Exercise 2, classify the statements in Exercise 5 as true or false.

(a) T
(b) T
(c) T

7. Consider the statement "If the sum of the digits of a number is nine, then the number is divisible by

9". Which of the following is logically equivalent to this statement?

(a) If the sum of the digits of a number is not nine, then the number is not divisible by 9.
(b) If a number is divisible by 9, then the sum of the digits of the number is nine.
(c) If a number is not divisible by 9, then the sum of its digits is not nine.

(c)

8. Write the following in "If ... then" notation.

(a) All dogs are loyal.
(b) Haste makes waste.
(c) No students are lazy.

(a) If an animal is a dog, then it is loyal.
(b) If a person is hasty, then he/she will be wasteful.
(c) If a person is a student, then he/she is not lazy.

9. Construct a truth table for each of the following.

(a) $p \to (p \vee q)$
(b) $\sim(p \to q)$
(c) $(p \vee q) \to (p \wedge q)$

(a)

p	q	$p \vee q$	$p \to (p \vee q)$
T	T	T	T
T	F	T	T
F	T	T	T
F	F	F	T

(b)

p	q	$p \to q$	$\sim(p \to q)$
T	T	T	F
T	F	F	T
F	T	T	F
F	F	T	F

(c)

p	q	$p \vee q$	$p \wedge q$	$(p \vee q) \to (p \wedge q)$
T	T	T	T	T
T	F	T	F	F
F	T	T	F	F
F	F	F	F	T

10. Using truth tables, determine whether the following pairs of statements are equivalent.

(a) $\sim(p \wedge q); p \to q$
(b) $p \to \sim q; q \to \sim p$

(a) Not equivalent

p	q	$p \wedge q$	$\sim(p \wedge q)$	$p \to q$
T	T	T	F	T
T	F	F	T	F
F	T	F	T	T
F	F	F	T	T

(b) Equivalent

p	q	$\sim p$	$\sim q$	$p \to \sim q$	$q \to \sim p$
T	T	F	F	F	F
T	F	F	T	T	T
F	T	T	F	T	T
F	F	T	T	T	T

11. Write the following as conditionals.

(a) John is tall, and Mary is short.
(b) Dogs are gentle or bears are not aggressive.

(a) It is false that if John is tall then Mary is short.
(b) If dogs are not gentle, then bears are not aggressive.

12. Use truth tables to prove the following are tautologies.

(a) $(p \to q) \to [(p \wedge r) \to q]$
(b) $[(p \to q) \wedge p] \to q$

(a)

p	q	r	$p \to q$	$p \wedge r$	$(p \wedge r) \to q$	$(p \to q) \to [(p \to r) \to q]$
T	T	T	T	T	T	T
T	T	F	T	F	T	T
T	F	T	F	T	F	T
T	F	F	F	F	T	T
F	T	T	T	F	T	T
F	T	F	T	F	T	T
F	F	T	T	F	T	T
F	F	F	T	F	T	T

(b)

p	q	$p \to q$	$(p \to q) \wedge p$	$[(p \to q) \wedge p] \to q$
T	T	T	T	T
T	F	F	F	T
F	T	T	F	T
F	F	T	F	T

13. Write one example for each of the following.

(a) Fallacy of experts
(b) Fallacy of compostion
(c) Fallacy of false cause

(a) Michael Jordan drinks Brand Y.
(b) You took too long to complete the test; therefore, you took too long on each question.
(c) I was in Dr. Cooper's class; I failed the course. Dr. Cooper is unfair.

14. Using a truth table, determine whether $p \wedge (q \vee r)$ and $(p \wedge q) \vee (p \wedge r)$ are logically equivalent.

Logically equivalent

p	q	r	$p \vee q$	$p \wedge q$	$p \wedge r$	$p \wedge (q \vee r)$	$(p \wedge q) \vee (p \wedge r)$
T	T	T	T	T	T	T	T
T	T	F	T	T	F	T	T
T	F	T	T	F	T	T	T
T	F	F	T	F	F	F	F
F	T	T	T	F	F	F	F
F	T	F	F	F	F	F	F
F	F	T	T	F	F	F	F
F	F	F	F	F	F	F	F

15. Write a statement logically equivalent to the statement "If a number is a multiple of 10, then it is a multiple of 5".

If a number is not a multiple of 5, then it is not a multiple of 10.

Section 5 Quantifiers

Cover the right side of the page and work on the left, then check your work

1. Make a Venn diagram for each of the following, showing:

    (a) the relationship among Students, Women, and Freshmen;
    (b) some A's are B's; all B's are C's, some A's are C's;
    (c) the relationship among surgeons (S), physicians (P), and men (M).

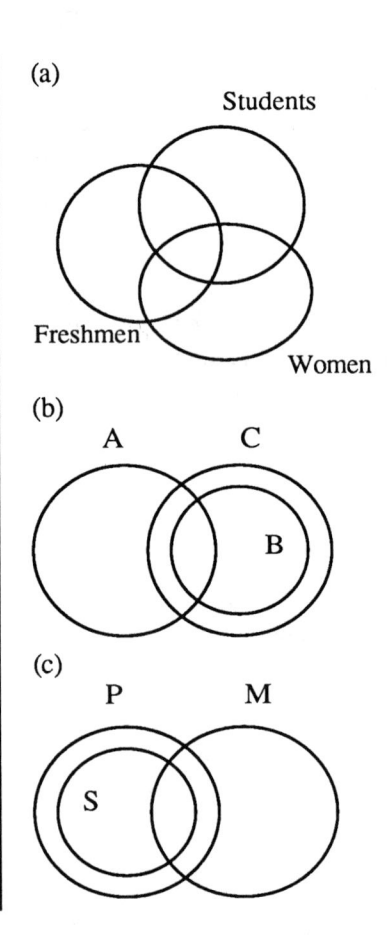

    (a) [Venn diagram: three overlapping circles labeled Students, Freshmen, Women]

    (b) [Venn diagram: circle A overlapping with circle C, with circle B inside C overlapping A]

    (c) [Venn diagram: circle S inside circle P, overlapping with circle M]

2. Make a Venn diagram showing each of the following.

    (a) Some Dweebs are Nerks.
    (b) No Kreels are Knarls.
    (c) All Bippys are Flatterips.
    (d) Some Singers are Men and all Singers are Music Lovers.

Section 5 Quantifiers

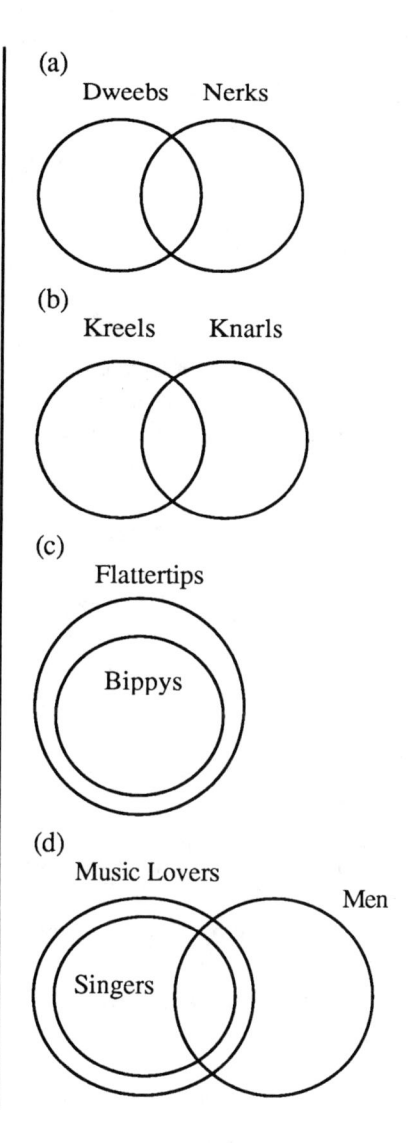

(a) Dweebs   Nerks

(b) Kreels   Knarls

(c) Flattertips / Bippys

(d) Music Lovers / Singers / Men

3. Use a quantifier to make each statement true.

(a) $x + 7 = 11$
(b) $x^2 = 9$
(c) $x - 5 = 13$
(d) $x + 0 = x$

(a) There exists a number x, such that $x + 7 = 11$.
(b) There is no counting number x, such that $x^2 = 7$.
(c) There exists a number x, such that $x - 5 = 13$.
(d) For every number x, $x + 0 = x$.

4. Use a quantifier to make each statement in Exercise 3 false.

(a) For all numbers x, $x + 7 = 11$.
(b) There exists a counting number x, such that $x^2 = 7$.
(c) For all numbers x, $x - 5 = 13$
(d) There exists a number x, such that $x + 0 \neq x$

5. Write the negation of each of the following.

(a) Some people have brown hair.
(b) All horses have four legs.
(c) No cats can fly.
(d) All squares are rectangles.

(a) No person has brown hair.
(b) Some horses do not have four legs.
(c) Some cats can fly.
(d) Some squares are not rectangles.

6. Write the negation of the statement below.

(a) There exists a counting number x such that $3x - 4 = 17$.
(b) Every counting number is divisible by 1.
(c) For all x, $x + 2 = 2 + x$.
(d) for all counting numbers x, $x + 3x = 4x$.

(a) There is no counting number x, such that $3x - 4 = 17$.
(b) There exists a counting number that is not divisible by 1.
(c) There exists a counting number x such that $x + 2 \neq 2 + x$.
(d) There exists a counting number x, such that $x + 3x \neq 4x$.

7. Write the negation of the following conditionals.

(a) If $x > 7$, then $x > 3$.
(b) If $x \neq 4$, then $y = 2$.

(a) $x > 7$ and $x > 3$
(b) $x \neq 4$ and $y \neq 2$

8. Write the negation of the following compound statement.

   Neither $x^2 = 3$ or $x^2 = 7$ has an integer solution.

   $\Big|$ $x^2 = 3$ has an integer solution or
   $x^2 = 7$ has an integer solution.

9. Describe figure X as completely as possible.

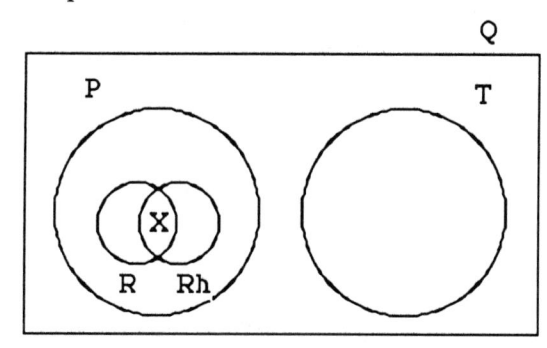

   Q - quadrilaterals
   P - parallelograms
   T - trapezoids
   R - rectangles
   Rh - rhombi

   $\Big|$ X is a quadrilateral
   X is a parallelogram
   X is a rectangle
   X is a rhombus
   X is not a trapezoid

10. Place ϕ (indicating empty) and $*$ (indicating at least one) in the following Venn diagram to indicate each of the following.

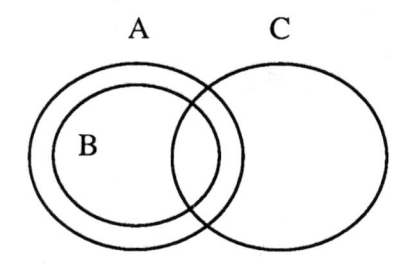

    (a) Some A's are C's
    (b) All C's are B's
    (c) All B's are C's
    (d) No C's are B's

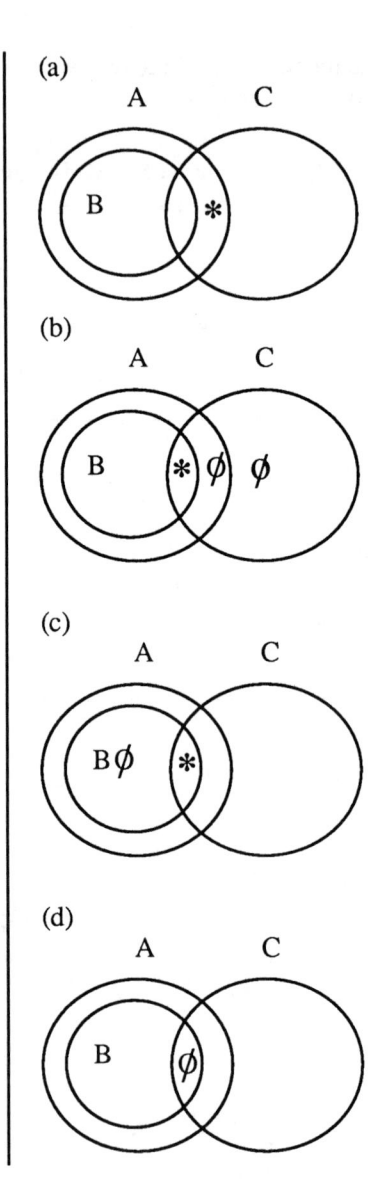

(a)

(b)

(c)

(d)

Section 6 Making Use of Deductive Logic

Cover the right side of the page and work on the left, then check your work

1. Rewrite each argument in symbolic form.

   (a) If tomorrow is Monday (*m*), then today is Sunday (*s*). Today is Sunday. Therefore, tomorrow is Monday.
   (b) If it snows (*s*), then the mail will be late (*l*). It is not snowing. Therefore, the mail will not be late.
   (c) If I am hungry (*h*), then I will not study (*s*). If I am worried (*w*), then I am hungry. Therefore, if I am worried, I will not study.
   (d) If it rains (*r*), the game will be cancelled (*c*). The game is not cancelled. Therefore, it does not rain.

   (a) $m \rightarrow s$
   s
   $\therefore m$
   (b) $s \rightarrow l$
   $\sim s$
   $\therefore \sim l$
   (c) $h \rightarrow s$
   $w \rightarrow h$
   $\therefore w \rightarrow s$
   (d) $r \rightarrow c$
   $\sim c$
   $\therefore \sim r$

2. Determine the validity of each of the arguments in Exercise 1.

   (a)

m	s	$m \rightarrow s$	$(m \rightarrow s) \wedge s$	$[(m \rightarrow s) \wedge s] \rightarrow m$
T	T	T	T	T
T	F	F	F	T
F	T	T	T	F
F	F	T	F	T

   ←Not Valid

   (b) Not Valid

(c) Valid

h	s	w	h → s	w → h	w → s	(h → s) ∧ (w → h)	[(h → s) ∧ (w → h)] → (w → s)
T	T	T	T	T	T	T	T
T	F	T	F	T	F	F	T
T	F	F	F	T	T	F	T
T	T	F	T	T	T	T	T
F	T	T	T	F	T	F	T
F	T	F	T	T	T	T	T
F	F	T	T	F	F	F	T
F	F	F	T	T	T	T	T

(d) Valid

3. Determine whether the following arguments are valid.

(a) $p \to q$
$\sim q \to p$
$\therefore \sim p$

(b) $(p \lor q) \to r$
p
$\therefore r$

(c) $p \land \sim q$
$\sim p \to q$
$\therefore q$

Answers follow

(a) Not Valid, because:

p	q	~p	~q	p → q	~q → p	(p → q) ∧ (~q → p)	[(p ∧ q) ∧ (~q → p)] → ~p
T	T	F	T	T	T	T	F
T	F	F	T	F	T	F	T
F	T	T	F	T	T	T	T
F	F	T	T	T	F	F	T

(b) Valid
(c) Not Valid, because:

p	q	$\sim p$	$\sim q$	$p \wedge \sim q$	$\sim p \to q$	$(p \wedge q) \wedge (\sim p \to q)$	$[(p \wedge q) \wedge (\sim p \to q)] \to q$
T	T	F	F	F	T	F	T
T	F	F	T	T	T	T	F
F	T	T	F	F	T	F	T
F	F	T	T	F	F	F	T

4. Use truth tables to determine the validity of each of the following.

   (a) If a woman is a professor (p), then she is smart (s). Rhonda is smart. Therefore, she is a professor.
   (b) If you work hard (w), then you will be successful (s). You do not work hard. Therefore, you will not be successful.
   (c) If you are an adult, then you have children (c). John has no children. Therefore, John is not an adult.
   (d) If a man is tall (t), then he is a basketball star (b). Chris is not a basketball star. Therefore, Chris is tall.

(a) Not Valid, because:

p	s	$p \to s$	$(p \to s) \wedge s$	$[(p \to s) \wedge s)] \to p$
T	T	T	T	T
T	F	F	F	T
F	T	T	T	F
F	F	T	F	T

(b) Not Valid
(c) Valid
(d) Not valid, because:

t	b	$\sim b$	$t \to b$	$(t \to b) \wedge \sim b$	$[(t \to b) \wedge \sim b] \to t$
T	T	F	T	F	T
T	F	T	F	F	T
F	T	F	T	F	T
F	F	T	T	T	F

5. Determine a valid conclusion that follows from each of the following statements.
   (a) If you are friendly, then you will be popular. You are friendly.

(b) If Sam is a fast runner, then he is a football star.
(c) All friends are loyal. All loyal people are trustworthy.

(a) Therefore, you are popular.
(b) Therefore, if Sam is not a football star, then he is not a fast runner.
(c) Therefore, all friends are trustworthy.

Use a Venn diagram to answer each question. Assume that the first two statements are true.

6. All men are mortal.
   Abraham Lincoln was a man.
   Can you be sure Lincoln was mortal?

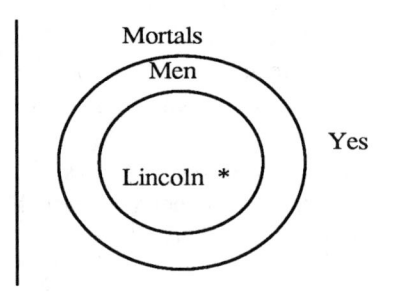

   Mortals
   Men
   Lincoln *
   Yes

7. All squares are parallelograms.
   All parallelograms are quadrilaterals.
   Can you be sure that all squares are quadrilaterals?

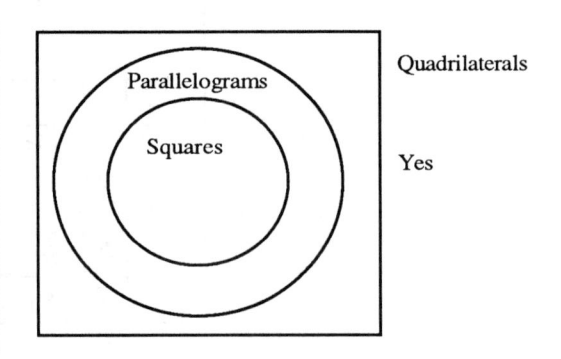

   Quadrilaterals
   Parallelograms
   Squares
   Yes

8. Some women are teachers.
   All teachers are college graduates.
   Can you be sure all women are college graduates?

   No.

Section 6   Making Use of Deductive Logic   **3 1**

Women   *College Graduates*

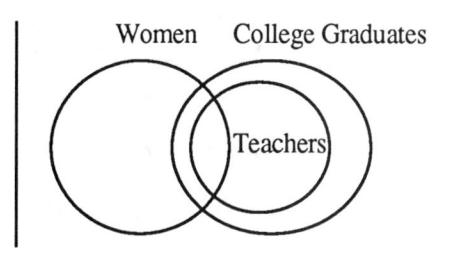

Teachers

Demonstrate the validity of each of the following arguments.

9. If it does not snow, then we will have school. We do not have school. Therefore, it snowed.

$\sim s \to h$ s = It snows
$\sim h$ h = We have school
$\therefore s$

$\sim s \to f \Leftrightarrow \sim h \to s$
Implication and its contrapositive
Then, $\sim h \to s$
         $\sim h$ Rule of Detachment
         $\therefore s$

10. If a penny is saved, then it is earned. A penny is saved. Therefore, the penny is earned.

Law of Detachment

11. Use the diagram to determine which of the following statements are true.

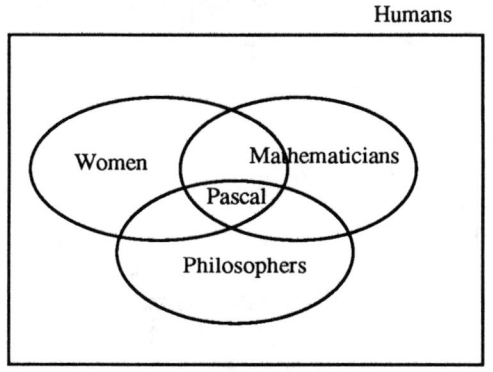

Humans

Women   Mathematicians

Pascal

Philosophers

(a) All mathematicians are women.
(b) Pascal was a woman.
(c) All philosophers are humans.
(d) Some mathematicians are philosophers.

(a) False
(b) False
(c) True
(d) True

Chapter 2
Tools for Problem Solving: Sets and Numbers

Section 1 An Introduction to Sets

Cover the right side of the page and work on the left, then check your work

1. Classify each statement as true or false.

    (a) $9 \in \{8, 9, 10\}$
    (b) $\{a, b, c\} \subset \{a, b, c\}$
    (c) $\{4, 2\} \subseteq \{1, 2, 3, 4\}$
    (d) $5 \subseteq \{4, 5, 6\}$

    (a) True
    (b) False (is not a proper subset)
    (c) True
    (d) False $5 \in \{4, 5, 6\}$

2. Let A = {Americans who have won the Olympic decathlon}, and C = {athletes who play on college athletic teams. Describe each of the following in words.

    (a) $A \cap C$
    (b) $C - A$

    (a) College athletes who have won the Olympic decathlon
    (b) College athletes who have not won the Olympic decathlon

3. Represent each of the following sets using the tabulation method.

    (a) the set of state names beginning with the letter T
    (b) the set of odd counting numbers less than 12
    (c) the set of even counting numbers between 1 and 10

    (a) {Tennessee, Texas}
    (b) {1, 3, 5, 7, 9, 11, 13}
    (c) {2, 4, 6, 8}

4. Represent each of the following sets using set-builder notation

    (a) the set of even counting numbers greater than 10
    (b) the set $\{1, 4, 9, 16, \ldots\}$
    (c) the set of months whose name begins with the letter A.

(a) $\{x \mid x$ is a counting number and $x > 10\}$
(b) $\{x \mid x$ is a perfect square$\}$
(c) $\{x \mid x$ is the name of a month and x begins with the letter $A\}$

5. List the subsets of the following set.

   $A = \{a, b, c\}$

   $\{a, b, c\}$ set itself
   $\{a\}, \{b\}, \{c\}$ all one element subsets
   $\{a, b\}, \{a, c\}, \{b, c\}$ all two element subsets
   ϕ empty set

6. Insert the symbol $\{\in, \notin, \subset, \subseteq\}$ to make each statement true. $A = \{1, 2, 3\}, B = \{3, 5, 7\}$ and $C = \{2, 3\}$.

   (a) C ___ A
   (b) 5 ___ A
   (c) 5 ___ B
   (d) B ___ B

   (a) $C \subseteq A$
   (b) $5 \notin A$
   (c) $5 \in B$
   (d) $B \subseteq B$

7. Given: $A = \{1, 2, 3, 4, 5\}, B = \{2, 4, 6\}$, and $C = \{1, 3, 5, 7, 9\}$. Find each of the following.

   (a) $A \cap B$
   (b) $B \cup C$
   (c) $(A \cup B) \cap C$
   (d) $(A \cap C) \cup (B \cap C)$

   (a) $\{2, 4\}$
   (b) $\{1, 2, 3, 4, 5, 6, 7, 9\}$
   (c) $\{1, 3, 5\}$
   (d) $\{1, 3, 5\}$

8. Given: $S = \{a, b, c, d, e\}, T = \{a, e, i\}$, and $U = \{a, b, c, d, e, f, g, h, i\}$ the universal set. Find each of the following.

   (a) $\overline{S}$
   (b) $S - T$
   (c) $\overline{S} \cap \overline{T}$
   (d) $\overline{S \cap T}$

34 CHAPTER 2 Tools for Problem Solving: Sets and Numbers

(a) $\{f, g, h, i\}$
(b) $\{b, c, d\}$
(c) $\{f, g, h\}$
(d) $\{b, c, d, f, g, h, i\}$

9. Shade the diagram to illustrate each of the following.

   (a) $\overline{A} \cup \overline{B}$
   (b) $(\overline{A \cup B}) \cap C$
   (c) $A - (B \cap C)$

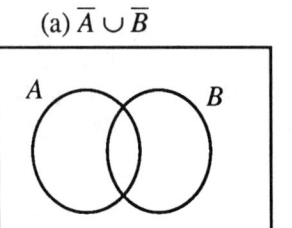

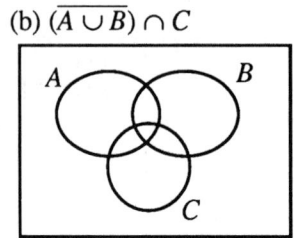

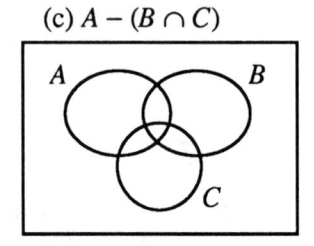

Answer to Exercise 9

   (a) $\overline{A} \cup \overline{B}$
   (b) $(\overline{A \cup B}) \cap C$
   (c) $A - (B \cap C)$

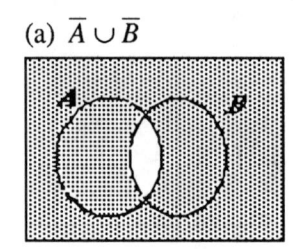

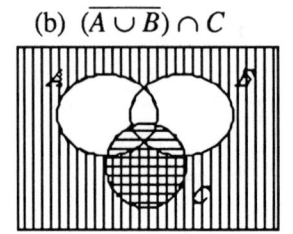

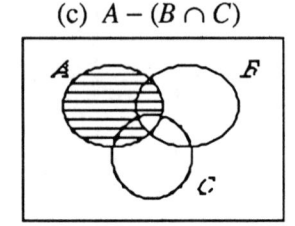

10. Draw a Venn diagram to illustrate each of the following.

    (a) In Math 1811, some of the students are mathematics minors, some are juniors, but none of the juniors are math minors.
    (b) In a survey, some people reported that they watch the news on Channel 9, some on Channel 4, and some on Channel 5. Some watch Channels 4 and 9, and some watch Channel 4 and 5. No one watches all three channels.

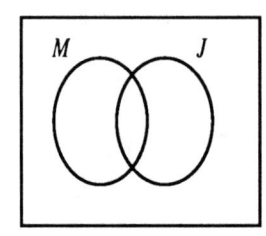

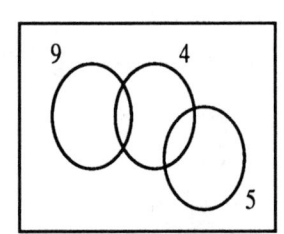

11. Use $R, S, T, \cup, \cap$, and $-$ to describe the shaded regions.

(a)

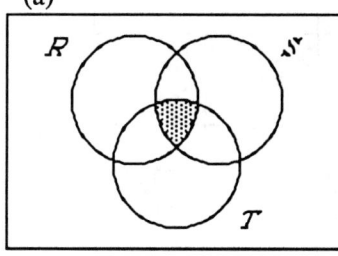

(b)

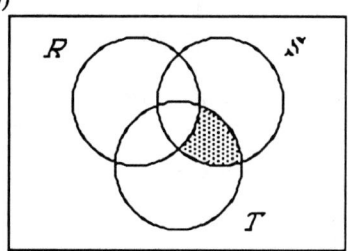

(a) $R \cap S \cap T$
(b) $(S \cap T) - R$ Intersection of S and T with elements of R removed

12. For each of the following, find $A - B$.

(a) $A \cap B = \emptyset$
(b) $A = B$
(c) $A \subseteq B$

(a) $A - B = A$ Look at the figure below.

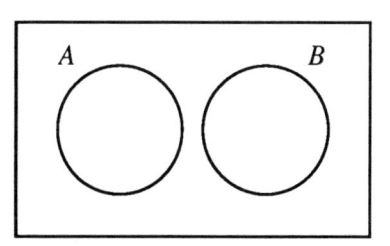

(b) ϕ A removed from itself
(c) ϕ

13. State the conditions under which each of the following is true.

(a) $A \cup B = A$
(b) $A \cap B = A$
(c) $A \cup B = A \cap B$

(a) When $B \subseteq A$
(b) When $A \subseteq B$
(c) When $A = B$

$\overline{A \cup B} = \overline{A} - \overline{B}$

$x \in A$	$x \in B$	$x \in (A \cup B)$	$x \in \overline{(A \cup B)}$	$x \in \overline{A}$	$x \in \overline{B}$	$x \in \overline{A} - \overline{B}$
T	T	T	F	F	F	F
F	T	T	F	T	F	F
T	F	T	F	F	T	F
F	F	F	T	T	T	T

15. If $a \in A \cup B$, is a in $A \cap B$? Why or why not?

No. Consider the figure below.

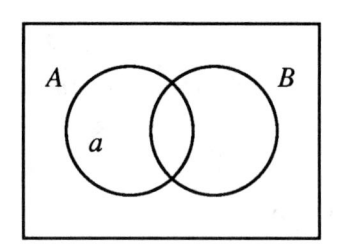

Section 2 Cartesian Products and Relations

Cover the right side of the page and work on the left, then check your work

1. If S has m elements and T has n elements, then

    (a) how many elements are in $S \times T$?
    (b) how many elements are in $T \times S$?
    (c) how many elements are in $T \times T$?

    > (a) $m \times n$ elements
    > (b) $n \times m$ elements
    > (c) $n \times n$ elements

2. Write three ordered pairs that satisfy the relation "is the capitol of".

    > (Austin, Texas), (Jefferson City, Missouri), (Albany, New York)
    > There would be many other examples.

3. If $S = \{1, 2, 3\}$ and $T = \{a, b\}$, tabulate each of the following.

    (a) $S \times T$
    (b) $T \times S$
    (c) $S \times S$

    > (a) $S \times T = \{(1, a), (1, b), (2, a), (2, b), (3, a), (3, b)\}$
    > (b) $T \times S = \{(a, 1), (a, 2), (a, 3), (b, 1), (b, 2), (b, 3)\}$
    > (c) $S \times S = \{(1, 1), (1, 2), (1, 3), (2, 1), (2, 2), (2, 3), (3, 1), (3, 2), (3, 3)\}$

4. For each relation below, determine if it is reflexive, symmetric, or transitive.

    (a) Has the same tens digit as
    (b) Is taller than
    (c) Is in the same mathematics class as

    > (a) Is reflexive. A number has the same tens digit as itself.
    > Is symmetric. If n has the same tens digit at m, then m has the same tens digit as n.
    > Example:

$(23, 29) \in R \Rightarrow (29, 23) \in R$
Is transitive. If n has the same tens digit as m and m has the same tens digit as p, then n has the same tens digit as p.
Example:
$(42, 46) \in R$ and $(46, 48) \in R \Rightarrow (42, 48) \in R$

(b) Is not reflexive. A person is not taller than himself/herself.
Is not symmetric. If a is taller than b, then b is not taller than a.
Is transitive. If a is taller than b and b is taller than c, then a is taller than c.

(c) Is reflexive. A person is in the same mathematics class as himself/herself.
Is symmetric. If a is in the same mathematics class as b, then b is in the same mathematics class as a.
Is symmetric. If a is in the same mathematics class as b and b is in the same mathematics class as c, then a is in the same mathematics class as c.

5. For each mathematical relation below, determine if it is reflexive, symmetric, or transitive.

   (a) Has the same perimeter as
   (b) Is parallel to
   (c) Is a proper subset of

6. Find sets S and T so that $S \times T$ has

   (a) 4 ordered pairs.
   (b) 6 ordered pairs

   (a) $S = \{1, 2\}$ and $T = \{x, y\}$
   (b) $S = \{4, 5, 6\}$ and $T = \{a, b\}$

7. The Cartesian product $S \times T$ is given below. Find S and T.

   (a) $\{(a, b), (b,b), (a, c), (b, c)\}$
   (b) $\{(1, 5), (1, 3), (2, 5), (2, 3), (3, 3), (3, 5)\}$

(a) $S = \{a, b\}$ Set of all first members of the ordered pairs.
$T = \{b, c\}$ Set of all second members of the ordered pairs.
(b) $S = \{1, 2, 3\}$
$T = \{3, 5\}$

8. A relation on $A = \{1, 2, 3, 4\}$ is defined as follows.

   $(a, b) \in R$ means a and b are equal or differ by 1

   For example, $(3, 4) \in R$ since 3 and 4 differ by 1.

   (a) Is this relation reflexive?
   (b) Is this relation symmetric?
   (c) Is this relation transitive?

(a) R is reflexive. For $a \in A$ we have $(a, a) \in R$ since $a = a$.
(b) R is symmetric. For $a, b \in A$, $(a, b) \in R$ means that a and b are equal or differ by 1. If a and b are equal, the $(b, a) \in R$. If a and b differ by 1, then b and a differ by 1 and $(b, a) \in R$.
(c) R is not transitive. $(1, 2) \in R$ and $(2, 3) \in R$, but $(1, 3) \notin R$

9. Each of the following sets of pairs are included in some relation. Describe each relation.

   (a) $(2, \frac{1}{2})$, $(3, \frac{1}{3})$, $(4, \frac{1}{4})$, $(5, \frac{1}{5})$
   (b) $(3, 8)$, $(4, 11)$, $(5, 14)$, $(6, 17)$

(a) $(a, b) \in R$ means that b is the reciprocal of a
(b) $(a, b) \in R$ means that $b = 3a - 1$

10. Determine if the relation R below is an equivalence relation on $A \times A$ where $A = \{5, 6, 7\}$.

    $R = \{(5, 5), (6, 6), (7, 7), (5, 6), (6, 5), (6, 7),$
    $(5, 7), (7, 6), (7, 5)\}$

R is an equivalence relation since it is reflexive, symmetric, and transitive.

11. If $A = \{a, b, c, d\}$ and a relation on A is illustrated in the diagram, determine if the relation is reflexive, symmetric, or transitive.

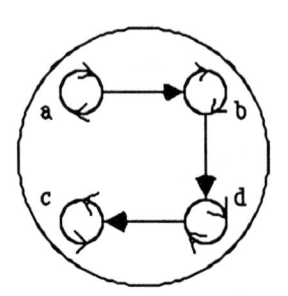

The relation is reflexive since each element is related to itself (notice that each element has an arrow from itself to itself).
The relation is not symmetric since there is an arrow from a to b but there is no arrow from b to a.
The relation is not transitive since there is an arrow from a to b and an arrow from b to c, but there is no arrow from a to c.

12. Define a relation R as follows, on the set $C = \{1, 2, 3, 4, \ldots\}$.

    $(a, b) \in R$ means a is a factor of b

    Is this an equivalence relation?

R is not an equivalence relation since it is not symmetric. For example, $(2, 4) \in R$ since 2 is a factor of 4. We cannot say, however, that $(4, 2) \in R$ since 4 is not a factor of 2.

13. Is the following true or false? Why or why not?

    $A \times B = B \times A$

$A \times B = B \times A$ is false. Consider the example below.

Let $A = \{1, 2\}$ and $B = \{r, s\}$
$A \times B = \{(1, r), (1, s), (2, r), (2, s)\}$

14. Let R be defined on the set $\{1, 2, 3, 4\}$. Give a relation R that is not reflexive, but is symmetric and is transitive.

15. Let $S = \{2, 3, 4, \ldots, 20\}$. Identify the equivalence classes for the relation R defined on S and described below.

    $(a, b) \in R$ means a and b have the same number of factors

and $B \times A =$
$\{(r, 1), (r, 2), (s, 1), (s, 2)\}$

We see that $A \times B \neq B \times A$

Example:
$R = \{(1, 1), (1, 2), (2, 1)\}$

$C_1 = \{2, 3, 5, 7, 11, 13, 17, 19\}$
All numbers with two factors
$C_2 = \{4, 9, 16\}$, All numbers with three factors
$C_3 = \{6, 8, 10, 12, 14, 15, 18, 20\}$, All numbers with four factors

Section 3 The Number of Elements in a Set

Cover the right side of the page and work on the left, then check your work

1. An automobile is available with 14 choices for exterior color and 19 choices for interior color. Use mental computation to determine the number of color combinations available.

   Number of combinations = 14×19
   = $14 \times (20 - 1) = 14 \times 20 - 14 \times 1$
   = $280 - 14 = 266$

2. $n(A) = 23$, $n(B) = 12$, and $n(A \cap B) = 3$. Find $n(A \cup B)$.

   $n(A \cup B) = 32$
   Recall that $n(A \cup B) = n(A) + n(B) - n(A \cap B)$
   In this example, we have $n(A \cup B) = 23 + 12 - 3 = 32$

3. Which of the following sets are equivalent to $A = \{1, 2, 3, 4\}$?

   (a) $\{4, 3, 2, 1\}$
   (b) $\{a, b, c, d, e\}$
   (c) $\{7, 8, 9, 10, \ldots\}$
   (d) $\{w, x, y, z\}$

   The following sets are equivalent to $A = \{1, 2, 3, 4\}$: $\{4, 3, 2, 1\}$ and $\{w, x, y, z\}$.

4. Classify each of the following as true or false.

   (a) $n(A \cup B) = n(A) + n(B)$ for all finite sets A and B
   (b) If set C is equivalent to set D, then $C = D$
   (c) The set $\{1, 2, 3, \ldots, 1000\}$ is a finite set
   (d) $n(A \times \phi) = n(A)$
   (e) $n(B - A) = n(A - B)$

   (a) False. Recall that the formula is
   $n(A \cup B) = n(A) + n(B) - n(A \cap B)$
   (b) False
   (c) True
   (d) True
   (e) False. Consider $A = \{1, 2, 3, 4\}$ and $B = \{2, 4, 7\}$. Then $A - B = \{1, 3\}$ and $n(A - B) = B - A = \{7\}$ and $n(B - A) = 1$

5. Use sets $A = \{c, a, t\}$ and $B = \{p, e, t\}$ to find each of the following.

   (a) $n(A \cup B)$
   (b) $n(B - A)$
   (c) $n(A \times B)$

   (a) $n(A \cup B) = 5$
   (b) $n(B - A) = 2$
   (c) $n(A \times B) = 9$

6. Consider sets $A = \{1, 2, 3\}$ and $B = \{x, y, z\}$.

   (a) Show a one-to-one correspondence between A and B.
   (b) How many different one-to-one correspondences are there between A and B?

   (a) One example of a one-to-one correspondence would be

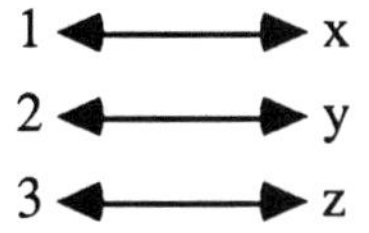

   1 ⟷ x
   2 ⟷ y
   3 ⟷ z

   (b) There are six different one-to-one correspodences between A and B.

7. Given: $A = \{1, 3, 5, 7, \ldots\}$ and $B = \{2, 4, 6, 8, \ldots\}$. Which of the following are true?

   (a) $A = B$
   (b) A is equivalent to B
   (c) A is equivalent to a proper subset of B

   (a) False. Sets A and B do not have the same elements.
   (b) True. A one-to-one Correspondence is shown below.

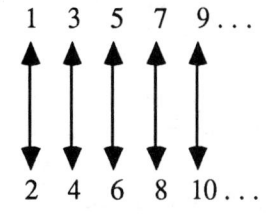

   1   3   5   7   9 ...
   ↕   ↕   ↕   ↕   ↕
   2   4   6   8   10 ...

(c) True. Consider the one-to-one correspondence between A and $C \subset B$.

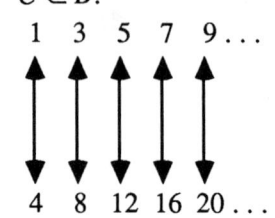

$$\begin{array}{ccccc} 1 & 3 & 5 & 7 & 9 \ldots \\ \updownarrow & \updownarrow & \updownarrow & \updownarrow & \updownarrow \\ 4 & 8 & 12 & 16 & 20 \ldots \end{array}$$

8. Define, if possible, sets R and S so that

   (a) $n(R) + n(S) > n(R \cup S)$
   (b) $n(R) + n(S) = n(R \cup S)$
   (c) $n(R) + n(S) < n(R \cup S)$

(a) $R = \{1, 2, 3, 4, 5\}$ and $S = \{4, 5, 6\}$

$n(R \cup S) = 6$
$n(R) = 5$
$n(S) = 3$

Thus, $n(R) + n(S) > n(R \cup S)$

(b) $R = \{1, 2, 3, 4, 5\}$ and $S = \{8, 9, 10\}$
$n(R \cup S) = 8$
$n(R) = 5$
$n(S) = 3$
Thus, $n(R) + n(S) = n(R \cup S)$

(c) not possible

9. In a student survey, it was found that 16 students liked history, 19 liked English, 18 like mathematics, 8 liked mathematics and English, 5 liked history and English, 7 liked history and mathematics, and 3 liked all three subjects. Every student liked at least one of the subjects. Draw a Venn diagram and use it to help answer the following questions.

   (a) How many students were in the survey?
   (b) How many students liked only mathematics?
   (c) How many students liked English and mathematics, but not history?

(a) 36
(b) 6
(c) 5

The Venn diagram is shown below.

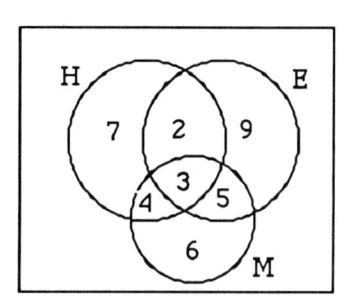

10. Given the following figure, find each of the following. The numbers represent the cardinal number of a region.

    (a) $n(R)$
    (b) $n(R \cup S)$
    (c) $n(R \cap S \cap T)$

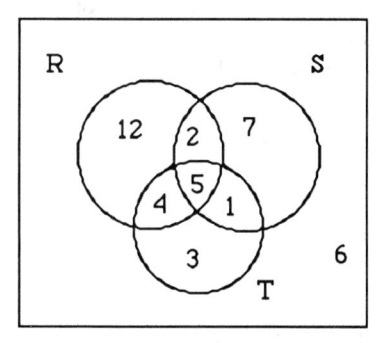

    (a) 23. Sum of the numbers in the circle R.
    (b) 31. Sum of the numbers in the circles R and S.
    (c) 5. Where the three circles intersect.

11. In a class of 45 students, 30 are men, 20 are juniors, and 7 are neither men nor juniors. How many junior men are in the class?

12. Consider the Venn diagram.

Men   **Juniors**

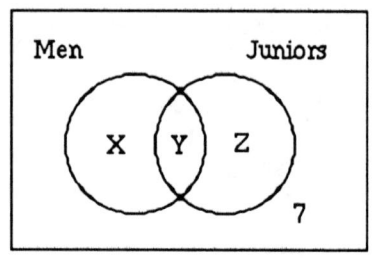

(Venn diagram with regions labeled x, y, z; 7 outside)

The letters x, y, and z represent the cardinal number of a given region. The following relationships can be found.

(1) $x + y + z = 45 - 7 = 38$
(2) $x + y = 30$
(3) $y + z = 20$

If we substitute 30 for $x + y$ in equation (1), we have
$30 + z = 38$ and we find $z = 8$.
We now substitute 8 for z in equation (3) and we have
$y + 8 = 20$ and we find $y = 12$.
Finally, we substitute 12 for y in equation (2) and we have $x + 12 = 30$ and we find $x = 18$.

12. If $n(A) = 7$ and $n(B) = 4$

    (a) What is the greatest number of elements that could be in $A \cup B$?
    (b) What is the greatest number of elements that could be in $A \cap B$?

(a) 11 $A \cap B = \phi$
(b) 4 $B \subset A$

13. Which of the following make use of cardinal numbers?

    (a) 4th of July
    (b) 16 ounces in a pound
    (c) 2nd semester
    (d) 7 swans a swimming
    (e) 47 home runs

Cardinal numbers are used in (b), (d), and (e)

14. Blood types are determined by the presence or absence of three types of antigens, A, B, and Rh. The various blood types are summarized in the table.

Antigen Present	Blood Type
A	A
B	B
Both A and B	AB
Neither A nor B	O

(a) In one sample of 100 people, 35 had type A antigen, 12 had type B antigen, and 6 had AB. How many were type O?

(b) In a second sample of 100 people, 11 were type A, 6 type B, and 87 type O. How many were type AB?

(a) 59 See the Venn diagram.

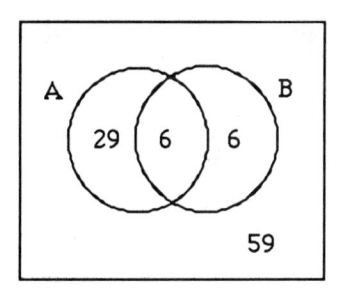

(b) 4

15. Every dork is a dweeb. Half of all zaps are dweebs. Half of all dweebs are dorks. There are 20 zaps and 12 dorks. How many dweebs are neither dorks nor zaps?

2. See the Venn diagram.

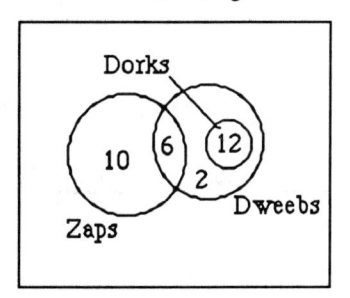

Section 4 Whole Number Addition, Subtraction, and Order

Cover the right side of the page and work on the left, then check your work

1. Use the associative property to help mentally calculate the following.

    (a) 37 + 98
    (b) 76 + 87

    (a) 37 + 98 = (35 + 2) + 98
    = 35 + (2 + 98)
    = 35 + 100
    = 135

    (b) 76 + 87 = 76 + (4 + 83)
    = (76 + 4) + 83
    = 80 + 83
    = 163

2. Without performing the computation, decide which sum below is larger.

    ```
       3472           6162
       5186           4356
    +  4213        +  2343
    ```

    ```
       3472                          6162
       5186  ←Larger sum             4356
    +  4213                       +  2343
    ```

    Note that the two problems have identical sums in the thousands colums (12), in the hundreds columns (7), and in the ones columns (11). The problem on the left, however, has a sum of 16 in the tens column compared to a sum of 15 in the tens column for the problem on the right.

3. Name the property of addition applied in each example below.

    (a) $2 + a = a + 2$
    (b) $b + 0 = 0 + b$
    (c) $(4 + 5) + 6 = (5 + 4) + 6$
    (d) $(2 + 3) + 7 = 2 + (3 + 7)$

    (a) commutative property
    (b) additive identity property

Section 4   Whole Number Addition, Subtraction and Order   49

(c) commutative property
(d) associative property

4. Show each of the following on a number line.

   (a) (3 + 5) + 6 = 3 + (5 + 6)
   (b) 9 − 6 = 3

(a)

```
                              (3 + 5) + 6 = 14
                     3 + 5 = 8              6
              3            5
         ├──┬──┬──┬──┬──┬──┬──┬──┬──┬──┬──┬──┬──┬──┤
         0  1  2  3  4  5  6  7  8  9 10 11 12 13 14
```

(b)

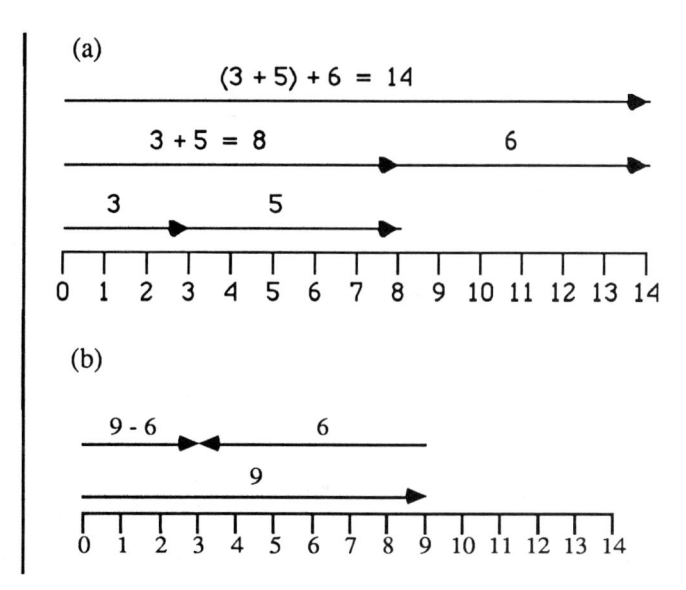

```
              9 - 6           6
                    9
         ├──┬──┬──┬──┬──┬──┬──┬──┬──┬──┬──┬──┬──┬──┤
         0  1  2  3  4  5  6  7  8  9 10 11 12 13 14
```

5. Use the definition of less than to classify the following as true or false.

   (a) 4 < 6
   (b) 3 + 7 > 12

   (a) True. Since 4 + 2 = 6 we can say that 4 < 6
   (b) False

6. For each of the following write two other related computation facts.

   (a) 8 + 9 = 17
   (b) 12 − 4 = 8

   (a) 17 − 9 = 8
       17 − 8 = 9
       9 + 8 = 17

   (b) 12 − 8 = 4
       8 + 4 = 12
       4 + 8 = 12

7. Determine whether each set is closed under addition.
   (a) $A = \{0, 4, 8, 12, \ldots\}$
   (b) $B = \{0, 2\}$
   (c) $C = \{x \mid x \text{ is a whole number less than } 10\}$

   (a) Closed
   (b) Not closed. $2 + 2 = 4$ and $4 \notin B$.
   (c) Not closed. $7 + 8 = 15$ and $15 \notin C$

8. Find the solution set for each of the following.
   (a) $a + 5 = 9$
   (b) $4 + a = 7$
   (c) $9 - 3 = a$
   (d) $a - 4 = 9$

   (a) $a = 4$
   (b) $a = 3$ If $4 + a = 7$, then $a = 7 - 4$
   (c) $a = 6$
   (d) $a = 13$

9. Find the solution set for each of the following.
   (a) $a > 6$
   (b) $a - 3 > 5$
   (c) $8 + a < 14$
   (d) $a - 0 < 4$

   (a) $\{7, 8, 9, \ldots\}$
   (b) $\{9, 10, 11, \ldots\}$
   (c) $a < 6$
   (d) $a < 0$

10. Illustrate $9 - 3 = 6$ using each of the following models.
    (a) Take-away model
    (b) Comparison model

    Answer to 10:
    (a)   (b)

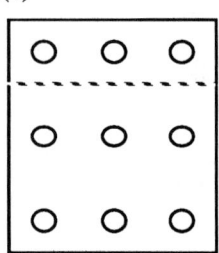

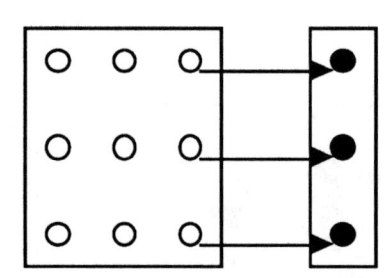

11. For each problem below, make an equation and then solve the equation.

    (a) Mary needs $4 more in order to purchase a $12 shirt. How much does she have now?
    (b) Bobby has 6 fewer baseball cards than Mark. If they combine their cards they have 34. How may cards does Mark have?

    (a) Let x represent the amount of money Mary has now.

    $$x + 4 = 12$$
    $$x = 8$$

    (b) Let x represent the number of cards Mark has.

    $$x + (x - 6) = 34$$
    $$2x - 6 = 34$$
    $$2x = 40$$
    $$x = 20$$

12. What property (or properties) justify the following computation.

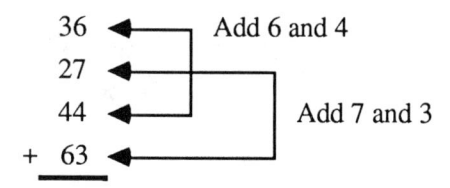

    ```
      36  ←┐  Add 6 and 4
      27  ←┤
      44  ←┤  Add 7 and 3
    + 63  ←┘
    ```

    The commutative and associative properties.

13. Suppose a set contains the number 2. What other numbers must it contain to be closed under addtion?

    The set must contain $2 + 2$, $4 + 2$, $6 + 2$, etc. Then the set must be $\{2, 4, 6, 8, \ldots\}$

14. A magic square is an arrangement of numbers in which the sum of every row, column, and diagonal is the same. Complete the following to make it a magic square. Use the numbers 10 through 18.

17		
		16
	18	

17	10	15
12	14	16
13	18	11

15. Let a, b, and c be whole numbers. For what values of a, b, and c will the following be true?

$b - c =$ a whole number and
$a - (b - c) =$ a whole number

We must have the following condition satisfied.

$b \geq c$ and $a \geq b - c$

Section 5 Whole Number Multiplication and Division

Cover the right side of the page and work on the left, then check your work

1. Use the distributive property to help perform the following computations mentally.

   (a) 39×8
   (b) $(374 \times 678{,}543) + (626 \times 678{,}543)$

   (a) $39 \times 8 = (40 - 1) \times 8$
   $= 40 \times 8 - 1 \times 8$
   $= 320 - 8$
   $= 312$

   (b) $(374 \times 678{,}543) + (626 \times 678{,}543) =$
   $(374 + 626) \times 678{,}543$
   $= (1000) \times 678{,}543$
   $= 678{,}543{,}000$

2. Estimate each of the following.

   (a) $16 (97 + 96)$
   (b) 26×59

   (a) $16 (97 + 96) \approx 16 \times 200 = 3200$
   (b) $26 \times 59 \approx 25 \times 60 = 25 \times 4 \times 15$
   $= 100 \times 15$
   $= 1500$

3. Name the property of multiplication applied in each example below.

   (a) $7 \times 8 = 8 \times 7$
   (b) $4 (3 + 7) = (3 + 7) 4$
   (c) $(9 \times 4) \times 6 = 9 \times (4 \times 6)$
   (d) $(x + y) 0 = 0$
   (e) $(a \times b) \times c = a \times (b \times c)$

   (a) commutative
   (b) commutative
   (c) associative
   (d) multiplicative property of zero
   (e) associative

4. Illustrate each of the following on a number line.

   (a) $4 \times 5 = 20$
   (b) $15 \div 5 = 3$

CHAPTER 2 Tools for Problem Solving: Sets and Numbers

(c) $3(2+4) = (3 \times 2) + (3 \times 4)$

Answer:
(a)

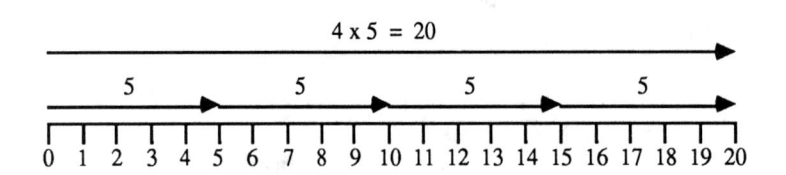

```
                            4 x 5 = 20
            5          5          5          5
   ┬─┬─┬─┬─┬─┬─┬─┬─┬─┬─┬─┬─┬─┬─┬─┬─┬─┬─┬─┬
   0 1 2 3 4 5 6 7 8 9 10 11 12 13 14 15 16 17 18 19 20
```

(b)

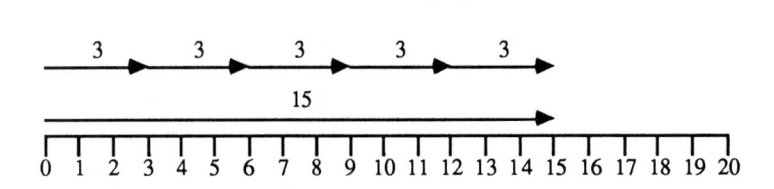

```
      3    3    3    3    3
              15
   0 1 2 3 4 5 6 7 8 9 10 11 12 13 14 15 16 17 18 19 20
```

(c)

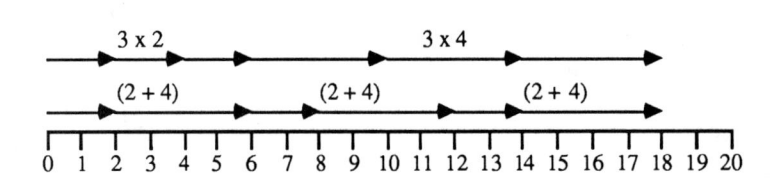

```
      3 x 2              3 x 4
      (2 + 4)    (2 + 4)    (2 + 4)
   0 1 2 3 4 5 6 7 8 9 10 11 12 13 14 15 16 17 18 19 20
```

5. For each of the following, write a related computation fact.

   (a) $9 \times 7 = 63$
   (b) $56 \div 8 = 7$
   (c) $a \times b = c$

   (a) $63 \div 7 = 9$ or $63 \div 9 = 7$
   (b) $7 \times 8 = 56$ or $8 \times 7 = 56$
   (c) $b \times a = c$ or $c \div a = b$ or $c \div b = a$

6. Compute each of the following.

   (a) $5 \times 3 - 6 \div 3$
   (b) $8(3 + 7 - 4)$

   (a) $5 \times 3 - 6 \div 3 = 15 - 2 = 13$
   (b) $8(3 + 7 - 4) = 8(10 - 4) = 8(6) = 48$

7. Find the solution set for each of the following.

Section 5   Whole Number Multiplication and Division   55

(a) $4x + 7 = 19$
(b) $3a - 4 = 17$
(c) $s \div 5 = 7$

> (a) $4x + 7 = 19$
> $4x = 19 - 7$
> (definition of subtraction)
> $4x = 12$
> $x = 12 \div 3$
> (definition of division)
> $x = 4$
>
> (b) $a = 7$
> (c) $s = 35$

8. Which of the following sets are closed under multiplication?

   (a) $\{0, 3, 6, 9, \ldots\}$
   (b) $\{x \mid x$ is a whole number less than 20$\}$
   (c) $\{0, 2\}$

> (a) closed (multiples of 3)
> (b) not closed
> $9 \times 7 = 63 \notin \{x \mid x$ is a whole number less than 20$\}$
> (c) not closed
> $2 \times 2 = 4 \notin \{0, 2\}$

9. Rewrite each of the following products as a sum.

   (a) $5(70 + 19)$
   (b) $4(2x + a)$
   (c) $7(30 + 45 + 2x)$

> (a) $5(70 + 19) =$
> $(5 \times 70) + (5 \times 19) = 350 + 95$
> (b) $4(2x + a) =$
> $(4 \times 2x) + (4 \times a) = 8x + 4a$
> (c) $7(30 + 45 + 2x) =$
> $(7 \times 30) + (7 \times 45) + (7 \times 2x) =$
> $210 + 315 + 14x$

10. Insert parentheses to make the following computation equal to the indicated value.

    $13 + 7 \times 12 - 6$
    (a) 234
    (b) 120

(c) 91

(a) $(13 + 7) \times 12 - 6$
(b) $(13 + 7) \times (12 - 6)$
(c) $13 + (7 \times 12) - 6$

11. For each of the following problems, write an equation that models the problem and then find the solution.

   (a) Becky can type 72 words per minute. How many words can she type in a five minute typing test?
   (b) Curtis played four rounds of golf. His scores were 90, 84, 87, and 83. What must he score on a fifth round to have an average for the five rounds of 85?

(a) $n = 5 \times 72$
$n = 360$
(b) $(90 + 84 + 87 + 83 + s) \div 5 = 85$
$(344 + s) \div 5 = 85$
$344 + s = 5 \times 85$
$344 + s = 425$
$s = 425 - 344 = 81$

12. Rewrite each of the following sums as a product.

   (a) $4x + 7x$
   (b) $3x^2 - 9x$
   (c) $4(a + 1) + 8(a + 1)$

(a) $4x + 7x = (4 + 7)x$
(b) $3x^2 - 9x = 3x(x - 3)$
(c) $4(a + 1) + 8(a + 1) = (4 + 8)(a + 1)$

13. A new operation * is defined as follows.

   $x * y = 2x + y$

   (a) Find $x * y$ for $x = 3$ and $y = 7$
   (b) Find $x * y$ for $x = 7$ and $y = 3$
   (c) Is * commutative?

(a) $3 * 7 = 2(3) + 7 = 14$
(b) $7 * 3 = 2(7) + 3 = 17$
(c) no. From (a) and (b) we see that $3 * 7 \neq 7 * 3$

Section 5   Whole Number Multiplication and Division   **57**

14. Name a set that is closed under multiplication but not closed under addition.

> One example would be the following set.
>
> $A = \{1, 3, 5, 7, \ldots\}$
> $3 + 7 = 10 \notin A$

15. Using the digits 1 through 9 (each once), fill in the boxes to make all the equations true.

$$\Box - \Box = \Box$$
$$\Box \div \Box = \Box$$
$$\Box + \Box = \Box$$

with the right-column of the three results linked by × and =.

> $9 - 5 = 4$
> $6 \div 3 = 2$
> $1 + 7 = 8$
>
> (with $4 \times 2 = 8$)

Chapter 3
Numeration Systems

Section 1 History of Numeration Systems

Cover the right side of the page and work on the left, then check your work

1. Ali delivers ⊃⊃ ||| newspapers each day.
   Estimate the number of papers he delivers in a month.

   > Ali delivers **approximately** 6000 newspapers per month.
   > The Egyptian numeral given would be 203 using Hindu-Arabic numerals. Since there are **about** 30 days in a month, we can estimate by thinking $200 \times 30 = 6000$.

2. About how many papers will Ali deliver in a year?

   > Ali delivers approximately 72,000 papers per year.

3. Write each of the following in expanded notation.

   (a) 3749
   (b) 123456
   (c) 40105

   > (a) $3(1000) + 7(100) + 4(10) + 9$
   > (b) $1(100000) + 2(10000) + 3(1000) + 4(100) + 5(10) + 6$
   > (c) $4(10000) + 1(100) + 5$

4. Write each of the following in standard notation.

   (a) $4(1000) + 7(100) + 6(10) + 8$
   (b) $7(10000) + 9(1000) + 7(10)$
   (c) $3(1000) + 4$

   > (a) 4768
   > (b) 79070
   > (c) 3001

5. Write the Hindu-Arabic numeral 5674 as an

   (a) Egyptian numeral
   (b) Babylonian numeral
   (c) Mayan numeral

Section 1   History of Numeration Systems   59

(a) 𝄞𝄞𝄞𝄞𝄞 999999 ∩∩∩∩∩∩ ||||

(b) ▼ ⟨⟨⟨⟨▼▼▼▼ ⟨⟨⟨⟨▼▼▼▼

(c)
=
•••
=
••••
=

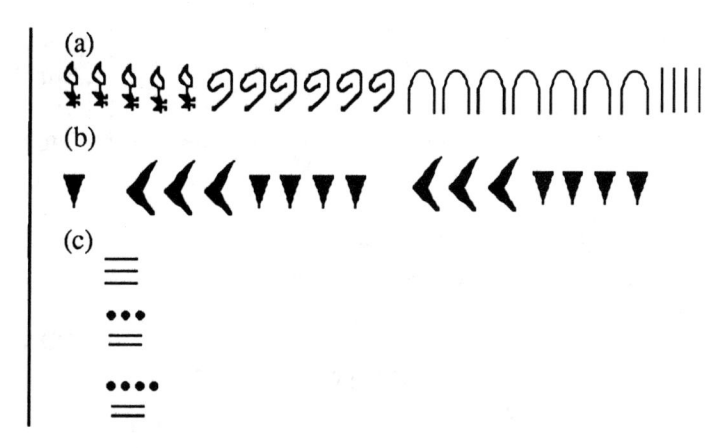

6. Rewrite each of the following using Hindu-Arabic numerals.

   (a) (𓀀 99 ∩∩∩ ||||

   (b) ⟨⟨▼ ▼▼▼

   (c) ••
       •••

(a) 11234
(b) 1263   We have $21 \cdot 60 + 3 = 1263$
(c) 828   We have $2(18)(20) = 720$
$5(20) = 100$
$8 = 8$
828

7. Arrange the following from largest to smallest.

   DCX   XI   MCXIV   CDXXIX

XI, MCXIV, DCX, CDXXIX
(largest to smallest)

8. What is the largest four-digit number that can be represented in base ten using the digits

   0, 2, 4, 9?

9429   Largest digits are assigned to the largest place values.

9. What is the smallest four-digit number that can be represented in base ten using the digits

   0, 2, 4, 9?

60 CHAPTER 3 Numeration Systems

10. What is the smallest number that can be represented in Roman notation using only the symbols C, D, X, V, I?

> 2049
> Smallest digits are assigned to the smallest place values. Cannot assign 0 to the thousand's place because you would not have a four-digit number.
>
> CDXIV

11. Write the numeral for the number that precedes each of the following.

    (a) MCMXC
    (b) ▼▼ ❮❮❮❮
    (c) 𓆓𓆓

> (a) MCMLXXXIX
> (b) ▼▼ ❮❮ ▼▼▼▼▼▼▼▼▼
> (c) 𓆓∩∩∩∩∩∩∩∩∩||||||||

12. Find the sum, the difference, and the product of XCVI and XXIV.

> sum - CXX
> difference - LXXII
> product - MMCCCIIII

13. Curtis used 1173 digits to number the pages of a book. How many pages are in the book.

> There are 427 pages.
>
> Consider the following table.
>
> | Pages | Number of Digits |
> |---|---|
> | 1 - 9 | 9 |
> | 10 - 99 | 180 |
> | 100 - ? | # pages × 3 digits |
>
> We could continue the table until we have used a total of 1173 digits. We could also develop an equation to represent the relationship between

Section 1 **History of Numeration Systems** **61**

the number of pages and number of digits. The equation would be:

$3n + 189 = 1173$ 3 times the number of pages (n) over 99 plus the 189 digits used for pages 1 through 99 equals 1173 (total number of digits used)

$3n = 984$
$n = 328$ pages over 99
Total number of pages = 99 + 328 = 427 pages

14. A number is equal to three times the product of its two digits. What is the number?

The number is 24.

We could solve this problem by trial and error - compare various two-digit numbers with the product of their digits. We can be a little more systematic, however, by doing the following.
Let a = the tens digit of the number
    b = the ones digit of the number
Then, $10a + b$ = the value of the number and we have
$10a + b = 3(ab)$ (three times the product of the digits)

We can set up a table such as the following and use it to help us make our trial and error more efficient.

a	b	$10a + b$	ab
3	2	32	6
3	5	35	15
3	4	34	12
2	6	26	12
2	4	24	8

3 x 6 < 32
3 x 15 > 35
3 x 12 > 34
3 x 12 > 26
3 x 8 = 24

15. The sum of the digits of a two digit number is 10. If the digits are reversed and the original number is subtracted from the number formed by this reversal, the difference is 18. Find the original number.

The original number is 46.
We know the following:
$a + b = 10$ where a is the tens digit and b is the ones digit of the original number

$10b + a - (10a + b) = 18$ original number subtracted from the reversed number. Then, we have

$10b + a - 10a - b = 18$
$9b - 9a = 18$
$b - a = 2$ or
$b = a + 2$

If we substitute $a + 2$ for b in the expression $a + b = 10$ we have

$a + (a + 2) = 10$
$2a + 2 = 10$
$2a = 8$
$a = 4$

Solving for b with $a = 4$, we find $b = 6$.

Section 2 Using Exponentials in Addition and Subtraction Algorithms

Cover the right side of the page and work on the left, then check your work

1. Find an estimate of each of the following by rounding to the nearest hundred.

   (a)   378
       + 921

   (b)   634
       − 289

   | (a)   378
   | + 921
   |
   | rounded to nearest 100
   | 400
   | + 900
   | 1300 estimate
   |
   | (b)   634
   | − 289
   |
   | rounded to nearest 100
   | 600
   | − 300
   | 300 estimate

2. Perform each of the following mentally.

   (a)   237
       − 79

   (b)   167
       + 438

   (c)   228
       + 126

   | (a) 237 − 80 + 1 = 157 + 1 = 158
   | (b) 167 + 3 = 170
   | 438 − 3 = 435
   |
   | 100 + 400 = 500   Add the hundreds
   | 70 + 30 = 100   Add the tens
   | 0 +   5 =   5   Add the ones
   | 605

64 CHAPTER 3 Numeration Systems

$$\begin{array}{rl} \text{(c)} & 228 + 2 = 230 \\ & + 126 - 2 = \underline{124} \end{array}$$

$$\begin{array}{ll} 200 + 100 = 300 & \text{Add the hundreds} \\ 30 + 20 = 50 & \text{Add the tens} \\ 0 + 4 = \underline{4} & \text{Add the ones} \\ 354 \end{array}$$

3. Simplify each of the following.

   (a) $3^4 \times 3^8$
   (b) $(2 \times 5)^5$
   (c) $[(7 \times 6)^{12}]^0$

   (a) 3^{12}
   (b) $2^5 \times 5^5 = 10^5$
   (c) 1 (<u>any</u> number with an exponent of 0 is equal to 1)

4. Write the following numerals in "expanded form," using powers of 10.

   (a) 4506
   (b) 70003
   (c) 5,897,087

   (a) $4(10^3) + 5(10^2) + 6(10^0)$
   (b) $7(10^4) + 3(10^0)$
   (c) $5(10^6) + 8(10^5) + 9(10^4) + 7(10^3) + 8(10^1) + 7(10^0)$

5. Write the following in standard form.

   (a) $5(10)^4 + 3(10)^3 + 8(10)^2 + 2(10)^1 + 9(10)^0$
   (b) $6(10)^3 + 3(10)^0$
   (c) $9(10)^7 + 7(10)^1 + 3(10)^0$

   (a) 53829
   (b) 6003
   (c) 90000073

6. Simplify the following by obtaining a common base.

   (a) $2^3 \times 64^7$
   (b) $4^9 \div 2^3$
   (c) $(5 \times 25)^2$

Section 2   Using Exponentials in Addition and Subtraction Algorithms   **65**

(a) $2^3 \times 64^7 = 2^3 \times (2^6)^7 = 2^3 \times 2^{6 \times 7}$
$= 2^3 \times 2^{42} = 2^{3+42} = 2^{45}$
(b) $4^9 \div 2^3 = (2^2)^9 \div 2^3 = 2^{18} \div 2^3 = 2^{18-3} = 2^{15}$
(c) $(5 \times 25)^2 = (5 \times 5^2)^2 = (5^{1+2})^2 = (5^3)^2 = 5^{2 \times 3} = 5^6$

7. Find a value for n to make each of the following true.

(a) $3^n = 9^2$
(b) $n^4 = 8n$
(c) $(4^n)^3 = 2^{12}$

(a) $n = 4$ $3^n = 9^2$
$3^n = (3^2)^2$
$3^n = 3^4$
$n = 4$
(b) $n = 2$ $n^4 = 8n$
$\dfrac{n^4}{n} = \dfrac{8n}{n}$
$n^3 = 8$
$n^3 = 2^3$
$n = 2$
(c) $n = 2$ $(4^n)^3 = 2^{12}$
$[((2^2)^n)^3] = 2^{12}$
$(2^{2n})^3 = 2^{12}$
$2^{6n} = 2^{12}$
$6n = 12$
$n = 2$

8. Find the missing digits in each of the problems below.

(a)   - - -
   + 9 2 8
     1 4 2 7

(b)   7 4 3
   - - - -
     2 6 7

(a) 479

$$\begin{array}{r} 1 \\ 9 \to \\ +\,928 \\ \hline 1427 \end{array} \quad \begin{array}{r} 1 \\ 79 \to \\ +\,928 \\ \hline 1427 \end{array} \quad \begin{array}{r} 1\,1 \\ 479 \\ +\,928 \\ \hline 1427 \end{array}$$

(b) 267

9. Do each of the following computations using expanded notation and a place value chart.

(a) 127
   + 185

(b) 235
   − 158

(a) Expanded Form:

$$\begin{array}{r} 100 + 20 + 7 \\ 100 + 80 + 5 \\ \hline 12 \end{array} \quad \begin{array}{r} 127 \\ +\,185 \\ \hline 100 \\ 200 \\ \hline 327 \end{array}$$

Place Value Chart

100	10	1
1	2	7
1	8	5
	10	12
	1	2
	11	
1	1	
3	1	2

(b) Expanded Form

$$\begin{array}{l} 200 + 30 + 5 \to \\ 100 + 50 + 8 \end{array} \quad \begin{array}{l} 200 + 20 + 15 \to \\ 100 + 50 + 8 \end{array} \quad \begin{array}{l} 100 + 120 + 15 \\ 100 + 50 + 8 \\ + 70 + 7 \end{array}$$

Section 2 Using Exponentials in Addition and Subtraction Algorithms 6 7

Place Value Chart

100	10	1
2	3	5
1	5	8
1	12	15
1	5	8
	7	7

10. Use the "equal additions method" to subtract.

(a) 854
   − 468

(b) 5003
   − 3728

(a) $854 \to 85\,{}^14 \to 8\,{}^15\,{}^14 \to 8\,{}^15\,{}^14$
− 468   47 8   5 7 8   5 7 8
          6   8 6   3 8 6

(b) $5003 \to 5\,{}^10\,{}^10\,{}^13$
− 3728   4 8 3 8
          1 2 7 5

11. Fill in the blanks for the following computation.

    589   500 + __ + 9
  + 345   300 + 40 + 5
               __ + __ + __

500 + 90 + 9
300 + 40 + 5
900 + 30 + 4

12. Multiply and write the answer as a decimal numeral.

(a) $10^3 [5\,(10^5) + 3\,(10^4) + 6\,(10^2) + 4\,(10^0)]$
(b) $10^2\,(3469)$

(a) 530604000
(b) 346900

CHAPTER 3 Numeration Systems

13. Solve for n.

    $(5 \times 3^2)^{2n-4} = 1$

    > $n = 2$
    > $2n - 4 = 0$ Only number with 0 exponent will give a result of 1.
    > $2n = 4$
    > $n = 2$

14. Place the digits 2, 3, 4, 5, 6, 8 in the boxes

    ```
      □ □ □
    - □ □ □
    ───────
    ```

    to obtain

    (a) the greatest difference
    (b) the least difference

    > (a) 865   Use largest and smallest
    >   − 234   digits for hundreds places
    >   631
    >
    > (b) 623
    >   − 584
    >   39
    >
    > Hundreds digits must differ by 1
    >   425
    >   − 386
    >   39

15. Using only addition and subtraction, arrange the digits 1, 2, 3, 4, 5, 6, 7, 8 in order to obtain a result of 100.
    (Example: 12 + 34 + 56 − 7 + 8 = 103)

    > 12 + 3 − 4 + 5 + 6 + 78 = 100 is one example

Section 3 Multiplication and Division Algorithms

Cover the right side of the page and work on the left, then check your work

1. Estimate each of the following.

   (a) 4362×2147
   (b) 879×4168
   (c) $79{,}367 \div 18$
   (d) $21{,}462 \div 737$

   (a) $4362 \times 2147 \approx 4000 \times 2000 = 8{,}000{,}000$ or
   $4362 \times 2147 \approx (4 \cdot 10^3)(2 \cdot 10^3) = 8 \cdot 10^6 = 8{,}000{,}000$
   (b) $879 \times 4168 \approx (9 \cdot 10^2)(4 \cdot 10^3) = 36 \cdot 10^5 = 3{,}600{,}000$
   (c) $79{,}367 \div 18 \approx (8 \cdot 10^4) \div (2 \cdot 10^1) = 4 \cdot 10^3 = 4000$
   (d) $21462 \div 737 \approx (21 \cdot 10^3) \div (7 \cdot 10^2) = 3 \cdot 10^1 = 30$

2. Perform each of the following mentally.

   (a) 342×5
   (b) 6×83
   (c) $312 \div 4$

   (a) $\begin{array}{r} 342 \\ \times\, 5 \end{array}$ $\begin{array}{r} 171 \\ \underline{10} \\ 1710 \end{array}$ Take half of 342
   Double 5

   (b) $6 \times 83 = (6 \times 80) + (6 \times 3) = 480 + 18 = 498$

   (c) $4\overline{)312}$ think $4\overline{)320-8}$ $80 - 2 = 78$

3. Consider the following.

   $24 \times 16 = (20 + 4)(20 - 4) =$
   $20^2 - 4^2 = 400 - 16 = 384$

   $37 \times 23 = (30 + 7)(30 - 7) =$
   $30^2 - 7^2 = 900 - 49 = 851$

Use this technique to do the following mentally.

(a) 53×47
(b) 72×88

$\quad$ (a) $53 \times 47 = (50 + 3)(50 - 3) =$
$\quad 50^2 - 3^2 = 2500 - 9 = 2491$
$\quad$ (b) $72 \times 88 = 80^2 - 8^2 = 6400 - 64$
$\quad\quad\quad = 6336$

4. Do each of the following computations using the horizontal format.

(a) 7×48
(b) 16×24

$\quad$ (a) $7 \times 48 = 7(40 + 8) =$
$\quad\quad 7 \cdot 40 + 7 \cdot 8 = 280 + 56 = 336$
$\quad$ (b) $16 \times 24 = 16(20 + 4) =$
$\quad\quad 16 \cdot 20 + 16 \cdot 4$
$\quad\quad = (10 + 6)20 + (10 + 6)4$
$\quad\quad = 10 \cdot 20 + 6 \cdot 20 + 10 \cdot 4 + 6 \cdot 4$
$\quad\quad = 200 + 120 + 40 + 24$
$\quad\quad = 384$

5. Do each of the following computations using the partial products format.

(a) 39×57
(b) 97×46

```
(a)    39        (b)    97
     × 57             × 46
       63              42
      210             540
      450             280
     1500            3600
     2223            4462
```

6. Do each of the following computations using the traditional algorithm.

(a) 83×48
(b) 256×37

```
(a)    83        (b)    256
     × 48             ×  37
      664             1792
      332              768
     3984             9472
```

7. Do each of the following computations using the scaffold format.

   (a) 187 ÷ 7
   (b) 8364 ÷ 32

```
(a)                        (b)
7 ⟌187                    32 ⟌8364
  140   20 (7's)             6400   200(32's)
   47                        1964
   42    6 (7's)             1920    60(32's)
    5   26                     44
                               32     1 (32)
                               12   261
```

8. Do each of the following computations using the intermediate algorithm.

   (a) 234 ÷ 6
   (b) 17,347 ÷ 47

```
(a)     9           (b)      9
       30                   60
   6 ⟌234                  300
      180          47 ⟌17347
       54                14100
       54                 3247
                          2820
                           447
                           423
                            24
```

9. Do each of the following computations using the traditional algorithm.

   (a) 764 ÷ 9
   (b) 23,471 ÷ 67

```
(a)   84          (b)    350
   9 ⟌764            67 ⟌23471
      72                  201
      44                  337
      36                  335
                           21
                            0
                           21
```

10. Fill in the missing numerals.

(a)
```
        3 9
      × 5 7
      2 □ □
    □ □ 5
    □ □ 2 □
```

(b)
```
              − 1
        23 ) 7 2 8
             □ □
             3 □
             □ 3
             1 □
```

(a)
```
        3 9
      × 5 7
      2 [7] 3
    [1][9] 5
  [2][2] 2 [3]
```

(b)
```
           [3] 1
      23 ) 7 2 8
           6 [9]
           3 [8]
           2 [3]
           1 [5]
```

11. Perform the following computations using the Russian Peasant Algorithm.

(a) 19×46
(b) 321×62

Halving	Doubling	Halving	Doubling
19	46	321	62
9	92	160	124
~~4~~	~~184~~	80	248
~~2~~	~~368~~	40	496
1	736	20	992
	874	10	1984
		5	3968
		2	7936
		1	15872
			19902

12. Find the q and r of the division algorithm for each of the following.

(a) $74 \div 12$
(b) $128 \div 15$
(c) $38 \div 42$

(a) $q = 6 \quad r = 2$
(b) $q = 8 \quad r = 8$
(c) $q = 0 \quad r = 42$

Section 3 Multiplication and Division Algorithms   73

13. In the division calculation shown below, box the digits that represent the difference
    50863 − (7000 × 7).

$$
\begin{array}{r}
7266 \\
7\,\overline{)50863} \\
\underline{49} \\
18 \\
\underline{14} \\
46 \\
\underline{42} \\
43 \\
\underline{42} \\
1
\end{array}
$$

$$
\begin{array}{r}
7266 \\
7\,\overline{)50863} \\
\underline{49} \\
\boxed{1}\,8 \\
\underline{14} \quad 50863 - (7000 \times 7) = 1863 \\
4\,\boxed{6} \\
\underline{42} \\
4\,\boxed{3} \\
\underline{42} \\
1
\end{array}
$$

14. If you could spend $1 per second, how long would it take you to spend

    (a) $1,000,000
    (b) $1,000,000,000

    (a) 1,000,000 seconds or approximately 11.6 days.
    (b) 100,000,000 seconds = 16,666.67 minutes = 277.77778 hours = 11.574074 days.

15. Find the quotient and remainder of 3627 ÷ 49 using a calculator.

    3627 ÷ 49 = 74.020408 (your calculator may give fewer or more decimal places). From the Division Algorithm, we know that 3627 = 49 · 74 + r (r is the remainder) We can rewrite this as
    r = 3627 − 49 · 74 or
    r = 3627 − 3626
    r = 1

Section 4 Patterns for Nondecimal Bases

Cover the right side of the page and work on the left, then check your work

1. Estimate each of the following by rounding.

   (a) 287_{ten} = _____ seven
   (b) 412_{five} = _____ ten

   (a) $287_{ten} \approx 600_{seven}$

   287 will round to 300. We then must consider the place values 1, 7, 49, and so on in base seven. Rounding 49 to 50, we can estimate that 300 ÷ 50 = 6. We than will have zeros in the 7's and 1's places.

   (b) $412_{five} \approx 100_{ten}$

   412_{five} will round to 400_{five}. We than have 4 in the 25's place and this would be 100_{ten}.

2. Write the first twenty counting numbers using base five notation.

   1 2 3 4 10 11 12 13 14 20
   21 22 23 24 30 31 32 33 34 40
   Base Five

3. For each of the following, write the next three numerals.

   (a) 45_{six}
   (b) 1111_{two}
   (c) 333_{four}

   (a) 45_{six}, 50_{six}, 51_{six}, 52_{six}
   45_{six} = 4 groups of six and 5 ones. If we add another one, we than have six ones and we can exchange this for one group of six which will give us 5 groups of six and no ones.
   (b) 1111_{two}, 10000_{two}, 10001_{two}, 10010_{two}
   (c) 333_{four}, 1000_{four}, 1001_{four}, 1002_{four}

Section 4   Patterns for Nondecimal Bases   75

4. For each of the following, write the number that precedes the given number.

   (a) 210_{three}
   (b) 1000_{seven}
   (c) 430_{five}

   (a) 202_{three}
   $210_{three} = 2(3)^2 + 1(3)^1 + 0$.
   If we take one away from this (to get the preceding number), we then have $2(3)^2 + 0(3)^1 + 2$
   (b) 666_{seven}
   (c) 424_{five}

5. Write each of the following as a base ten numeral.

   (a) $3ET_{twelve}$
   (b) 432_{five}
   (c) 1221_{three}

   (a) $3ET_{twelve} = 574_{ten}$
   $3ET_{twelve} = 3(12)^2 + 11(12)^1 + 10$
   $= 3(144) + 132 + 10$
   $= 574.$
   (b) 117_{ten}
   (c) 52_{ten}

6. Do each of the following.

   (a) $378_{ten} = \underline{\quad}_{six}$
   (b) $1426_{ten} = \underline{\quad}_{twelve}$
   (c) $127_{ten} = \underline{\quad}_{two}$

   (a) $378_{ten} = 1430_{six}$
   The place values of base six expressed in base ten notation are 1, 6, 36, 216, 1296 and so on. Since 378 is less than 1296, we begin with deciding how many 216's are in 378.
   ```
   216| 378 |1
        216
    36| 162 |4
        144
     6|  18 |3
         18
   ```
   (b) $9TT_{twelve}$
   (c) 111111_{two}

7. Find the missing digit for each of the following.

   (a) $4 \_ 3_{\text{five}} = 113_{\text{ten}}$
   (b) $\_27_{\text{eight}} = 215_{\text{ten}}$
   (c) $323_{\text{four}} = 5 \_\_\_\__{\text{ten}}$

   (a) $4 \underline{\ 2\ } 3_{\text{five}} = 113_{\text{ten}}$

   $4\ \underline{n}\ 3_{\text{five}} = 4(5)^2 + n(5) + 3 = 113$
   $100 + 5n + 3 = 113$
   $103 + 5n = 113$
   $5n = 10$
   $n = 2$

   (b) $\underline{\ 3\ } 27_{\text{eight}} = 215_{\text{ten}}$
   (c) $323_{\text{four}} = 5\ \underline{\ 9\ }_{\text{ten}}$

8. Place the following in order from smallest to largest.

   (a) 27_{ten}, 56_{seven}, 24_{twelve}
   (b) 21_{three}, 110111_{two}, 15_{twelve}

   (a) 27_{ten}, 24_{twelve}, 56_{seven}
   $24_{\text{twelve}} = 2(12) + 4 = 24 + 4 = 28$
   $56_{\text{seven}} = 5(7) + 6 = 35 + 6 = 41$

   (b) 21_{three}, 15_{twelve}, 110111_{two}

9. Find the base in each of the following.

   (a) $53_{\text{ten}} = 104_b$
   (b) $392_{\text{ten}} = 327_b$

   (a) $53_{\text{ten}} = 104_{\text{seven}}$
   $104_b = 1(b)^2 + 0(b) + 4 = 53$
   $b^2 + 4 = 53$
   $b^2 = 49$
   $b = 7$

   (b) $392_{\text{ten}} = 327_{\text{eleven}}$
   Hint: The base b must be a number larger than 10.

10. Write each of the following.

    (a) smallest three-digit numeral in base nine.
    (b) smallest three-digit numeral in base three.
    (c) largest four-digit numeral in base seven.

    (a) 100_{nine}
    (b) 100_{three}
    (c) 6666_{seven}

Section 4 Patterns for Nondecimal Bases 77

11. Change each of the following to a base eleven numeral.

(a) 376
(b) 9321
(c) 1000

(a) 376 = 312$_{eleven}$
$$121 | \overline{376} \; |3$$
$$\underline{363}$$
$$11 | \overline{13} \; |1$$
$$\underline{11}$$
$$1 | \overline{2} \; |2$$
$$2$$

(b) 9321 = 7004$_{eleven}$
(c) 1000 = 82T$_{eleven}$
We need a symbol for ten and we cannot use 10 since this will represent eleven in base eleven.

12. Complete the following table.

Base	Place Values				
2	16	8	4	2	1
5			25		1
7		343			
9					
12				12	

Base	Place Values				
2	16	8	4	2	1
5	625	125	25	5	1
7	2401	343	49	7	1
9	6481	729	81	9	1
12	20736	1728	144	12	1

13. How many different symbols would be necessary for a base twenty-eight system?

28 symbols would be needed. We would need symbols for numerals zero through twenty-seven. The symbol for twenty-eight would be 10.

14. Find the smallest values for a and b so that $21_b = 25_a$

$a = 4$
$b = 6$
$21_b = 25_a$
This would mean that
$2(b) + 1 = 2(a) + 5$
$2b = 2a + 4$
$b = a + 2$
We then make the following table.

a	b
0	2
1	3
2	4
3	5
4	6

Why can't we use these?

15. What can you say about the ones digit of every even number (0, 2, 4, 6, ...) in

   (a) base ten
   (b) base four
   (c) base five

(a) The ones digits of even numbers will be 0, 2, 4, 6, or 8.
(b) The ones digits of even numbers will be 0 or 2.
(c) The ones digits of even numbers will be 0, 1, 2, 3, or 4.

Consider the following table.

Base Ten	Base Four	Base Five
0	0	0
2	2	2
4	10	4
6	12	11
8	20	13
10	22	20
12	30	22
14	32	24
16	100	31
18	102	33

Section 5 Computations in Different Bases

Cover the right side of the page and work on the left, then check your work

1. Estimate each of the following.

   (a) 435_{six}
        $+\ 243_{six}$

   (b) $T3_{twelve}$
        $\times\ 59_{twelve}$

   (c) $34_{five} \overline{)2412_{five}}$

   (a) 435_{six} Use rounding 400_{six}
       $+\ 243_{six}$ $+\ 200_{six}$
                                    1000_{six}

   (b) 500_{twelve}

   (c) $34_{five} \overline{)2412_{five}}$ Use rounding:
       $\phantom{34_{five}\ \ \ }80_{five}$
       $30_{five} \overline{)2400_{five}}$

2. Perform the following mentally.

   (a) 326_{eight}
        $-\ \ 67_{eight}$

   (b) 637_{nine}
        $+\ 243_{nine}$

   (c) 34_{five}
        $\times\ \ 4_{five}$

   (a) 326_{eight}
       $-\ \ 67_{eight}$

   $326_{eight} - 70_{eight} + 1_{eight} =$
   $236_{eight} + 1_{eight} = 237_{eight}$

   (b) 637_{nine} $637_{nine} + 2_{nine} = 640_{nine}$
       $+\ 243_{nine}$ $243_{nine} - 2_{nine} = 241_{nine}$
       $\phantom{+\ 243_{nine}}\ 681_{nine}$

80 CHAPTER 3 Numeration Systems

$$\begin{array}{r} \text{(c)} \quad 34_{six} \\ \times \; 3_{six} \\ \hline 150_{six} \end{array} \quad \begin{array}{l} 15_{six} \quad \text{Take half of } 34_{six} \\ \underline{10_{six}} \quad \text{Double } 3_{six} \end{array}$$

3. Make an addition table for base six.

+	0	1	2	3	4	5
0	0	1	2	3	4	5
1	1	2	3	4	5	10
2	2	3	4	5	10	11
3	3	4	5	10	11	12
4	4	5	10	11	12	13
5	5	10	11	12	13	14

4. Make a multiplication table for base six.

×	0	1	2	3	4	5
0	0	0	0	0	0	0
1	0	1	2	3	4	5
2	0	2	4	10	12	14
3	0	3	10	13	20	23
4	0	4	12	20	23	32
5	0	5	14	23	32	41

5. Perform the following additions.

    (a) 221_{three}
          $+ \; 122_{three}$

    (b) $4E78_{twelve}$
          $+ \; 6946_{twelve}$

$$\begin{array}{r} \text{(a)} \quad 221_{three} \\ + \; 122_{three} \\ 10 \\ 110 \\ \underline{1000} \\ 1120_{three} \end{array}$$

$$\begin{array}{r} \text{(b)} \quad 4E78_{twelve} \\ + \; \underline{6946_{twelve}} \\ E802_{twelve} \end{array}$$

6. Perform the following subtractions.

(a) 732_{eight}
    $- 426_{eight}$

(b) 4002_{five}
    $- 1233_{five}$

| (a) 732_{eight} $7\,2\ ^{1}2_{eight}$ |
| $- 426_{eight}$ $-\,4\,2\ \ 6_{eight}$ |
| $3\,0\ \ 4_{eight}$ |

(b) 4002_{five}
    $-\ 1233_{five}$
      2214_{five}

7. Perform the following multiplications.

   (a) 33_{four}
       $\times\ 3_{four}$

   (b) 47_{eight}
       $\times\ 35_{eight}$

(a) 33_{four}
    $\times\ 3_{four}$
      21
     210
     231_{four}

(b) 47_{eight}
    $\times\ 35_{eight}$
     303
    165
    2153_{eight}

8. (a) $35_{six} \overline{)4342_{six}}$
   Use the scaffold method.

   (b) $7_{twelve} \overline{)3TE_{twelve}}$

82 **CHAPTER 3** Numeration Systems

(a)
$$\begin{array}{r} 1 \\ 10 \\ 100 \\ 35_{six} \overline{)4342_{six}} \\ \underline{3500} \\ 442 \\ \underline{350} \\ 52 \\ \underline{35} \\ 13 \end{array}$$

(b)
$$\begin{array}{r} 68 \\ 7_{twelve} \overline{)3TE_{twelve}} \\ \underline{36} \\ 4E \\ \underline{48} \\ 3 \end{array}$$

9. What base is being used in each of the following?

(a) 53 − 26 = 24
(b) 432 + 213 = 1200
(c) $\quad$ 18
$\quad\times\;\underline{\;7\;}$
$\quad$ 111

(a) 53 − 26 = 24   base seven

$\quad$ 53   We note that we had to
− 26   regroup (borrow) in
$\quad$ 24   order to subtract in the ones
place. We borrowed one group of the
base, say b. Thus, we have:
$$b + 3 - 6 = 4$$
$$b - 3 = 4$$
$$b = 7$$

(b) 432 + 213 = 1200   base five
Note the 0 in the ones place of the answer (2 + 3 = 0 and we must have carried one group of five).

(c) $\quad$ 18   base eleven
$\quad\times\;\underline{\;7\;}$
$\quad$ 111

10. For what base is the following correct?

$$\begin{array}{r} 121 \\ 42\overline{\smash{)}5432} \\ \underline{42} \\ 123 \\ \underline{114} \\ 62 \\ \underline{42} \\ 20 \end{array}$$

Correct for base seven. Note the multiplication of $2 \times 42 = 114$

11. Express the fact that "five times six equals thirty" in each of the following bases.

(a) Base two
(b) Base five
(c) Base nine

(a) $101_{two} \times 110_{two} = 11110_{two}$
(b) $10_{five} \times 11_{five} = 110_{five}$
(c) $5_{nine} \times 6_{nine} = 33_{nine}$

12. Solve the following using the lattice method.

$$\begin{array}{r} 314_{six} \\ \times 45_{six} \end{array}$$

13. Mr. Braden has a pan balance that he wants to use to weigh any amount from 1 gram through 63 grams. What weights will he need if he wants to use the least number of weights?

Mr. Braden will need the following weights: 1 gram, 2 grams, 4 grams, 8 grams, 16 grams, and 32 grams. These weights are also the place values in base two. Note that in base two we always need only "one" of

84   CHAPTER 3   Numeration Systems

each place value. We may even want to look at some examples.

1	1
2	10
3	11
4	100
5	101
10	1010
30	11110
63	111111

14. What is the minimum number of pennies, nickels, and quarters needed to make 88¢?

3 quarters, 2 nickels, 3 pennies
$323_{\text{five}} = 3(25) + 2(5) + 3 = 88$

15. The concession stand has 10 gross, 4 dozen, and 9 single paper cups. Express the number of cups using base twelve notation. Express the number of cups using base ten notation. A gross is twelve dozen.

1497 cups

10 gross = 10 (12) (12) = 1440
4 dozen = 4 (12) = 48
1440 + 48 + 9 = 1497

Chapter 4
The System of Integers & Elementary Number Theory

Section 1 The System of Integers

Cover the right side of the page and work on the left, then check your work

1. Perform the following computations mentally.

   (a) $278 + {}^-140$
   (b) $84 + {}^-67 + {}^-34$

   > (a) $278 + {}^-140 = 138 + (140 + {}^-140)$
   > $\phantom{(a) 278 + {}^-140} = 138 + 0 = 138$
   > (b) $84 + {}^-67 + {}^-34 = (50 + 34) + {}^-67 + {}^-34$
   > $= 50 + (34 + {}^-34) + {}^-67$
   > $= 50 + 0 + {}^-67$
   > $= 50 + {}^-67 = {}^-17$

2. Estimate each of the following.

   (a) ${}^-119 + 351 + {}^-463$
   (b) $637 - {}^-352$

   > (a) ${}^-119 + 351 + {}^-463 \approx {}^-120 + 350 + {}^-460$
   > $= {}^-580 + 350 = {}^-230$
   > (b) $637 - {}^-352 \approx 640 + 350 = 990$

3. Write the additive inverse for each of the following.

   (a) 9
   (b) ${}^-63$
   (c) s
   (d) ${}^-r$

   > (a) ${}^-9$
   > (b) 63
   > (c) ${}^-s$
   > (d) r

4. Calculate each of the following.

   (a) $14 - 26$
   (b) ${}^-21 + 33$

(c) ⁻31 − ⁻15
(d) $(x - y) - {}^-y$

(a) ⁻12
$14 - 26 = 14 + {}^-26 = {}^-12$
(b) 12
(c) ⁻16
${}^-31 - {}^-15 = {}^-31 + 15 = {}^-16$
(d) x
$(x - y) - {}^-y = (x + {}^-y) + y$
$= x + ({}^-y + y)$
$= x + 0 = x$

5. Find the solution for each of the following.

(a) $x + {}^-3 = {}^-10$
(b) $5 - x = {}^-4$
(c) $3 + {}^-x = 7$

(a) $x = {}^-7$
(b) $x = 9$
$\quad 5 - x = {}^-4 \Rightarrow 5 = x + {}^-4$
$\quad x = 9$
(c) $x = {}^-4$
$\quad 3 + {}^-x = 7$
$\quad {}^-x = 7 - 3$
$\quad {}^-x = 4$
$\quad x = {}^-4$

6. Illustrate each of the following on a number line.

(a) ⁻3 + ⁻7
(b) 3 + ⁻6
(c) 3 + (⁻2 + ⁻5)

Answers to 6:

(a)

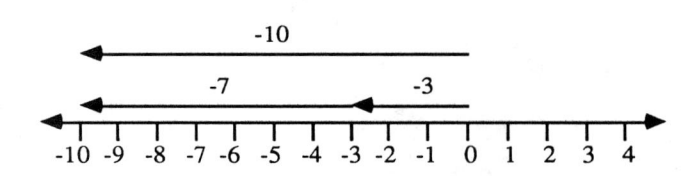

(b)

```
                              3 + -6 = -3
                           ←─────────────
                                   -6
                           ←───────────────────
                                              3
                                          ──────→
        ├──┼──┼──┼──┼──┼──┼──┼──┼──┼──┼──┼──┼──┼──→
       -10 -9 -8 -7 -6 -5 -4 -3 -2 -1  0  1  2  3  4
```

(c)

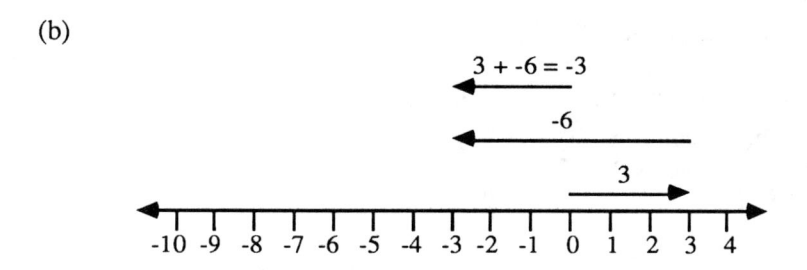

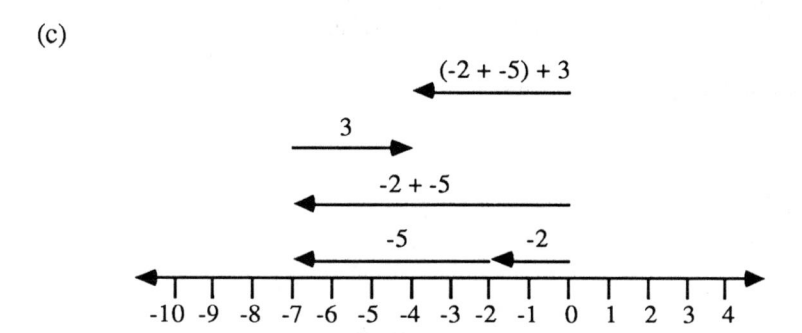

```
                                 (-2 + -5) + 3
                           ←─────────────────
                         3
                      ────→
                              -2 + -5
                           ←───────────
                              -5
                           ←───────
                                        -2
                                     ←─────
        ├──┼──┼──┼──┼──┼──┼──┼──┼──┼──┼──┼──┼──┼──→
       -10 -9 -8 -7 -6 -5 -4 -3 -2 -1  0  1  2  3  4
```

7. Make a model for each of the following. Solve each.

   (a) At the beginning of a week, Gerry Levine Inc. stock sold for $12 per share. For the next five days the changes in the stock selling price were gained $4, lost $3, gained $12, gained $9, and lost $15. What was the selling price at the end of the week?

   (b) At noon the temperature was $30°$ C. By 4:00 p.m., the temperature had increased by $4°$. A cold front arrived and by 7:00 p.m. the temperature had dropped by $14°$ from the 4:00 reading. What was the change in temperature from noon to 7:00 p.m.?

(a) $12 + 4 - 3 + 12 + 9 - 15 = n$
$(12 + 4 + 12 + 9) - 3 - 15 = n$
$37 - 3 - 15 = n$
$19 = n$

(b) $t = 30 - (30 + 4 - 14)$
$t = 30 - 20$
$t = 10$

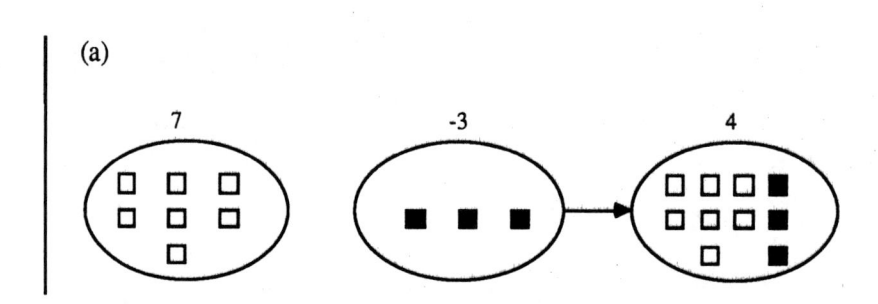

8. Answer true or false for each of the following.

   (a) Every whole number is an integer.
   (b) $x \neq {}^-x$ for every whole number x
   (c) The smallest positive integer is 0.
   (d) There is no smallest integer.

   | (a) True
   | (b) False $0 = {}^-0$
   | (c) False
   | (d) True

9. Calculate each of the following.

   (a) $|6|$
   (b) $|{}^-9|$
   (c) $|4-6|$
   (d) $|4| - |6|$

   | (a) 6
   | (b) 9
   | (c) 2 $|4-6| = |{}^-2| = 2$
   | (d) $^-2$ $|4| - |6| = 4 - 6 = {}^-2$

10. Fill in the missing number.

    (a) $7 - {}^-3 = \square$ because ${}^-3 + 10 = 7$
    (b) ${}^-3 - {}^-5 = 2$ because ${}^-5 + \square = {}^-3$
    (c) $9 - {}^-3 = 9 + \square$

    | (a) $\square = 10$
    | (b) $\square = 2$
    | (c) $\square = 3$

11. Use colored blocks to illustrate each of the following.

    (a) $7 + {}^-3$
    (b) ${}^-3 + {}^-4$
    (c) $9 - {}^-3$

    (a)

(b)

⁻3   -4   -7

(c)

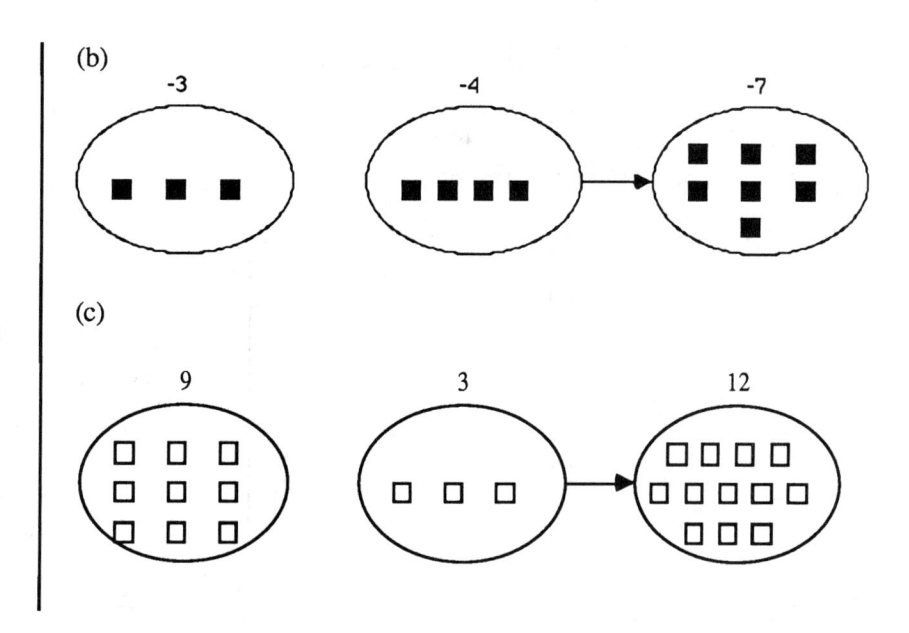

9   3   12

12. Perform the following in each of the three ways given.

$$^-4 - {}^-2$$

(a) Missing addend
(b) Adding the opposite
(c) Colored blocks model

(a) $^-4 - {}^-2 = \Box \Rightarrow {}^-2 + \Box = {}^-4$
$\Box = {}^-2$

(b) $^-4 - {}^-2 = {}^-4 + 2 = {}^-2$

(c)

-4   2   -2

13. Fill in each empty box with the sum of the pair of numbers beneath that box.

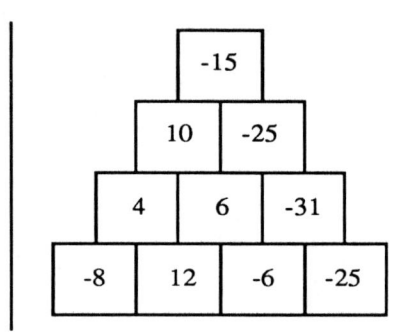

CHAPTER 4 The System of Integers and Elementary Number Theory

```
            [-15]
          [   ][   ]
        [   ][   ][   ]
      [-8][ 12][-6][   ]
```

```
              [-15]
           [10][-25]
        [ 4][ 6][-31]
     [-8][12][-6][-25]
```

14. For each of the following, find a pair of integers that make the statement true.

(a) $|x+y| \neq |x| + |y|$
(b) $|x+y| < |x| + |y|$
(c) $|x+y| > |x| + |y|$

(a) $x = 4 \quad y = {}^-5$
(b) $x = {}^-2 \quad y = 7$
(c) not possible

15. Let S be a set closed under subtraction. Suppose 5 and 9 are elements of S. Show that each of the following must also be elements of S.

(a) 4 (b) ${}^-4$ (c) 0 (d) 14 (e) 1

(a) Since $9 \in S$ and $5 \in S$, then $9 - 5 = 4 \in S$
(b) Since $5 \in S$ and $9 \in S$, then $5 - 9 = {}^-4 \in S$
(c) Since $5 \in S$ and $5 \in S$, then $5 - 5 = 0 \in S$
(d) Since $4 \in S$ and $9 \in S$, then $4 - 9 = {}^-5 \in S$. Since $9 \in S$ and ${}^-5 \in S$, then $9 - {}^-5 = 14 \in S$
(e) Since $5 \in S$ and $4 \in S$, then $5 - 4 = 1 \in S$

Section 2 Integer Multiplication and Division

Cover the right side of the page and work on the left, then check your work

1. Use mental computation to calculate each of the following.

   (a) $35 \times {}^-14$
   (b) $({}^-26 \times 37) + ({}^-26 \times 63)$

   (a) $35 \times {}^-14 = 35 \times ({}^-10 + {}^-4)$
   $= 35 \times {}^-10 + 35 \times {}^-4$
   $(35 \times {}^-4 = 30 \times {}^-4 + 5 \times {}^-4)$
   $= {}^-350 + {}^-140 = {}^-490$.
   (b) $({}^-26 \times 37) + ({}^-26 \times 63) =$
   ${}^-26 \times (37 + 63)$
   $= {}^-26 \times 100 = {}^-2600$.

2. Find a pattern for each of the following and write the next three terms.

   (a) 8, 3, ${}^-2$, ${}^-7$, ${}^-12$, ___, ___, ___
   (b) ${}^-20$, ${}^-17$, ${}^-14$, ${}^-11$, ${}^-8$, ___, ___, ___

   (a) 8, 3, ${}^-2$, ${}^-7$, ${}^-12$, ${}^-17$, ${}^-22$, ${}^-27$
   (b) ${}^-20$, ${}^-17$, ${}^-14$, ${}^-11$, ${}^-8$, ${}^-5$, ${}^-2$, 1

3. Perform the following computations.

   (a) ${}^-({}^-3) \times {}^-7$
   (b) ${}^-3(5 + {}^-6)$
   (c) $4[({}^-2)(12) + (3)({}^-5)]$

   (a) ${}^-21$
   (b) 3
   (c) ${}^-156$ $4[({}^-2)(12) + (3)({}^-5)] =$
   $4({}^-24 + {}^-15) = 4({}^-39) = {}^-156$.

4. Solve each of the following.

   (a) ${}^-5x = 20$
   (b) $\frac{x}{7} = {}^-4$
   (c) ${}^-4x + 3 = {}^-5$

92 CHAPTER 4 The System of Integers and Elementary Number Theory

(a) $x = {}^-4$
(b) $x = {}^-28$
(c) $x = 2$ $\quad {}^-4x + 3 = {}^-5$
$\qquad\qquad\quad {}^-4x = {}^-8$
$\qquad\qquad\quad\;\; x = 2$

5. Make a model for each of the following. Solve the model.

   (a) The Amazon rain forest is being destroyed at the rate of 75,000 acres per year. How much acreage will be lost by the end of the year 2000 if this pattern continues?

   (b) The temperature has been rising 7^0 C every hour. If the temperature is 15^0 C now, what was the temperature 3 hours ago?

(a) $a = 9 \times 75{,}000$
$\quad a = 675{,}000$ acres

(b) $t = 15 - (3 \times 7)$
$\quad t = 15 - 21$
$\quad t = {}^-6^0$ C

6. Simplify each of the following.

   (a) $({}^-3)^5 \times ({}^-3)^2$
   (b) $\dfrac{4^3 \times 4^2}{4}$
   (c) $\dfrac{a^8}{a^3} \times \dfrac{({}^-a)^4}{-a^2}$

(a) $({}^-3)^7$
(b) $4^4 \qquad \dfrac{4^3 \times 4^2}{4} = \dfrac{4^5}{4} = 4^{5-1} = 4^4$
(c) ${}^-a^7$

7. Show each of the following on a number line.

   (a) $3 \times {}^-4$
   (b) $5 \times {}^-2$

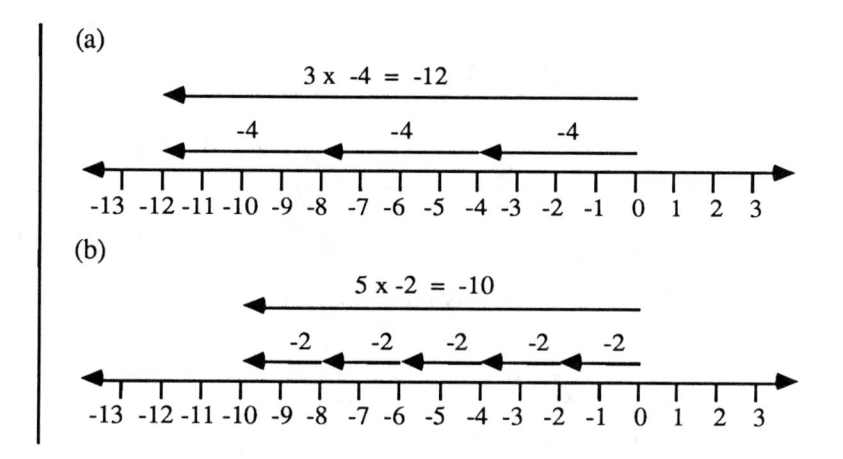

Section 2 Integer Multiplication and Division 93

(a)
$3 \times -4 = -12$

(b)
$5 \times -2 = -10$

8. Expand each of the following.

   (a) $(x - 3)(x + 2)$
   (b) $x^2(x - 3)$
   (c) $(2x - 5)(x + 4)$

   | (a) $x^2 - x - 6$
   | (b) $x^3 - 3x^2$
   | (c) $2x^2 + 3x - 20$
   | $(2x - 5)(x + 4) =$
   | $(2x - 5)x + (2x - 5)4$
   | $= 2x \cdot x + {}^-5 \cdot x + 2x \cdot 4 + {}^-5 \cdot 4$
   | $= 2x^2 - 5x + 8x - 20$
   | $= 2x^2 + 3x - 20$

9. Solve each of the following.

   (a) $4x < {}^-20$
   (b) $2x + 7 > 9$
   (c) $5x > {}^-15$

   | (a) $x < {}^-5$
   | (b) $x > 1$ $2x + 7 > 9$
   | $2x > 2$
   | $x > 1$
   | (c) $x > {}^-3$

10. Complete the following pattern.

    $5 \times 4 = 20$
    $5 \times 3 = 15$
    $5 \times 2 = 10$
    $5 \times 1 = 5$
    $5 \times 0 = 0$

94 CHAPTER 4 The System of Integers and Elementary Number Theory

$$5 \times {}^-1 = {}^-5$$
$$5 \times {}^-2 = {}^-10$$
$$5 \times {}^-3 = {}^-15$$

11. Write each of the following lists from least to greatest.

   (a) 3, ⁻4, ⁻3, 0, 4
   (b) ⁻4, ⁻7, ⁻9, ⁻2, ⁻1

   (a) ⁻4, ⁻3, 0, 3, 4
   (b) ⁻9, ⁻7, ⁻4, ⁻2, ⁻1

12. Fill in the blanks with the appropriate symbols (<, >, or =).

   (a) ⁻2 ___ ⁻1
   (b) 3 + (⁻4) ___ 2 × ⁻4
   (c) If $x < 7$, then $x + 4$ ___ 11
   (d) If $x > 3$, then ⁻3x ___ ⁻9

   (a) ⁻2 < ⁻1
   (b) 3 + (⁻4) > 2 × ⁻4
   (c) If $x < 7$, then $x + 4 \leq 11$
   (d) If $x > 3$, then ⁻3$x \leq$ ⁻9

13. Complete the following multiplication magic square.

-2		
3	-4	-18

-2	-9	12
-36	6	0
3	-4	-18

14. The population of a worm ranch doubles every month. At the end of twelve months the population is 2,000,000. What was the population after the eighth month?

Population after 8th month = 125,000. Consider the following table.

END OF MONTH	POPULATION
12	2,000,000
11	1,000,000
10	500,000
9	250,000
8	125,000

15. Find three integers for which the following is true.

$$(x + y) \div z = (x \div z) + (y \div z)$$

$x = {}^-9 \quad y = 21 \quad z = 3$

Section 3 Divisibility

Cover the right side of the page and work on the left, then check your work

1. List four proper divisors for each of the following.

    (a) 399
    (b) 1001

    (a) Divisors: 1, 2, 7, 19, 21, 57, 133
    (b) Divisors: 1, 7, 11, 13, 77, 91, 143

2. Classify each as true or false.

    (a) 33 is a factor of 333
    (b) 9 is a multiple of 45
    (c) 111 is a prime number

    (a) False   33 does not divide 333
    (b) False   9 is a divisor of 45
    (c) False $3 \mid 111$

3. List all the divisors of each of the following numbers.

    (a) 63
    (b) 120
    (c) 199

    (a) 1, 3, 9, 21, 63
    (b) 1, 2, 3, 4, 5, 6, 8, 10, 12, 15, 20, 24, 30, 60
    (c) 1, 199

4. Test each of the following for divisibility by 2, 3, 4, 5, 6, 7, 8, 9, 10, 11, 12.

    (a) 11,223,344
    (b) $2^3 \cdot 3^4 \cdot 5^6$

    (a) Divisible by: 2, 4, 8, 11
    (b) Divisible by: 2, 3, 4, 5, 6, 8, 9, 10, 12

5. Classify each of the following as true or false.

    (a) If a number is divisible by 3, then every digit of the number is divisible by 3.
    (b) If a number is divisible by 15, then it is divisible by 3 and by 5.

Section 3   Divisibility   97

(c) If a number is divisible by 3 and by 5, then it is divisible by 15.
(d) If a number is divisible by 5 and 10, then it is divisible by 50.

> (a) False $3|45$ but $3 \nmid 4$ and $3 \nmid 5$
> (b) True
> (c) True
> (d) False   Consider $5|20$ and $10|20$ but $50 \nmid 20$

6. Fill in the box with the largest digit that makes the statement true.

   (a) $4|64\square$
   (b) $9|74\square25$
   (c) $11|94\square1$

> (a) $\square = 8$
> (b) $\square = 9$
> (c) $\square = 7$

7. A mathematics class has an enrollment of 247. Can the students be seated in 9 rows with an equal number of seats in each row?

> No $9 \nmid 247$

8. Classify each of the following as true or false. If false, give a counterexample.

   (a) If $a|(b+c)$, then $a|b$ or $a|c$.
   (b) If $a|b$ and $a|c$, then $a|bc$.
   (c) $1|a$.

> (a) False
> $5|(7+3)$ but $5 \nmid 7$ and $5 \nmid 3$
> (b) True
> (c) True

9. Mr. Hampton owes $2916 on a used car. Can this be paid in 12 equal monthly payments?

> Yes $12|2916$

10. A palindrome is a number that reads the same forward or backwards. Which of the following 4-digit palindromes are divisible by 11?

(a) 5225
(b) 7117
(c) 1331

> (a) $11 \mid 5225$
> (b) $11 \mid 7117$
> (c) $11 \mid 1331$

11. Replace the two missing digits in the number below so that it will be divisible by 88.

$$22 \_\_ \_\_ 2$$

> $88 \mid 22352$ If $8 \mid a$ and $11 \mid a$, then $8 \cdot 11 \mid a$

12. State the property that justifies each of the following.

(a) $4 \mid 20$; therefore $4 \mid 27 \cdot 20$.
(b) $5 \mid 100$ and $5 \mid 27$; therefore $5 \mid (100 + 27)$.
(c) $5 \mid a$; therefore $5 \mid a^2$

> (a) If $x \mid y$ or $x \mid z$, then $x \mid yz$
> (b) If $x \mid y$ and $x \nmid z$, then $x \nmid (y + z)$
> (c) If $x \mid y$ or $x \mid z$, then $x \mid yz$

13. Make a conjecture about the following.

$ab - ba = ?$ [Example: $54 - 45$]
where $a = b + 1$

> $ab - ba = 9$

14. Make a conjecture about the following.

$ab - ba = ?$ [Example: $53 - 35$]
where $a = b + 2$

> $ab - ba = 18$

15. Generalize the results from 13 and 14.

$ab - ba = ?$

> $ab - ba = 9 \cdot (a - b)$
> Consider the table below
>
> | a | b | ab − ba |
> |---|---|---------|
> | 6 | 5 | 9 |
> | 6 | 4 | 18 |
> | 6 | 3 | 27 |
> | 6 | 2 | 36 |

Section 4 Primes, Composites, and Factorization

Cover the right side of the page and work on the left, then check your work

1. Which of the following are prime numbers.

   (a) 149
   (b) 433
   (c) 379

   | (a) Prime
   | (b) Prime
   | (c) Prime

2. Use a factor tree to find the prime factorization for each of the following.

   (a) 504
   (b) 828
   (c) 1840

   (a)
   ```
   504
   / \
   2   252
       / \
       2   126
           / \
           2   63
               / \
               3   21
                   / \
                   3   7
   ```

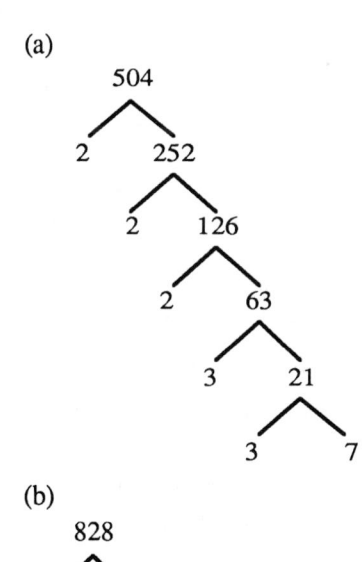

   (b)
   ```
   828
   / \
   2   414
       / \
       2   207
           / \
           3   69
               / \
               3   23
   ```

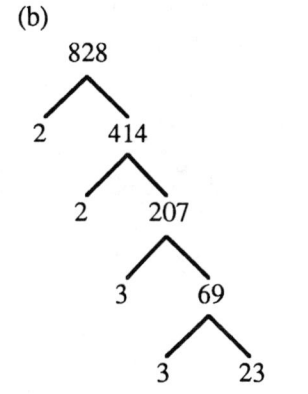

(c)
```
       1840
      /    \
     2     920
          /   \
         2   460
             /   \
            2   230
                /   \
               2   215
                   /  \
                  5   23
```

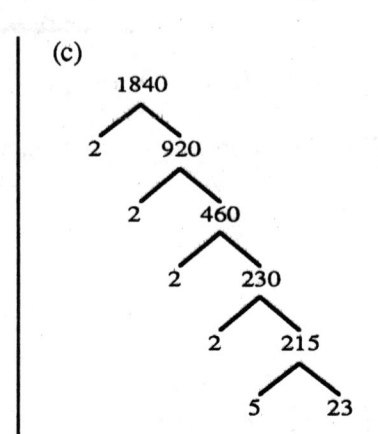

3. Use the compact form of the division technique to find the prime factorization for each of the following.

   (a) 326
   (b) 540
   (c) 1404

(a)
```
  2 | 326
163 | 163
    |   1
```

(b)
```
2 | 540
2 | 270
3 | 135
3 |  45
3 |  15
5 |   5
  |   1
```

(c)
```
2 | 1404
2 |  702
3 |  351
3 |  117
3 |   39
1 |   13
  |    1
```

Section 4 Primes, Composites, and Factorization   101

4. Classify each of the following as perfect, deficient, or abundant.

   (a) 500
   (b) 326
   (c) 120

   > (a) Deficient
   > $1 + 2 + 4 + 10 + 20 + 25 + 50 + 100 + 250 = 462 < 500$
   > (b) Deficient
   > $1 + 2 + 163 = 166 < 326$
   > (c) Abundant
   > $1 + 2 + 4 + 6 + 10 + 12 + 15 + 20 + 24 + 30 + 40 + 60 = 232 > 120$

5. What is the greatest prime you must conider to test whether 4879 is prime?

   > 69 $69^2 = 4761$
   > $70^2 = 4900$

6. What is the smallest number that has exactly eight positive proper divisors?

   > 840 Has factors = $\{1, 2, 3, 4, 5, 6, 7, 8\}$
   > $840 = 2^3 \cdot 3 \cdot 5 \cdot 7$

7. (a) Find all the positive divisors of 3^3 and 2^2.
   (b) How many divisors does $3^3 \cdot 2^2$ have?

   > (a) Divisors of $3^3 = \{1, 3, 9, 27\}$
   >   Divisors of $2^2 = \{1, 2, 4\}$
   > (b) Divisors of $3^3 \cdot 2^2 = \{1, 2, 3, 4, 6, 9, 12, 18, 27, 36, 54, 108\}$

8. Fill in the boxes in the following table.

Number of Divisors	Numbers with Given Number of Divisors
1	1
2	2 3 5 7 11 13 ☐ ☐ ☐
3	4 9 ☐ ☐
4	6 8 10 14 15 21 ☐ ☐
5	☐
6	12 18 20 ☐ ☐
8	24 ☐
9	☐

Number of Divisors	Numbers with Given Number of Divisors
1	1
2	2 3 5 7 11 13 17 19 23
3	4 9 25 49
4	6 8 10 14 15 21 22 26
5	16
6	12 18 20 28 32
8	24 30
9	36

9. Find a pair of twin primes between 95 and 105.

> 101 and 103

10. A Niven number is a number that is divisible by the sum of its digits. Find five Niven numbers.

> 12, 18, 20, 21, 24.
> Other answers are possible.

11. Show that 1184 and 1210 are ammicable numbers.

> Proper divisors of 1184 – 1, 2, 4, 8, 16, 32, 37, 74, 148, 296, and 592
> Sum of proper divisors = 1210
>
> Proper divisors of 1210 – 1, 2, 5, 10, 11, 22, 55, 110, 121, 242, and 605
> Sum of proper divisors = 1184

12. Show that each of the following satisfy Part I of Goldbach's conjecture.

    (a) 58
    (b) 94

> (a) $58 = 37 + 19$
> Other answers are possible
> (b) $94 = 5 + 89$
> Other answers are possible

Section 4 Primes, Composites, and Factorization 103

13. The United States flag had the stars arranged in a 6 × 8 rectangular arrangement when there were 48 states. What other rectangular arrangement could have been used?

$$\begin{array}{l} 1 \times 48 \\ 2 \times 24 \\ 3 \times 16 \\ 4 \times 12 \end{array}$$ Consider divisors of 48

14. Find the smallest number that has factors 2, 3, 4, 5, 6, 7, 8, 9, and 10.

2520 $2520 = 2^3 \cdot 3^2 \cdot 5 \cdot 7$

15. The Applesauce company always plant their apple trees in square arrays, like those illustrated. This year they planted 31 more apple trees in the square orchard. If the orchard is still square, how many apple trees are there in the orchard this year?

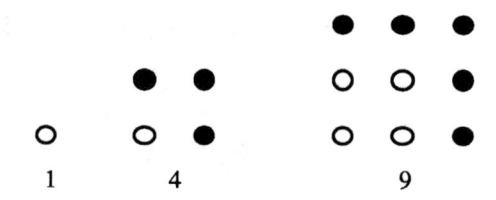

```
                            • • •
          • •               ○ ○ •
  ○    ○  ○  •              ○ ○ •
  1       4                    9
```

256   Consider the following table

Number of Trees	# added from previous year
1	1
4	3
9	5
16	7
25	9
⋮	⋮
256	31

Section 5 Greatest Common Divisor and Least Common Multiple

Cover the right side of the page and work on the left, then check your work

1. Find the greatest common divisor and the least common multiple for each of the following by using intersection of sets.

    (a) 18 and 30
    (b) 24 and 36
    (c) 8, 12, 36

    (a) g.c.d. (18, 30) = 6
    l.c.m. (18, 30) = 90

    Divisors of 18 = {1, 2, 3, 6, 9, 18}
    Divisors of 30 = {1, 2, 3, 5, 6, 10, 15, 30}

    Multiples of 18 = {18, 36, 54, 72, 90, 108, ...}
    Multiples of 30 = {30, 60, 90, 120,

    (b) g.c.d. (24, 36) = 12
    l.c.m. (8, 12, 36) = 72

2. Find the greatest common divisor and the least common multiple by using the Fundamental Theorem of Arithmetic.

    (a) 132 and 404
    (b) 85 and 1580
    (c) 800, 180, 530

    (a) g.c.d. (132, 404) = 2^2 = 4
    l.c.m. (132, 404) = $2^2 \cdot 3 \cdot 11 \cdot 101$ = 13, 332

    $132 = 2^2 \cdot 3 \cdot 11$
    $404 = 2^2 \cdot 101$

    (b) g.c.d. (85, 1580) = 5
    l.c.m. (85, 1580) = $2^2 \cdot 5 \cdot 17 \cdot 79$ = 26, 860

    $85 = 5 \cdot 17$
    $1580 = 2^2 \cdot 5 \cdot 79$

(c) g.c.d.(800, 180, 530) = 2 · 5 = 10
l.c.m. (800, 180, 530)
= $2^5 \cdot 3^2 \cdot 53$ = 381,600
800 = $2^5 \cdot 5^2$
180 = $2^2 \cdot 3^2 \cdot 5$
530 = 2 · 5 · 53

3. Find the greatest common divisor using the Euclidean algorithm.

   (a) 220 and 1928
   (b) 13985 and 9474

(a) g.c.d. (220, 1928) = 4

```
        8                    1
 220 ⌐1928       168 ⌐220
       1760              168
       ────              ────
        168               52

         3            4            3
  52 ⌐168     12 ⌐52     4 ⌐12
      156          48          12
      ────         ───         ──
       12           4           0
```

(b) g.c.d. (13985, 9474) = 1

4. Classify each statement as true or false.

   (a) If g.c.d. (a, b) = 1 then a and b cannot both be even.
   (b) If g.c.d. (a, b) = 3 then both a and b are multiples of 3.
   (c) If a and b are both even, then g.c.d. (a, b) = 2.
   (d) g.c.d. (a, b) ≥ a

(a) True
(b) True
(c) False   g.c.d. (4, 12) = 4
(d) True

5. Which of the following are relatively prime?

   (a) 4363 and 9
   (b) 951 and 3
   (c) 181, 345, 913 and 11

(a) Relatively prime
(b) Not relatively prime
(c) Relatively prime

6. Find the least common multiple for each pair of numbers using the formula for least common multiple.

    (a) 220 and 1928
    (b) 44,200 and 4810

    (a) l.c.m. (220, 1928) = $\frac{(220)(1928)}{4 \leftarrow \text{g.c.d.}}$
    $= \frac{424160}{4} = 106040$
    (b) l.c.m. (44200, 4810) = 1,635,400

7. Find all counting numbers such that g.c.d. (30, b) = 1 and $b \leq 25$.

    1, 7, 11, 13, 17, 19, 23

8. What is the relationship between a and b if g.c.d. $(a, b) = a$

    $a | b$

9. Marvin mows the lawn every ten days and washes the car every fourteen days. He did both today. How many days will it be before he will do both jobs on the same day?

    70 days   l.c.m. (10, 14) = 70

10. Carmela has 144 foreign stamps and 360 United States stamps. She wants to put them in books so that there are an equal number of stamps in each book. What is the greatest number of stamps that she can put in each book?

    72 stamps in a book
    g.c.d. (360, 144) = 72

11. Find the g.c.d. and l.c.m. for the following and express these in prime factored form.

    $2^3 \cdot 3^5 \cdot 5^7$ and $2^7 \cdot 3^4 \cdot 5^3$

    g.c.d. = $2^3 \cdot 3^4 \cdot 5^3$
    l.c.m. = $2^7 \cdot 3^5 \cdot 5^7$

12. Use a calculator to find two integers a and b such that $a \times b = 100,000$ and neither x nor y contain any zeros as digits.

    $a = 32$ $100,000 = 2^5 \cdot 5^5$
    $b = 3,125$

13. If g.c.d. $(a, b) = 1$, then what can we say about g.c.d. (a^2, b^2)?

    g.c.d. $(a^2, b^2) = 1$

Section 6 Modular Arithmetic

Cover the right side of the page and work on the left, then check your work

1. Perform, if possible, each of the following operations on a 12-hour clock.

    (a) 7 ⊕ 8
    (b) 3 ⊖ 9
    (c) 5 ⊗ 6
    (d) 2 ⊘ 5

    > (a) 3
    > (b) 6
    > (c) 6
    > (d) 10

2. Construct an addition table for a 7-hour clock.

    > | + | 0 | 1 | 2 | 3 | 4 | 5 | 6 |
    > |---|---|---|---|---|---|---|---|
    > | 0 | 0 | 1 | 2 | 3 | 4 | 5 | 6 |
    > | 1 | 1 | 2 | 3 | 4 | 5 | 6 | 0 |
    > | 2 | 2 | 3 | 4 | 5 | 6 | 0 | 1 |
    > | 3 | 3 | 4 | 5 | 6 | 0 | 1 | 2 |
    > | 4 | 4 | 5 | 6 | 0 | 1 | 2 | 3 |
    > | 5 | 5 | 6 | 0 | 1 | 2 | 3 | 4 |
    > | 6 | 6 | 0 | 1 | 2 | 3 | 4 | 5 |

3. Construct a multiplication table for a 7-hour clock.

    > | x | 0 | 1 | 2 | 3 | 4 | 5 | 6 |
    > |---|---|---|---|---|---|---|---|
    > | 0 | 0 | 0 | 0 | 0 | 0 | 0 | 0 |
    > | 1 | 0 | 1 | 2 | 3 | 4 | 5 | 6 |
    > | 2 | 0 | 2 | 4 | 6 | 1 | 3 | 5 |
    > | 3 | 0 | 3 | 6 | 2 | 5 | 1 | 4 |
    > | 4 | 0 | 4 | 1 | 5 | 2 | 6 | 3 |
    > | 5 | 0 | 5 | 3 | 1 | 6 | 4 | 2 |
    > | 6 | 0 | 6 | 5 | 4 | 3 | 2 | 1 |

4. Perform, if possible, each of the following operations on a 7-hour clock.

    (a) 3 ⊕ 4
    (b) 6 ⊗ 5
    (c) 4 ⊘ 6

108 CHAPTER 4 The System of Integers and Elementary Number Theory

(d) $4 \bigcirc 3$

(a) 0
(b) 2
(c) 5
$4 \bigcirc 6 = \square \Rightarrow \square \oplus 6 = 4 \Rightarrow \square = 5$
(d) 6
$4 \bigcirc 3 = \square \Rightarrow \square \otimes 3 = 4 \Rightarrow \square = 6$

5. Find t for the following, where the answer is some number on a 7-hour clock.

   (a) $t = 6 \oplus 5$
   (b) $3 = t \oplus 6$
   (c) $t = 4 \otimes 4$
   (d) $1 = 3 \otimes t$

(a) $t = 4$
(b) $t = 4$
(c) $t = 2$
(d) $t = 5$

6. Using a 7-hour clock, two numbers a and b are additive inverses if $a \oplus b \equiv 0 \pmod{7}$.

   (a) What is the additive inverse for 3?
   (b) What is the additive inverse for 5?
   (c) What is the additive inverse for 0?

(a) 4 $3 \oplus 4 \equiv 0 \pmod{7}$
(b) 2 $5 \oplus 2 \equiv 0 \pmod{7}$
(c) 0 $0 \oplus 0 \equiv 0 \pmod{7}$

7. Using a 7-hour clock, two numbers a and b are multiplicative inverse if $a \otimes b \equiv 1 \pmod{7}$.

   (a) What is the multiplicative inverse for 3?
   (b) What is the multiplicative inverse for 4?
   (c) What is the multicplicative inverse for 0?

(a) 5 $3 \otimes 5 \equiv 1 \pmod{7}$
(b) 2 $4 \otimes 2 \equiv 1 \pmod{7}$
(c) there is none

8. Label each of the following as either true or false.

   (a) $81 \equiv 1 \pmod{8}$
   (b) $33 + 3 \equiv 3 \pmod{9}$
   (c) $3^5 \equiv 2 \pmod{6}$

(a) True $81 \div 8$ has a remainder of 1

Section 6   Modular Arithmetic   109

(b) False   36 ÷ 9 has a remainder of 0
(c) False $3^5 ÷ 6 = 243 ÷ 6$ has a remainder of 3.

9. Perform the following operations and in each case reduce the answer to a whole number less than the modulus.

   (a) $473 \equiv x \pmod 5$
   (b) $36 \cdot 97 \equiv \pmod 7$
   (c) $4x + 3 \equiv \pmod 9$

(a) $x = 3$
(b) $x = 6$ $36 \cdot 97 \equiv 1 \cdot 6 \pmod 7$
   $\equiv 6 \pmod 7$
(c) $x = 7$ $4x + 3 \equiv 4 \pmod 9$
   $4x \equiv 1 \pmod 9$ subtract 3
   $x = 7$ use a table

10. Find all a such that $a \equiv 2 \pmod 5$.

$a = \{..., -13, -8, -3, 2, 7, 12, ...\}$

11. Translate each of the following into a statement using congruence.

   (a) $5 \mid 35$
   (b) $4 \mid 28$
   (c) $3 \mid -30$

(a) $35 \equiv 0 \pmod 5$ $35 ÷ 5$ has a remainder of 0
(b) $28 \equiv 0 \pmod 4$
(c) $-30 \equiv 0 \pmod 3$

12. If October 2 falls on Tuesday, on what day of the week will it fall

   (a) next year?
   (b) next year, if next year is a leap year?

(a) Wednesday $365 \equiv 1 \pmod 7$
(b) Thursday $366 \equiv 2 \pmod 7$

13. It can be shown that if $a \equiv b \pmod m$ and k is a natural number, then $a^k \equiv b^k$. We can use this property to help us with some problems.

Example: Find the remainder when 2^{100} is divided by 5.

$2^3 \equiv 3 \pmod 5$
$2^4 \equiv 1 \pmod 5$
$2^{100} = (2^4)^{25} \equiv 1^{25} \pmod 5$
or $2^{100} \equiv 1 \pmod 5$ remainder = 1

Solve each of the following:

(a) Find the remainder when 4^{100} is divided by 7.

(b) Find the remainder when 10^{99} is divided by 11.

(a) remainder = 4
$4^2 \equiv 2 \pmod 7$
$4^3 \equiv 1 \pmod 7$
$(4^3)^{33} \equiv 1^{33} \pmod 7$
$4^{99} \equiv 1 \pmod 7$
$4^{99} \cdot 4 \equiv 1 \cdot 4^1 \pmod 7$
$4^{100} \equiv 4 \pmod 7$

(b) remainder = 10
$10^1 \equiv 10 \pmod{11}$
$10^2 \equiv 1 \pmod{11}$
$(10^2)^{50} \equiv 1^{50} \pmod{11}$
$10^{100} \equiv 1 \pmod{11}$
$10^{100} \div 10^1 \equiv 1 \div 10 \pmod{11}$
$10^{99} \equiv 1 \div 10 \pmod{11}$
$1 \div 10 = \square \pmod{11} \Rightarrow \square \otimes 10 \equiv 1 \pmod{11} \Rightarrow \square = 10$

14. True or false: $10^{84} \equiv 1 \pmod 9$

True
$10^1 \equiv 1 \pmod 9$
$10^2 \equiv 1 \pmod 9$
$10^3 \equiv 1 \pmod 9$

15. Find an example to show the following.

$ac \equiv bc \pmod m$ does not always imply $a \equiv b \pmod m$

$3 \cdot 2 \equiv 3 \cdot 4 \pmod 6$ but $2 \neq 4 \pmod 4$

Chapter 5
Introduction to the Rational Numbers

Section 1 The Set of Rational Numbers

Cover the right side of the page and work on the left, then check your work

1. Estimate a fraction to represent the shaded portion of each diagram.

   (a)

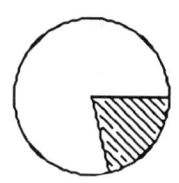

   (b)

   (c)

   (a) $\dfrac{1}{5}$

   (b) $\dfrac{1}{6}$

   (c) $\dfrac{1}{3}$

2. Use estimation to decide which of each pair of fractions is larger.

   (a) $\dfrac{24}{75}$ and $\dfrac{1}{3}$

   (b) $\dfrac{49}{64}$ and $\dfrac{3}{4}$

   (a) $\dfrac{24}{75} < \dfrac{1}{3}$ (75 is more than 3 × 24)

   (b) $\dfrac{49}{64} > \dfrac{3}{4}$

112   CHAPTER 5   Introduction to the Rational Numbers

3. What fraction does the shaded portion of each diagram illustrate?

   (a)

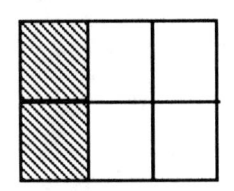

   (b)

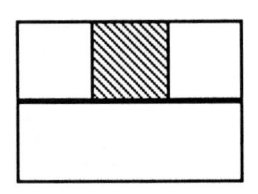

   (c)

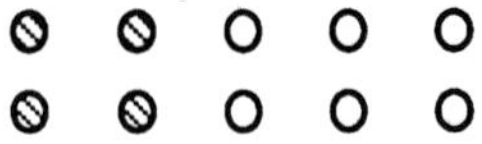

   (a) $\dfrac{2}{6}$

   (b) $\dfrac{1}{6}$

   (c) $\dfrac{4}{10}$

4. Write three fractions equivalent to each of the following fractions.

   (a) $\dfrac{4}{9}$

   (b) $\dfrac{1}{3}$

   (c) $\dfrac{a}{4}$

   (a) $\dfrac{8}{18}, \dfrac{12}{27}, \dfrac{16}{36}$ (others are possible).

   (b) $\dfrac{2}{6}, \dfrac{3}{9}, \dfrac{4}{12}$

   (c) $\dfrac{2a}{8}, \dfrac{3a}{12}, \dfrac{4a}{16}$

5. Find the simplest form for each of the following.

(a) $\dfrac{156}{93}$

(b) $\dfrac{27}{45}$

(c) $\dfrac{6543}{37{,}467}$

(d) $\dfrac{x^2y + xy}{x^3y + xy^2}$

(a) $\dfrac{52}{31}$

$\left(\dfrac{156}{93} = \dfrac{3 \cdot 4 \cdot 13}{3 \cdot 31} = \dfrac{4 \cdot 13}{31} = \dfrac{52}{31} \right)$

(b) $\dfrac{3}{5}$

(c) $\dfrac{717}{4163}$ $\left(\dfrac{6453}{37467} = \dfrac{3 \cdot 3 \cdot 3 \cdot 239}{3 \cdot 3 \cdot 23 \cdot 181} \right.$

$\left. = \dfrac{3 \cdot 239}{23 \cdot 181} = \dfrac{717}{4163} \right)$

(d) $\dfrac{x+1}{x^2+y}$ $\left(\dfrac{x^2y + xy}{x^3y + xy^2} = \right.$

$\left. \dfrac{xy(x+1)}{xy(x^2+y)} = \dfrac{x+1}{x^2+y} \right)$

6. Decide whether the following pairs are equivalent by writing each in simplest form.

(a) $\dfrac{3}{8}$ and $\dfrac{450}{1200}$

(b) $\dfrac{17}{27}$ and $\dfrac{25}{45}$

(c) $\dfrac{24}{36}$ and $\dfrac{6}{9}$

(a) $\dfrac{3}{8} = \dfrac{450}{1200}$ $\left(\dfrac{450}{1200} = \right.$

$\left. \dfrac{2 \cdot 3 \cdot 3 \cdot 5 \cdot 5}{2 \cdot 2 \cdot 2 \cdot 2 \cdot 3 \cdot 5 \cdot 5} = \dfrac{3}{8} \right)$

(b) $\dfrac{17}{27} \neq \dfrac{25}{45}$ $\left(\dfrac{25}{45} = \dfrac{5 \cdot 5}{5 \cdot 9} = \dfrac{5}{9} \right)$

(c) $\dfrac{24}{36} = \dfrac{6}{9}$ $\left(\dfrac{24}{36} = \dfrac{2 \cdot 12}{3 \cdot 2} = \dfrac{2}{3}, \right.$

$\left. \dfrac{6}{9} = \dfrac{2 \cdot 3}{3 \cdot 3} = \dfrac{2}{3} \right)$

7. Decide whether the following pairs are equivalent by changing both to the same denominator.

114 CHAPTER 5 Introduction to the Rational Numbers

(a) $\dfrac{10}{16}$ and $\dfrac{14}{20}$

(b) $\dfrac{4}{12}$ and $\dfrac{30}{90}$

(c) $\dfrac{2a+2b}{4}$ and $\dfrac{a+b}{2}$

> (a) $\dfrac{10}{16} \neq \dfrac{14}{20}$
> $\left(\dfrac{10}{16} = \dfrac{50}{80} \text{ and } \dfrac{14}{20} = \dfrac{76}{80} \right)$
> (b) $\dfrac{4}{12} = \dfrac{30}{90}$
> (c) $\dfrac{2a+2b}{4} = \dfrac{a+b}{2}$
> $\left(\dfrac{a+b}{2} = \dfrac{2(a+b)}{2 \cdot 2} = \dfrac{2a+2b}{4} \right)$

8. Solve for x.

   (a) $\dfrac{3}{4} = \dfrac{x}{20}$

   (b) $\dfrac{7}{x} = \dfrac{9}{63}$

   (c) $\dfrac{9}{2} = \dfrac{x^2}{8}$

> (a) $x = 15$
> (b) $x = 49$ $9x = 7 \cdot 63$
> $9x = 441$
> $x = 49$
> (c) $x = 6, \,^-6$ $2x^2 = 9 \cdot 8$
> $2x^2 = 72$
> $x^2 = 36$
> $x = 6, \,^-6$

9. Insert >, =, or < to make each of the following true.

   (a) $\dfrac{7}{9}, \dfrac{15}{17}$

   (b) $\dfrac{33}{90}, \dfrac{15}{40}$

   (c) $\dfrac{a}{b}, \dfrac{2a}{2b}$

> (a) $\dfrac{7}{9} < \dfrac{15}{17}$
> $(7 \cdot 17 = 119 < 9 \times 15 = 135)$
> (b) $\dfrac{33}{90} < \dfrac{15}{40}$
> (c) $\dfrac{a}{b} = \dfrac{2a}{2b}$

Section 1   The Set of Rational Numbers   115

10. Solve for x.

   (a) $\dfrac{x}{9} < \dfrac{2}{3}$

   (b) $\dfrac{2}{3} = \dfrac{x}{6}$

   (c) $\dfrac{3}{4} > \dfrac{9}{x}$

$(a \cdot 2b = 2ab = b \cdot 2a = 2ab)$

(a) $x < 6$

$\left(\dfrac{x}{9} < \dfrac{2}{3} \Rightarrow 3x < 18 \Rightarrow x < 6 \right)$

(b) $x = 4$

(c) $x > 12$ $\dfrac{3}{4} > \dfrac{9}{x}$

$3x > 36$

$x > 12$

11. Demonstrate the fraction $\dfrac{3}{8}$ using:

   (a) sets
   (b) fraction bar
   (c) number line

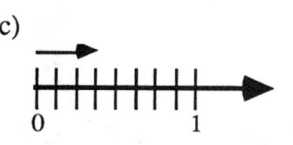

12. Classify each of the following as true or false
    (W = whole numbers; I = integers; Q = rationals).

   (a) $Q \cap I = W$
   (b) $(I \cup W) \subset Q$
   (c) $Q \cap W = W$

(a) False
(b) False
(c) True

116 CHAPTER 5 Introduction to the Rational Numbers

13. On her first mathematics test Lori answered 20 out of 25 questions correctly. On the second test she answered 24 out of 30 questions correctly. Did she do equally well on the two tests?

$$\text{Yes} \quad \frac{20}{25} = \frac{24}{30}$$
$$20 \cdot 30 = 600 = 25 \cdot 24$$

14. Which of the following are models for the same fraction?

    (a)

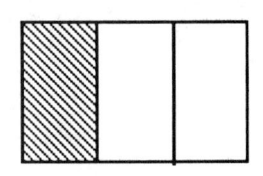

    (b)

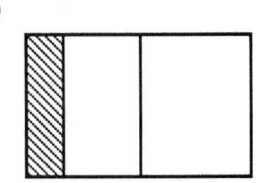

    (c)

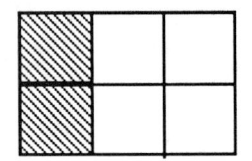

    (d)

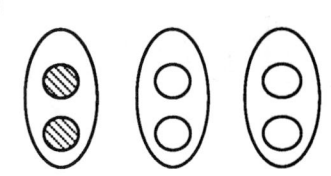

    (a) and (d)

Section 2 Addition and Subtraction of Rational Numbers

Cover the right side of the page and work on the left, then check your work

1. Calculate mentally each of the following.

   (a) $\dfrac{3}{8} + \dfrac{2}{3} + \dfrac{5}{8}$

   (b) $2\dfrac{3}{5} + 3\dfrac{4}{5}$

   (c) $4\dfrac{5}{9} - 3\dfrac{8}{9}$

   (d) $3\dfrac{3}{4} + \dfrac{1}{2}$

   (a) $1\dfrac{2}{3}$

   $\left(\dfrac{3}{8} + \dfrac{2}{3} + \dfrac{5}{8} = \left(\dfrac{3}{8} + \dfrac{5}{8}\right) + \dfrac{2}{3} = \dfrac{8}{8} + \dfrac{2}{3} = 1\dfrac{2}{3}\right)$

   (b) $6\dfrac{2}{5}$

   $\left(2\dfrac{3}{5} + 3\dfrac{4}{5} = \left(2\dfrac{3}{5} - \dfrac{1}{5}\right) + \left(3\dfrac{4}{5} + \dfrac{1}{5}\right)\right.$
   $\left. = 2\dfrac{2}{5} + 4\right)$

   (c) $\dfrac{6}{9}$

   $\left(4\dfrac{5}{9} - 3\dfrac{8}{9} = \left(4\dfrac{5}{9} + \dfrac{1}{9}\right) - \left(3\dfrac{8}{9} + \dfrac{1}{9}\right)\right.$
   $\left. = 4\dfrac{6}{9} - 4\right)$

   (d) $1\dfrac{1}{4}$

2. Estimate each answer by rounding to the nearest whole number.

   (a) $3\dfrac{4}{7} + 12\dfrac{7}{9} + 4\dfrac{2}{5}$

   (b) $13\dfrac{2}{3} - 7\dfrac{1}{4}$

   (c) $4\dfrac{3}{5} + 19\dfrac{1}{8} - 7\dfrac{7}{11}$

   (a) 21   (4 + 13 + 4)
   (b) 7   (14 − 7)

118　CHAPTER 5　Introduction to the Rational Numbers

3. Estimate each answer by rounding to the neartest one half.

   (a) $3\frac{4}{7} + 12\frac{7}{9} + 4\frac{2}{5}$

   (b) $13\frac{2}{3} - 7\frac{1}{4}$

   (c) $4\frac{3}{5} + 19\frac{1}{8} - 7\frac{7}{11}$

   > (a) 21　$\left(3\frac{1}{2} + 13 + 4\frac{1}{2}\right)$
   >
   > (b) 6　$\left(13\frac{1}{2} - 7\frac{1}{2}\right)$
   >
   > (c) 16　$\left(4\frac{1}{2} + 19 - 7\frac{1}{2}\right)$

4. Perform the following additions.

   (a) $\dfrac{4}{5} + \dfrac{6}{7}$

   (b) $\dfrac{7}{8} + \dfrac{2}{5}$

   (c) $\dfrac{3}{y} + \dfrac{5}{x}$

   (d) $\dfrac{d}{b} + \dfrac{a}{bc}$

   > (a) $\dfrac{58}{35}$
   >
   > $\left(\dfrac{4}{5} + \dfrac{6}{7} = \dfrac{7 \cdot 7 + 5 \cdot 6}{35} = \dfrac{28 + 30}{35} = \dfrac{58}{35}\right)$
   >
   > (b) $\dfrac{51}{40}$
   >
   > $\left(\dfrac{7}{8} + \dfrac{2}{5} = \dfrac{7 \cdot 5 + 8 \cdot 2}{40} = \dfrac{35 + 16}{40} = \dfrac{51}{40}\right)$
   >
   > (c) $\dfrac{3x + 5y}{xy}$
   >
   > $\left(\dfrac{3}{y} + \dfrac{5}{x} = \dfrac{3x + 5y}{xy}\right)$
   >
   > (d) $\dfrac{dc + a}{bc}$
   >
   > $\left(\dfrac{d}{b} + \dfrac{a}{bc} = \dfrac{dc}{bc} + \dfrac{a}{bc} = \dfrac{dc + a}{bc}\right)$

Section 2   Addition and Subtraction of Rational Numbers   **119**

5. Perform the following subtractions.

   (a) $\dfrac{4}{12} - \dfrac{1}{3}$

   (b) $11 - (\dfrac{3}{5} - \dfrac{7}{45})$

   (c) $\dfrac{x}{5} - \dfrac{y}{10}$

   > (a) $0 \;\left(\dfrac{4}{12} - \dfrac{1}{3} = \dfrac{4}{12} - \dfrac{4}{12} = 0\right)$
   >
   > (b) $\dfrac{95}{9}$
   >
   > $\left(11 - \left(\dfrac{3}{5} - \dfrac{7}{45}\right) = 11 - \left(\dfrac{27}{45} - \dfrac{7}{45}\right) = \right.$
   > $\left. 11 - \dfrac{20}{45} = \dfrac{495}{45} - \dfrac{20}{45} = \dfrac{475}{45} = \dfrac{95}{9}\right)$
   >
   > (c) $\dfrac{2x - y}{10}$
   >
   > $\left(\dfrac{x}{5} - \dfrac{y}{10} = \dfrac{2x}{10} - \dfrac{y}{10} = \dfrac{2x - y}{10}\right)$

6. Change each improper fraction to a mixed number.

   (a) $\dfrac{37}{6}$

   (b) $\dfrac{19}{5}$

   (c) $\dfrac{41}{8}$

   > (a) $6\dfrac{1}{6}$
   >
   > $\dfrac{37}{6} = \dfrac{36 + 1}{6} = \dfrac{36}{6} + \dfrac{1}{6} = 6 + \dfrac{1}{6} = 6\dfrac{1}{6}$
   >
   > (b) $3\dfrac{4}{5}$
   >
   > (c) $5\dfrac{1}{8}$

7. Change each mixed number to an improper fraction.

   (a) $6\dfrac{3}{4}$

CHAPTER 5 Introduction to the Rational Numbers

(b) $14\frac{3}{7}$

(c) $10\frac{4}{5}$

(a) $\frac{27}{4}$ $\left(6\frac{3}{4} = 6 + \frac{3}{4} = \frac{6}{1} + \frac{3}{4} = \frac{24}{4} + \frac{3}{4} = \frac{27}{4}\right)$

(b) $\frac{101}{7}$

$\left(14\frac{3}{7} = 14 + \frac{3}{7} = \frac{14}{1} + \frac{3}{7} = \frac{98}{7} + \frac{3}{7} = \frac{101}{7}\right)$

(c) $\frac{54}{5}$

8. Carry out the indicated operations.

(a) $3\frac{1}{3} + 1\frac{3}{4}$

(b) $3\frac{1}{3} - 1\frac{3}{4}$

(c) $26\frac{5}{8} - 19\frac{9}{10}$

(a) $4\frac{7}{12}$ $\quad 3\frac{1}{3} = 3\frac{4}{12}$
$\quad\quad\quad\quad + 1\frac{3}{4} = 1\frac{3}{12}$
$\quad\quad\quad\quad\quad\quad\quad = 4\frac{7}{12}$

(b) $\frac{7}{12}$ $\quad 3\frac{1}{3} = 3\frac{4}{12} = 2\frac{16}{12}$
$\quad\quad\quad - 1\frac{3}{4} = 1\frac{9}{12} = 1\frac{9}{12}$
$\quad\quad\quad\quad\quad\quad\quad\quad\quad\quad = \frac{7}{12}$

(c) $6\frac{29}{40}$

9. Solve each of the following for x.

(a) $x - \frac{1}{3} = \frac{3}{4}$

(b) $x + 3\frac{1}{2} = 5\frac{3}{4}$

(c) $\frac{1}{3} - x = \frac{1}{5}$

(a) $x = 1\frac{1}{12}$ $x - \frac{1}{3} = \frac{3}{4}$

$x = \frac{3}{4} + \frac{1}{3} = \frac{9+4}{12} = \frac{13}{12} = 1\frac{1}{12}$

(b) $x = 2\frac{1}{4}$

(c) $x = \frac{2}{15}$ $\frac{1}{3} - x = \frac{1}{5}$

$\frac{1}{3} = \frac{1}{5} + x$

$x = \frac{1}{3} - \frac{1}{5} = \frac{5-3}{15} = \frac{2}{15}$

10. Illustrate the addition $\frac{1}{3} + \frac{1}{4}$ using each of the following models.

   (a) fraction bars
   (b) number line
   (c) set model

Answers follow.

(a)

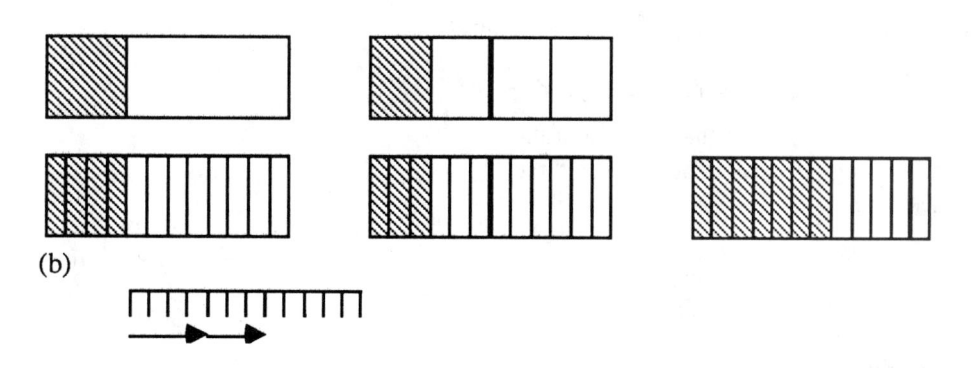

(b)

(c)

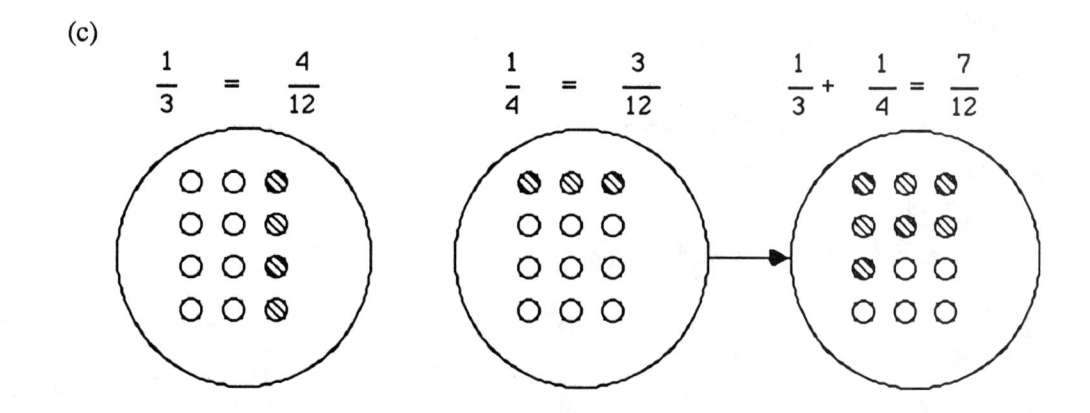

$\frac{1}{3} = \frac{4}{12}$ $\frac{1}{4} = \frac{3}{12}$ $\frac{1}{3} + \frac{1}{4} = \frac{7}{12}$

11. Perform each of the following.

(a) $\dfrac{2xy}{x^2y} + \dfrac{5x^2y}{xy^2}$

(b) $\dfrac{5}{a+b} + \dfrac{3}{a-b}$

(a) $\dfrac{2xy^2 + 5x^3y}{x^2y^2}$

$\left(\dfrac{2xy}{x^2y} + \dfrac{5x^2y}{xy^2} = \dfrac{2xy \cdot y}{x^2y^2} + \dfrac{5x^2y \cdot x}{x^2y^2}\right.$

$\left.= \dfrac{2xy^2 + 5x^3y}{x^2y^2}\right)$

(b) $\dfrac{8a - 2b}{(a+b)(a-b)}$

$\left(\dfrac{5}{a+b} + \dfrac{3}{a-b} = \dfrac{5(a-b)}{(a+b)(a-b)} + \dfrac{3(a+b)}{(a+b)(a-b)} = \dfrac{5a - 5b + 3a + 3b}{(a+b)(a-b)} = \dfrac{8a - 2b}{(a+b)(a-b)}\right)$

12. In a certain region of Europe, $\dfrac{2}{3}$ of the people speak French, $\dfrac{1}{5}$ of the people speak German, and $\dfrac{1}{10}$ of the people speak English. What fraction of the people in this region speak none of these three languages?

$\dfrac{1}{30}$

$x = 1 - \left(\dfrac{2}{3} + \dfrac{1}{5} + \dfrac{1}{10}\right)$

$= 1 - \left(\dfrac{20}{30} + \dfrac{6}{30} + \dfrac{3}{10}\right) = 1 - \dfrac{29}{30} = \dfrac{1}{30}$

13. Hester worked $7\dfrac{1}{2}$ hours last week grading mathematics assignments. This week she worked $13\dfrac{2}{3}$ hours.

(a) How many hours did she work for these two weeks?

(b) How many more hours did she work this week than last week?

(a) $21\frac{1}{6}$

(b) $6\frac{1}{6}$

14. Carry out the computations to verify each of the following.

   (a) $\frac{1}{2} = \frac{1}{3} + \frac{1}{2 \cdot 3}$

   (b) $\frac{1}{3} = \frac{1}{4} + \frac{1}{3 \cdot 4}$

   (c) $\frac{1}{4} = \frac{1}{5} + \frac{1}{4 \cdot 5}$

   (d) Write $\frac{1}{n}$ as a sum of 2 unit fractions. (That is, fractions with numerator equal to 1.)

   (e) Prove your answer for (d).

(a) $\frac{1}{3} + \frac{1}{2 \cdot 3} = \frac{1}{3} + \frac{1}{6} = \frac{2}{6} + \frac{1}{6} = \frac{3}{6}$
$= \frac{1}{2}$

(b) $\frac{1}{4} + \frac{1}{3 \cdot 4} = \frac{1}{4} + \frac{1}{12} = \frac{3}{12} + \frac{1}{12}$
$= \frac{4}{12} = \frac{1}{3}$

(c) $\frac{1}{5} + \frac{1}{4 \cdot 5} = \frac{1}{5} + \frac{1}{20} = \frac{4}{20} + \frac{1}{20}$
$= \frac{5}{20} = \frac{1}{4}$

(d) $\frac{1}{n} = \frac{1}{n+1} + \frac{1}{n(n+1)}$

(e) $\frac{1}{n+1} + \frac{1}{n(n+1)} =$
$\frac{n}{n(n+1)} + \frac{1}{n(n+1)} = \frac{n+1}{n(n+1)} = \frac{1}{n}$

Section 3 Multiplication and Division of Rational Numbers

Cover the right side of the page and work on the left, then check your work

1. Calculate mentally each of the following.

   (a) $5\frac{3}{4} \cdot 12$

   (b) $14 \cdot \left(\frac{4}{5} \cdot \frac{1}{7}\right)$

   (c) $4\frac{1}{2} \cdot \frac{1}{2}$

   (d) $21\frac{2}{5} \div 3$

   (a) 69

   $\left(5\frac{3}{4} \cdot 12 = \left(5 + \frac{3}{4}\right) \cdot 12 = 5 \cdot 12 + \frac{3}{4} \cdot 12 = 60 + 9 = 69\right)$

   (b) $\frac{8}{5}$

   $\left(14 \cdot \left(\frac{4}{5} \cdot \frac{1}{7}\right) = \left(14 \cdot \frac{1}{7}\right) \cdot \frac{4}{5} = 2 \cdot \frac{4}{5} = \frac{8}{5}\right)$

   (c) $2\frac{1}{4}$

   $\left(4\frac{1}{2} \cdot \frac{1}{2} = \left(4 + \frac{1}{2}\right)\frac{1}{2} = 4 \cdot \frac{1}{2} + \frac{1}{2} \cdot \frac{1}{2} = 2 + \frac{1}{4} = 2\frac{1}{4}\right)$

   (d) $7\frac{2}{15}$

   $\left(21\frac{2}{5} \div 3 = 21\frac{2}{5} \cdot \frac{1}{3} = \left(21 + \frac{2}{5}\right) \cdot \frac{1}{3} = 21 \cdot \frac{1}{3} + \frac{2}{5} \cdot \frac{1}{3} = 7 + \frac{2}{15} = 7\frac{2}{15}\right)$

2. Estimate each of the following in an appropriate range.

   (a) $3\frac{2}{5} \cdot 5\frac{9}{10}$

   (b) $8\frac{3}{4} \cdot 7\frac{1}{3}$

(c) $9\frac{5}{7} \cdot 3\frac{1}{2}$

(a) $15 - 24$
$$\left(3 < 3\frac{2}{5} < 4 \text{ and } 5 < 5\frac{9}{10} < 6\right)$$
(b) $56 - 72$
(c) $27 - 40$

3. Find an estimate for each of the following by rounding one factor to the nearest one-half and the other to the nearest whole number.

(a) $3\frac{2}{5} \cdot 5\frac{9}{10}$

(b) $8\frac{3}{4} \cdot 7\frac{1}{3}$

(c) $9\frac{5}{7} \cdot 3\frac{1}{2}$

(a) 21
$$\left(3\frac{2}{5} \cdot 5\frac{9}{10} \approx 3\frac{1}{2} \cdot 6 = 21\right)$$
(b) $67\frac{1}{2}$
$$\left(8\frac{3}{4} \cdot 7\frac{1}{3} \approx 9 \cdot 7\frac{1}{2} = 67\frac{1}{2}\right)$$
(c) 35
$$\left(9\frac{5}{7} \cdot 3\frac{1}{2} \approx 10 \cdot 3\frac{1}{2} = 35\right)$$

4. Decide whether each of the following is greater than or less than 5.

(a) $\frac{14}{13} \cdot \frac{51}{10}$

(b) $10\frac{3}{4} \div 2\frac{1}{3}$

(c) $10 \times \frac{1}{3}$

(a) greater than 5
$$\left(\frac{14}{13} \cdot \frac{51}{10} > 5 \text{ since } \frac{14}{13} > 1 \text{ and } \frac{51}{10} > 5\right)$$
(b) greater than 5
(c) less than 5

126 CHAPTER 5 Introduction to the Rational Numbers

$$\left(10 \cdot \frac{1}{3} < 5 \text{ since } \frac{1}{3} < \frac{1}{2}\right)$$

5. In the following figures, a unit rectangle is used to illustrate the product of two fractions. Write the fractions and their product.

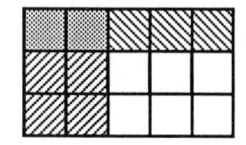

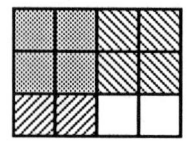

   (a)   (b)

(a) $\frac{1}{3} \times \frac{2}{5} = \frac{2}{15}$

(b) $\frac{2}{4} \cdot \frac{2}{3} = \frac{4}{12}$

6. Find each product. Write each answer as a reduced fraction.

   (a) $\frac{10}{9} \cdot \frac{17}{40}$

   (b) $\frac{22}{7} \cdot \frac{14}{3}$

   (c) $1\frac{1}{2} \cdot 1\frac{1}{3} \cdot 1\frac{1}{4}$

   (d) $\frac{xy}{z} \cdot \frac{xz^2}{x^3y^2}$

(a) $\frac{3}{4}$

$$\left(\frac{10}{9} \cdot \frac{27}{40} = \frac{\cancel{10}^{1}}{\cancel{9}_{1}} \cdot \frac{\cancel{27}^{3}}{\cancel{40}_{4}} = \frac{3}{4}\right)$$

(b) $\frac{44}{3}$

(c) $\frac{5}{2}$

$$\left(1\frac{1}{2} \cdot 1\frac{1}{3} \cdot 1\frac{1}{4} = \frac{3}{2} \cdot \frac{\cancel{4}^{1}}{\cancel{3}_{1}} \cdot \frac{5}{4} = \frac{5}{2}\right)$$

(d) $\frac{2}{xy}$

$$\left(\frac{xy}{z} \cdot \frac{xz^2}{x^3y^2} = \frac{xy}{z} \cdot \frac{xz^2}{x^3y^2} = \frac{z}{xy}\right)$$

Section 3 Multiplication and Division of Rational Numbers 127

7. Perform each of the following divisions and write the answer as a reduced fraction.

(a) $5 \div \dfrac{1}{8}$

(b) $\dfrac{15}{16} \div \dfrac{3}{8}$

(c) $2\dfrac{3}{4} \div 1\dfrac{1}{4}$

(d) $\dfrac{x}{y} \div \dfrac{x}{z}$

(a) 40 $\left(5 \div \dfrac{1}{8} = \dfrac{5}{1} \cdot \dfrac{8}{1}\right)$

(b) $\dfrac{5}{2}$ $\left(\dfrac{15}{16} \div \dfrac{3}{8} = \dfrac{\cancel{15}^5}{\cancel{16}_2} \cdot \dfrac{\cancel{8}^1}{\cancel{3}_1} = \dfrac{5}{2}\right)$

(c) $\dfrac{11}{5}$

$\left(2\dfrac{3}{4} \div 1\dfrac{1}{4} = \dfrac{11}{4} \div \dfrac{5}{4} = \dfrac{11}{4} \cdot \dfrac{4}{5} = \dfrac{11}{5}\right)$

8. Express each of the following in simplest form.

(a) $\dfrac{\frac{8}{7}}{\frac{3}{4}}$

(b) $\dfrac{\frac{1}{2} + \frac{1}{3}}{\frac{1}{2} - \frac{1}{3}}$

(c) $\dfrac{2\frac{1}{2} - 1\frac{1}{3}}{\frac{1}{6} + \frac{2}{3}}$

(a) $\dfrac{32}{21}$ $\left(\dfrac{\frac{8}{7}}{\frac{3}{4}} = \dfrac{8}{7} \cdot \dfrac{4}{3} = \dfrac{32}{21}\right)$

(b) 54

128 CHAPTER 5 Introduction to the Rational Numbers

(c) $\dfrac{7}{5} \left(\dfrac{2\frac{1}{2} - 1\frac{1}{3}}{\frac{1}{6} + \frac{2}{3}} = \dfrac{2\frac{3}{6} - 1\frac{2}{6}}{\frac{1}{6} + \frac{4}{6}} = \dfrac{\frac{7}{6}}{\frac{5}{6}} = \dfrac{7}{6} \cdot \dfrac{6}{5} = \dfrac{7}{5} \right)$

9. Solve each of the following for x.

    (a) $\dfrac{1}{3}x = \dfrac{7}{8}$

    (b) $3x = \dfrac{4}{5}$

    (c) $\dfrac{2}{3}\left(\dfrac{1}{2}x - 7\right) = \dfrac{3}{4}x$

(a) $x = \dfrac{21}{8}$ $\quad \dfrac{1}{3}x = \dfrac{7}{8}$

$\qquad\qquad\qquad x = \dfrac{7}{8} \cdot \dfrac{3}{1} = \dfrac{21}{8}$

(b) $x = \dfrac{4}{15}$

(c) $x = \dfrac{-56}{5}$ $\quad \dfrac{2}{3}\left(\dfrac{1}{2}x - 7\right) = \dfrac{3}{4}x$

$\qquad\qquad\qquad \dfrac{2}{6}x - \dfrac{14}{3} = \dfrac{3}{4}x$

$\qquad\qquad\qquad \dfrac{-14}{3} = \dfrac{3}{4}x - \dfrac{2}{6}x$

$\qquad\qquad\qquad \dfrac{-14}{3} = \dfrac{9}{12}x - \dfrac{4}{12}x$

$\qquad\qquad\qquad \dfrac{-14}{3} = \dfrac{5}{12}x$

$\qquad\qquad\qquad \dfrac{-14}{3} \cdot \dfrac{12}{5} = x = \dfrac{-56}{5}$

10. Find the multiplicative inverse for each of the following.

    (a) $\dfrac{3}{5}$

    (b) $2\dfrac{1}{4}$

    (c) $\dfrac{a}{b+c}$

(a) $\dfrac{5}{3}$

(b) $\dfrac{4}{9}$ $2\dfrac{1}{4} = \dfrac{9}{4}$

(c) $\dfrac{b+c}{a}$

11. Two weeks is:

(a) What part of a year?
(b) What part of a decade?
(c) What part of a century?

(a) $\dfrac{2}{52} = \dfrac{1}{26}$

(b) $\dfrac{2}{520} = \dfrac{1}{260}$

(c) $\dfrac{2}{5200} = \dfrac{1}{2600}$

12. Which of the following are geometric sequences? If a sequence is a geometric sequence, find the sum of its first 5 terms.

(a) $\dfrac{4}{3}, 1, \dfrac{3}{4}, \dfrac{9}{16}, \dfrac{27}{64}, \ldots$

(b) $\dfrac{1}{3}, \dfrac{2}{3^2}, \dfrac{3}{3^3}, \dfrac{4}{3^4}, \ldots$

(a) Geometric Sequence $a = \dfrac{4}{3}$ $r = \dfrac{3}{4}$

$$S_5 = \dfrac{\dfrac{4}{3}\left(\left(\dfrac{3}{4}\right)^5 - 1\right)}{\dfrac{3}{4} - 1} = \dfrac{\dfrac{4}{3}\left(\dfrac{243}{1024} - 1\right)}{\dfrac{3}{4} - 1}$$

$$= \dfrac{\dfrac{4}{3}\left(-\dfrac{781}{1024}\right)}{-\dfrac{1}{4}} = -\dfrac{16}{3}\left(-\dfrac{781}{1024}\right) = \dfrac{781}{192}$$

(b) Not a geometric sequence

$\left(\dfrac{2}{3^2} \div \dfrac{1}{3} \ne \dfrac{3}{3^3} \div \dfrac{2}{3^2}\right)$

130 CHAPTER 5 Introduction to the Rational Numbers

13. A school had a decrease in enrollment and lost one fifth of its students. If there were 280 students left after the decrease, how many students were there before the decrease?

(a) 350 $x - \dfrac{1}{5}x = 280$
(x = enrollment before decrease.)
$$\dfrac{4}{5}x = 280$$
$$x = \dfrac{280}{1} \cdot \dfrac{5}{4} = 350$$

14. Jonelle has $31\dfrac{1}{2}$ yards of material to use to make blouses. Each blouse requires $\dfrac{2}{3}$ yards of material.

    (a) How many blouses can be made?
    (b) How much material will be left over?

(a) 47 $x = 31\dfrac{1}{2} \div \dfrac{2}{3}$
$$x = \dfrac{63}{2} \cdot \dfrac{3}{2} = \dfrac{189}{4} = 47\dfrac{1}{4}$$
(b) $\dfrac{1}{4}$ of $\dfrac{2}{3}$ of a yard = $\dfrac{1}{6}$ yd.

15. Vince spent $\dfrac{1}{4}$ of his life as a boy, $\dfrac{1}{10}$ of his life in college, and $\dfrac{1}{2}$ his life as a teacher. He spent 12 years in retirement before he died. How old was he when he died?

80 years old
Let x = age when he died. Then,
$$\dfrac{1}{4}x + \dfrac{1}{10}x + \dfrac{1}{2}x + 12 = x$$
$$\left(\dfrac{1}{4} + \dfrac{1}{10} + \dfrac{1}{2}\right)x + 12 = x$$
$$\left(\dfrac{5}{20} + \dfrac{2}{20} + \dfrac{10}{20}\right)x + 12 = x$$
$$\dfrac{17}{20}x + 12 = x$$
$$12 = x - \dfrac{17}{20}x$$
$$12 = \dfrac{3}{20}x$$
$$\dfrac{12}{1} \cdot \dfrac{20}{3} = 80 = x$$

Section 4 A Return to Problem Solving

Cover the right side of the page and work on the left, then check your work

1. Solve for the variable on the domain of rational numbers.

   (a) $5x - 6 = 14$
   (b) $14 - 3x = 2x - 4$
   (c) $\dfrac{x}{3} - 4 = {}^-1$
   (d) $\dfrac{x}{3} - \dfrac{1}{6} = \dfrac{x}{6} + \dfrac{1}{3}$

   (a) $x = 4$ $5x - 6 = 14$
                  $5x = 20$ add 6 to both sides
                  $x = 4$ divide both sides by 5
   (b) $x = \dfrac{18}{5}$
   (c) $x = 9$ $\dfrac{x}{3} - 4 = {}^-1$
                  $\dfrac{x}{3} = 3$ add 4 to both sides
                  $x = 9$ multiply both sides by 3
   (d) $x = 3$ $\dfrac{x}{3} - \dfrac{1}{6} = \dfrac{x}{6} + \dfrac{1}{3}$
                  $2x - 1 = x + 2$ multiply both sides by 6
                  $2x = x + 3$ add 3 to both sides
                  $x = 3$ subtract x from both sides

2. State the property that is use in each.

   (a) $4x - 3 = 9$ so $(4x - 3) + 3 = 9 + 3$
   (b) $5x = 30$ so $\dfrac{5x}{5} = \dfrac{30}{5}$
   (c) $2x + 7 = {}^-2$ so $(2x + 7) - 7 = {}^-2 - 7$
   (d) $\dfrac{2x}{3} = 7$ so $3 \cdot \dfrac{2x}{3} = 3 \cdot 7$

   (a) Property 1
   (b) Property 3
   (c) Property 1
   (d) Property 2

3. Solve for the variable indicated.

   (a) $P = 2l + 2w$; w
   (b) $V = lwh$; l

(c) $V = \frac{1}{3}\pi r^2 h$; h

(a) $w = \frac{P}{2} - l \quad P = 2l + 2w$
$P - 2l = 2w$
$\frac{P}{2} - l = w$

(b) $l = \frac{V}{wh}$

(c) $h = \frac{3v}{\pi r^2}$

For exercises 4 - 12 use a guess and make a table to set up the model. Then solve and check the equation.

4. Mr. Delmar mixed 5 pounds of peanuts worth $.89 per pound with 3 pounds of cashews worth $1.37 per pound. How much should he charge per pound for the 8 pounds of mix in order to make the same amount of money compared to selling each type of nut separately.

   $1.07 per pound

   Guess that he charges $1.00 per pound for the mixture. Then complete a table as shown below.

Price/lb.	Mixture	Peanuts	Cashews
Guess	$1.00	$0.89	$1.37
Value of	($1.00)8 =	($0.89)5 +	($1.37)3
	$8.00 =	$4.45 +	$4.11
Replace guess			
Value of	$8x =$	5($0.89) +	3($1.87)

   $8x = 5(0.89) + 3(1.37)$ Check:
   $8x = 4.45 + 4.11$ $5(0.89) = 4.45$
   $8x = 8.56$ $3(1.37) = \underline{4.11}$
   $x = \$1.07$ $8(1.07) = 8.56$

5. The sum of 2 numbers is 22. If 3 times the smaller is 1 more than twice the larger, what are the numbers?

> smaller number = 9
> larger number = 13

6. Winston has 3 dollars in nickels, dimes, and quarters. There are 2 more dimes than nickels, but the number of quarters is the same as the number of nickels. How many coins does he have?

> number of quarters = 7
> number of dimes = 9
> number of nickels = 7
> TOTAL = 23

	Quarters	Dimes	Nickels	Total
Guess	8	10	9	26
Value of	$(0.25)8 =$ $8.00 =$	$(0.10)10 +$ $4.45 +$	$(0.05)8$ 4.11	= \$3.40 (too large)
Assign Variables	x	$x + 2$	x	
Value of	$(0.25)x +$	$(0.10)(x + 2) +$	$(0.05)x =$	\$3.00

> $.25x + (.10)(x + 2) + .05x = 3.00$
> $.25x + .10x + .20 + .05x = 3.00$
> $.40x = 2.80$
> $x = 7$
> $x + (x + 2) + x = 7 + 9 + 7 = 23$

7. Mrs. Burnam purchased 6 greeting cards for a cost of \$3.75. Some were \$.50 each, and some \$1.25 each. How many cards of each price did she buy?

> number of 50¢ cards = 1
> number of \$1.25 cards = 5

8. Thor broke a meter stick into 2 pieces. The longer piece was 8 cm short of being 3 times the other. Find the length of each piece.

> length of short piece = 27 cm
> length of long piece = 73 cm

Length	Short Piece	Long Piece	Total
Guess	$20 +$	$3(20) - 8 = 52$	72 (too short)
Assign Variables	$x +$	$3x - 8 =$	100

134 CHAPTER 5 Introduction to the Rational Numbers

$$x + 3x - 8 = 100$$
$$4x = 108$$
$$x = 27$$
$$3x - 8 = 73$$

9. Mr. Podstepny invested part of $10,000 in a savings certificate at 7% interest. He invested the remainder in a bond fund at 12% interest. If he received a total of $900 in interest for the year, how much money was invested in the bond fund?

$4,000 invested in bond fund

10. Find 3 consecutive integers who sum is 48.

first integer = 15
second integer = 16
third integer = 17

Integer	1st	2nd	3rd	Total
Guess	12 +	13 +	14 =	39
Assign Variables	$x +$	$x + 1 +$	$x + 1 =$	48

$$x + (x + 1) + (x + 2) = 48$$
$$3x + 3 = 48$$
$$3x = 45$$
$$x = 15$$

11. The longest side of a triangle is 3 cm longer than the shortest side. The third side is 2 cm less than twice the length of the shortest side. If the perimeter is 13 cm, how long is each side?

shortest side = 3
longest side = 6
other side = 4

12. In 5 years Mark Francisco will be 3 times as old as he was 7 years ago. How old is Mark?

Mark's age = 13

Mark's Age	Now	7 years ago	in 5 years
Guess	10	3	3(3) = 9 (Too small)
Assign Variables	x	$x - 7$	$x + 5$

Relationship $x + 5 = 3(x - 7)$
$x + 5 = 3(x - 7)$

$$x + 5 = 3x - 21$$
$$26 = 2x$$
$$x = 13$$

13. When 6 gallons of gasoline are put into a car, the indicator goes from $\frac{1}{4}$ to $\frac{5}{8}$. What is the total capacity of the gasoline tank?

   | 16 gallons

14. Connie has $40,000 to invest. If she invests $16,000 at 12% and $14,000 at 8%, at what rate should she invest the remainder in order to have a yearly income of $4000 from her investments.

   | Invest at a rate of 9.6%

Investment	$16,000	$14,000	$10,000	Total
Guess	12%	8%	10%	
Interest	16,000(0.12) +	14,000(0.08) +	10,000(0.10) =	$4,090 Too large
Assign Variables	16,000(0.12) +	14,000(0.08) +	10,000(x) =	4,000

$$16000(.12) + 14000(.08) + 10000x = 4000$$
$$1920 + 1120 + 10000x = 4000$$
$$10000x = 960 \qquad x = 9.6\%$$

15. Mr. Thornton invests x dollars, while Mr. Greenberg and Mr. Roberts each invest y dollars to form a corporation. What part of $100 profit should Mr. Thornton receive if each shares in the profits in proportion to the amount invested?

   | $\dfrac{100x}{x + 2y}$
   | $x \leftarrow$ Mr. Thornton investment
   | $x + 2y \leftarrow$ total investment
   | $\dfrac{x}{x + 2y} \cdot 100 = \dfrac{100x}{x + 2y}$

Section 5 An Introduction to Decimals

Cover the right side of the page and work on the left, then check your work

1. Do each of the following using mental mathematics.

   (a) 5.96 + 4.25
   (b) 23.24 − 13.97
   (c) 7.08 + 5.60
   (d) 18.95 − 11.37

   (a) 10.21   5.96 + 4.25 =
   (5.96 − .04) + (4.25 − .04) =
   6.00 + 4.21 = 10.21
   (b) 9.27
   23.24 − 13.97 =
   (23.24 + .03) − (13.97 + .03) =
   23.27 − 14.00 = 9.27
   (c) 12.68
   7.08 + 5.60 =
   (7.08 + .02) + (5.60 − .02) =
   7.10 + 5.58 = 12.68
   (d) 7.58

2. Find the whole number range for each of the following.

   (a) 3.96 + 7.58
   (b) 12.82 + 7.26
   (c) 4.123 − 3.432
   (d) 37.675 − 14.412

   (a) 10 – 12
   3 < 3.96 < 4 and 7 < 7.58 < 8
   (b) 19 – 21
   (c) 0 – 2
   (d) 22 – 24

3. Estimate each of the following by rounding to the nearest whole number.

   (a) 3.96 + 7.58
   (b) 12.82 + 7.26
   (c) 4.123 − 3.432
   (d) 37.675 − 14.412

   (a) 12   3.96 ≈ 4 and 7.58 ≈ 8
   (b) 20
   (c) 1   4.123 ≈ 4 and 3.432 ≈ 3
   (d) 24

Section 5   An Introduction to Decimals   137

4. Write each of the following in expanded form.

   (a) 10.362
   (b) 341.63
   (c) 0.317

   > (a) 1.0362×10^1
   > (b) 3.4163×10^2
   > (c) 3.17×10^{-1}

5. Express each of the following as a fraction.

   (a) 0.436
   (b) 25.31
   (c) 18.9102

   > (a) $\dfrac{436}{1000}$
   > (b) $\dfrac{2531}{100}$
   > (c) $\dfrac{189,102}{10000}$

6. Express each of the following as a decimal fraction.

   (a) $\dfrac{48}{10}$
   (b) $\dfrac{7}{16}$
   (c) $\dfrac{74}{25}$

   > (a) 4.8 $\dfrac{48}{10} = \dfrac{40}{10} + \dfrac{8}{10} = 4\dfrac{8}{10}$
   > (b) 0.4375
   > $\dfrac{7}{16} = \dfrac{7}{16} \cdot \dfrac{625}{625} = \dfrac{4375}{10000} =$
   > $\dfrac{4000}{10000} + \dfrac{3}{10000} + \dfrac{70}{10000} + \dfrac{5}{10000}$
   > $= \dfrac{4}{10} + \dfrac{3}{100} + \dfrac{7}{1000} + \dfrac{5}{1000} =$
   > .4375
   > (c) 2.96

7. Which of the following can be expressed as a terminating decimal.

   (a) $\dfrac{19}{60}$

138 CHAPTER 5 Introduction to the Rational Numbers

(b) $\dfrac{107}{625}$

(c) $\dfrac{18}{90}$

> (a) not terminating $60 = 2 \cdot 2 \cdot 3 \cdot 5$
> (b) terminating $625 = 5^4$
> (c) terminating $\dfrac{18}{90} = \dfrac{18 \cdot 1}{18 \cdot 5} = \dfrac{1}{5}$

8. Round each number below as indicated.

   (a) 367.483 (nearest hundredth)
   (b) 86.392 (nearest tenth)
   (c) 0.1347 (nearest thousandth)

> (a) $367.483 \approx 367.48$
> (b) $86.392 \approx 86.4$
> (c) $0.1347 \approx 0.135$

9. Write each of the following numbers in scientific notation.

   (a) 3 billion
   (b) 4967.37
   (c) 0.12345

> (a) 3×10^9
> 3 billion = 3,000,000,000
> (b) 4.96737×10^3
> (c) 1.2345×10^{-5}

10. Write each of the following in standard notation

    (a) 27.3×10^{-3}
    (b) 7×10^5
    (c) 19.36×10^{-3}

> (a) 0.273
> (b) 700,000
> (c) 0.01936

11. Perform the following operations.

    (a) $36.812 + 1.96$
    (b) $126.431 + 237.478$
    (c) $132.36 - 63.48$
    (d) $200.01 - 32.007$

	(a) 38.772 36.812
	             + 1.960
	              38.772
	 (b) 363.909
	 (c) 68.88 132.36
	            − 63.48
	             68.88
	 (d) 168.003 200.010
	             − 32.007
	              168.003

12. Perform each of the following.

    (a) $10.000_{two} + 1.11_{two}$
    (b) $10.100_{two} - 1.11_{two}$
    (c) $36.25_{seven} + 143.44_{seven}$

	(a) 100011_{two} 10.100_{two}
	                       $+ 1.111_{two}$
	                          1
	                        10
	                      100
	                    1000
	                 10000
	              100011_{two}
	
	 (b) 0.101_{two} 10.100_{two}
	                    $- 1.111_{two}$
	                     0.101_{two}
	
	 (c) 213.02 36.25_{seven}
	                  $+ 143.44_{seven}$
	                     12
	                    60
	                 1200
	               10000
	               10000
	               213.02

13. If the rainfall in August was 1.34 inches and in September was 5.17 inches, how much more rain was there in September?

	3.83 inches $x = 5.17 - 1.34$ $x = 3.83$

14. Leo bought a shirt for $13.98, a pair of pants for $21.49, and some shoes for $45.69. How much change will he receive if he pays with a $100 bill?

$18.84
$x = 100 - (13.98 + 21.49 + 45.69)$
$= 100 - (81.16)$
$= \$18.84$

15. Stacy bought 20 pounds of one type of candy for $2.00 per pound and 30 pounds of another type of candy for $4.00 per pound. She wants to buy 10 pounds of another type of candy so that the average price per pound of all the candy is $3.00. What price must she pay for the third type of candy?

$2.00

Price/lb.	Type 1	Type 2	Type 3
Guess	$2	$4	$3
Value	2(20) +	4(30) +	3(10) = 190
Average	190 ÷ 60 = $3.17	(too high)	
Assign Variables	2(20)	4(30)	x(10)

$[2(20) + 4(30) + x(10)] \div 60 = 3$
$x(10) = 20$
$x = \$2$

Section 6 The Arithmetic of Decimals

Cover the right side of the page and work on the left, then check your work

1. Use mental arithmetic to answer each of the following.

   (a) (.875) (40)
   (b) (.15) (80)
   (c) 7 (8.5)
   (d) 4.2 (3.65 + 6.35)

   > (a) 35
   > $(.875) \cdot (40) = \dfrac{7}{8} \cdot \dfrac{40}{1} = 35$
   > (b) 12
   > (c) 59.5
   > $7 \cdot (8.5) = (7 \cdot 8) + (7 \cdot 0.5)$
   > $= 56 + 3.5 = 59.5$
   > (d) 42
   > $4.2 (3.65 + 6.35) = 4.2 (10) = 42$

2. Find the range for each of the following.

   (a) (12.321) · (2.674)
   (b) (8.57) · (7.49)
   (c) 15.37 ÷ 4.63
   (d) 7.83 ÷ 2.43

   > (a) 24 – 39
   > $12 < 12.321 < 13$ and $2 < 2.674 < 3$
   > (b) 56 – 72
   > (c) 3 – 4
   > $15 < 15.37 < 16$ and $4 < 4.63 < 5$
   > (d) $2\dfrac{1}{3} - 4$

3. Estimate each of the following by rounding to one significant digit.

   (a) (12.321) · (2.674)
   (b) (8.57) · (7.49)
   (c) 15.37 ÷ 4.63
   (d) 7.83 ÷ 2.43

   > (a) 30
   > $12.32 \approx 10$ and $2.674 \approx 3$
   > (b) 72
   > (c) 4
   > $15.37 \approx 20$ and $4.63 \approx 5$
   > (d) 4

4. Place the decimal point in each by estimating.

142 CHAPTER 5 Introduction to the Rational Numbers

(a) $(19.7) \cdot (6.4) = 12608$
(b) $(0.32) \cdot (27.362) = 875584$
(c) $432.5 \div 19.4 = 222938$

<blockquote>
(a) 126.08
$19.7 \approx 20$ and $6.4 \approx 6$
$20 \times 6 = 120$
(b) 8.75584
(c) 22.2938
$432.5 \approx 400$ and $19.4 \approx 20$
$400 \div 20 = 20$
</blockquote>

5. For each of the following, change each decimal to a fraction, perform the computation, and then change the answer to a fraction.

   (a) $(1.36)(0.02)$
   (b) $1.36 \div 0.02$
   (c) $(12.25) \cdot (1.31)$

<blockquote>
(a) 0.0272
$(1.36)(0.02) = \dfrac{136}{100} \times \dfrac{2}{100} = \dfrac{272}{10000}$
$= 0.0272$
(b) 68
(c) 16.0475
$(12.25)(1.31) = \dfrac{1225}{100} \times \dfrac{131}{100} =$
$\dfrac{160475}{10000} = 16.0475$
</blockquote>

6. Perform each division below by first writing each division as a fraction having a whole number denominator.

   (a) $0.45\overline{)2.835}$
   (b) $2.6\overline{)24.336}$
   (c) $0.28\overline{)1.176}$

<blockquote>
(a) 6.3
$0.45\overline{)2.835} = \dfrac{283.5}{45} = 6.3$
(b) 9.36
$2.6\overline{)24.336} = \dfrac{243.36}{26} = 9.36$
(c) 4.2
</blockquote>

7. Arrange the following numbers from smallest to largest.

   (a) 4.3, 4.03, 4.33, 4.031
   (b) 0.345, 0.3451, 0.3449, 0.346

(c) 2.47, ⁻2.47, ⁻2.475, 2.475

8. List 3 decimal numbers that are between the two given numbers.

(a) 4.3 and 4.4
(b) 0.345 and 0.3451
(c) $\frac{2}{4}$ and $\frac{3}{8}$

(a) 4.03, 4.031, 4.3, 4.33
(b) 0.3449, 0.345, 0.3451, 0.346
(c) ⁻2.475, ⁻2.47, 2.47, 2.475

(a) 4.31, 4.32, 4.33 (others are possible)
(b) 0.34505, 0.34506, 0.34507
(c) 0.30, 0.32, 0.35
$\frac{2}{4} = 0.25$ and $\frac{3}{8} = 0.375$

9. Change each factor to scientific notation and then multiply or divide the approximate numbers.

(a) (375.21) · (0.35)
(b) (426,370) · (0.0084)
(c) 573.89 ÷ 43.2

(a) 1.3×10^2
(375.21) (0.35) =
$(3752 \times 10^2)(3.5 \times 10^{-1})$
$= 13.13235 \times 10^1 = 1.313235 \times 10^2$
(round to 2 significant digits)
(b) 3582
(c) 13.3   (573.89) ÷ (43.2) =
$(5.7389 \times 10^2) \div (4.32 \times 10^1)$
$= 1.3283391 \times 10^1$
(round to 3 significant digits)

10. Perform each of the following computations using approximate numbers. Leave the answers in scientific notation.

(a) $(8 \cdot 10^{12}) \cdot (6 \cdot 10^{15})$
(b) $(5.2 \times 10^3) \cdot (3.7 \cdot 10^4)$
(c) $(16 \cdot 10^{12}) \div (4 \cdot 10^5)$
(d) $(9.3 \times 10^8) \div (5.67 \times 10^3)$

(a) 4.8×10^{28}
$(8 \cdot 10^{12}) \cdot (6 \cdot 10^{15}) = 48 \times 10^{27} = 4.8 \times 10^{28}$

144   CHAPTER 5   Introduction to the Rational Numbers

11. Rusty's car travels 347 miles on 13 gallons of gas. How many miles to the gallon does his car get? (Round to the nearest tenth of a mile).

12. Yesterday, 2.675 inches of rain fell in a 6 hour period. What was the average amount of rain per hour? (to the nearest hundredth of an inch).

13. One quart of water weighs 2.082 pounds. One cubic foot of water is 29.922 quarts. How much does the cubic foot of water weigh? (to the nearest hundredth of a pound).

14. A local bank has the following formula for determining the service charge on a checking account. Plan 1 charges 10¢ per check written. Plan 2 has a 75¢ service charge per month plus 5¢ per check written.

    (a) Which plan is most economical if 14 checks per month are written?
    (b) Which plan is most economical if 40 checks per month are written?
    (c) For what number of checks would the two plans charge the same amount?

15. Write a story problem in which the required operation is $30 \div 4.59$.

(b) 1.9×10^8
(c) 4×10^7
(d) 1.6×10^5
$(9.3 \times 10^8) \div (5.67 \times 10^3) = 1.6402116 \times 10^5$

| 26.7 miles $x = 347 \div 13$

| 0.45 inch $x = 2.675 \div 6$

| 62.3 pounds $x = 2.082 \times 29.922$

| (a) Plan 1 Plan 1   Plan 2
         $14 \times \$0.10 = \$1.40$ $\$0.75 + 14 \times \$0.05 = \$1.45$
  (b) Plan 2 Plan 1   Plan 2
         $40 \times \$0.10 = \$4.00$ $\$0.75 + 40 \times \$0.05 = \$2.75$
  (c) 15 checks
      $0.10x = 0.75 + 0.05x$
      $0.05x = 0.754$
      $x = 15$

| Possible answer: Cassette tapes are on sale for $4.59 each. How many could be purchased with $30?

Chapter 6
From Rational Numbers to Real Numbers

Section 1 Ratio and Proportion

Cover the right side of the page and work on the left, then check your work

1. Express each ratio as a fraction in simplest terms.

   (a) 7 : 14
   (b) 20 : 5
   (c) $4\frac{1}{2} : 9\frac{1}{2}$
   (d) 10.5 : 15

   (a) $\frac{1}{2}$ $\frac{7}{14} = \frac{7 \cdot 1}{7 \cdot 2} = \frac{1}{2}$

   (b) $\frac{4}{1}$

   (c) $\frac{9}{19}$ $\frac{4\frac{1}{2}}{9\frac{1}{2}} = \frac{\frac{9}{2}}{\frac{19}{2}} = \frac{9}{2} \cdot \frac{2}{19} = \frac{9}{19}$

   (d) $\frac{7}{10}$ $\frac{10 \cdot 5}{15} = \frac{10\frac{1}{2}}{15} = \frac{\frac{21}{2}}{15} = \frac{21}{2} \cdot \frac{1}{15} = \frac{7}{10}$

2. Solve for x in each proportion.

   (a) $\frac{12}{x} = \frac{18}{45}$
   (b) $\frac{x}{15} = \frac{7}{3}$
   (c) $\frac{20}{80} = \frac{14}{x}$
   (d) $5 : 7 = 3x : 98$

   (a) $x = 30$ $\frac{12}{x} = \frac{18}{45} \Rightarrow 18x = 540$
              $x = 30$

   (b) $x = 35$
   (c) $x = 56$
   (d) $x = 21\frac{1}{3}$ $\frac{5}{7} = \frac{3x}{98} \Rightarrow 21x = 98$
              $21x = 490$
              $x = 21\frac{1}{3}$

146   CHAPTER 6   From Rational Numbers to Real Numbers

3. In Mo's family, there are 5 boys and 7 girls.

    (a) What is the ratio of boys to girls?
    (b) What is the ratio of girls to boys?
    (c) What is the ratio of boys to all children in Mo's family?

    | (a) 5 : 7
    | (b) 7 : 5
    | (c) 5 : 12   5 boys + 7 girls = 12 children in all

4. If a 4-ounce can of salt costs 79¢, what is the cost per ounce?

    | 19.75¢ $\dfrac{79}{4} = \dfrac{x}{1}$
    | $4x = 79$
    | $x = 19.75$

5. If 2 watermelons cost $1.79, how much do 9 watermelons cost?

    | $8.06
    |
    | | # of Watermelons | 2 | 9 |
    | |------------------|------|---|
    | | Cost | $1.79 | x |
    |
    | $\dfrac{2}{1.79} = \dfrac{9}{x}$
    | $2x = 16.11$
    | $x = 8.055$

6. On a road map, $\dfrac{1}{2}$ inch represents 40 miles. If Kansas City and St. Louis are 3 inches apart on the map, what is the actual distance between these cities?

    | 240 miles

7. If it takes 25 gallons of gasoline to drive 450 miles, how much gasoline should it take to drive 950 miles?

    | 53 gallons
    |
    | | Distance | 450 | 954 |
    | |-------------|-----|-----|
    | | # of gallons | 25 | x |

$$\frac{450}{25} = \frac{954}{x}$$
$$450x = 25 \cdot 954$$
$$450x = 23850$$
$$x = 53$$

8. In a certain school district the student-teacher ratio is 27 : 1. If there are 137 teachers in the district, how many students are there?

3699 students

# of Students	27	x
# of Teachers	1	137

$$\frac{27}{1} = \frac{x}{137}$$
$$x = 27 \cdot 137$$
$$x = 3699$$

9. Hazel types at a rate of 67 words per minute. At that rate, how long will it take her to type a paper of 1541 words?

23 minutes

10. Which is the better buy, a 16-ounce can of corn that costs 65¢ or a 12-ounce can that costs 53¢?

The 16-ounce can

Cost	65	53
# of Ounces	16	12
Cost per Ounce	$\frac{65}{16} = 4.07¢$	$\frac{53}{12} = 4.42¢$

11. An airliner travels at a speed of 620 miles per hour.

   (a) How long will it take the airliner to travel 3410 miles?
   (b) How far can it travel in 7 hours?

(a) 5.5 hours
(b) 4340 miles

12. Find x in each of the following.

148 CHAPTER 6 From Rational Numbers to Real Numbers

(a)

(b)

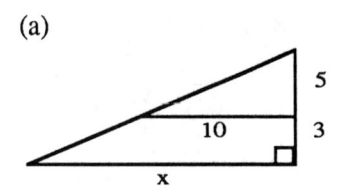

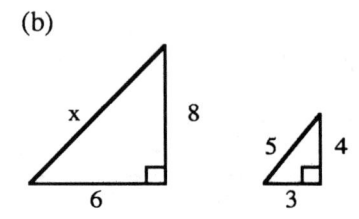

(a) $x = 16$	$\dfrac{x}{10} = \dfrac{8}{5}$
	$5x = 80$
	$x = 16$
(b) $x = 10$	$\dfrac{x}{5} = \dfrac{8}{4}$
	$4x = 40$
	$x = 10$

13. Charlie's recipe calls for the following: 3 eggs, $2\frac{1}{2}$ cups of flour, $1\frac{1}{4}$ cups of sugar, and $1\frac{1}{2}$ cups of milk. What amount of each ingredient will be needed in order to triple the recipe?

6 eggs
5 cups of flour
$2\frac{1}{2}$ cups of sugar
3 cups of milk

14. A school has 1200 students and teacher-student ratio of 1: 30.

   (a) How many additional teachers must be hired to reduce the ratio to 1 : 20?
   (b) If beginning teachers are paid $22,000, how much will it cost the school to hire these additional teachers?

(a) 20 teachers

Section 1   Ratio and Proportion   149

Number of Students	30	1200	20	1200
Number of Teachers	1	x	1	y

$\dfrac{30}{1} = \dfrac{1200}{x}$ $\dfrac{20}{1} = \dfrac{1200}{y}$
$30x = 1200$ $20y = 1200$
$x = 40$ teachers $y = 60$ teachers
additional teachers $= 60 - 40 = 20$

(b) $440,000
$\dfrac{22000}{1} = \dfrac{x}{20}$
$x = 20 \cdot 22000$
$x = \$440,000$

15. Jay is going to drive to Chicago to visit a friend. If he travels at 65 mph, he will be 1 hour early in arriving. If he travels at 55 mph, he will be 1 hour late in arriving. How far is it to Chicago?

715 miles
Let d = distance traveled
    t = number of hours to arrive on time

Rate	Time	Distance
65	t - 1	65(t - 1)
55	t - 1	55(t - 1)

$65(t - 1) = 55(t + 1)$
$65t - 65 = 55t + 55$
$10t = 120,\ t = 12$
$d = 65(t - 1) = 65(11) = 715$

Section 2 The Language of Percent

Cover the right side of the page and work on the left, then check your work

1. Estimate each of the following.

   (a) The lunch bill is $15.60. What will be the amount of the tip if it is 15% of the bill?
   (b) Howard pays 21% interest on a car loan of $4,000. How much interest does he pay?
   (c) 25% of $358
   (d) 40% markdown on a dress that sells for $97.99.

   (a) $2.40
   15% of $15.60 =
   10% of $15.60 + 5% of $15.60
   $\approx 1.60 + \frac{1}{2}$ of 10% = 1.60 + 0.80
   = $2.40
   (b) $800
   (c) 90
   25% of 358 ≈ 25% of 360
   = $\frac{1}{4}$ of 360 = 90
   (d) $40

2. Do each of the following using mental computation.

   (a) 25% × 56 = _____
   (b) 36 is 25% of _____
   (c) 38 is _____% of 50
   (d) 48% × 25 = _____

   (a) 14
   25% of 56 = $\frac{1}{4}$ of 56 = 16
   (b) 144
   25% of □ = 36
   $\frac{1}{4}$ of □ = 36
   □ = 4 · 36
   □ = 144
   (c) 76%
   38 is 38% of 100
   38 is (38% × 2) of 50
   (d) 12
   48% × 25 = 48 × 25%
   = 48 × $\frac{1}{4}$ = 12

3. Convert each to decimals

   (a) 19.3%
   (b) 219%
   (c) 0.34%
   (d) 7.9%

   | (a) 0.193
   | (b) 2.19
   | (c) 0.0034
   | (d) 0.079

4. Convert each to percents.

   (a) $\dfrac{7}{25}$

   (b) $\dfrac{13}{20}$

   (c) $\dfrac{128}{8}$

   (d) $\dfrac{37}{80}$

   | (a) 28%
   | (b) 65%
   | (c) 1600%
   | (d) 46.25%

5. Complete the follwing chart.

FRACTION	DECIMAL	PERCENT
$\dfrac{1}{5}$	_____	_____
_____	0.35	_____
_____	_____	110%
$\dfrac{7}{40}$	_____	_____
_____	0.0004	_____

Fraction	Decimal	Percent
$\dfrac{1}{5}$	0.2	20%
$\dfrac{35}{100}$	0.35	35%
$\dfrac{110}{100}$	1.1	110%
$\dfrac{7}{40}$	0.175	17.5%
$\dfrac{4}{1000}$	0.004	0.4%

152 CHAPTER 6 From Rational Numbers to Real Numbers

6. Find the markdown and sale price for each of the following.
   (a) original price = $5.40 markdown = 20%
   (b) original price = $10.00 markdown = 15%
   (c) original price = $7.60 markdown = 40%

	Markdown	Sale Price
(a)	$1.08	$4.32
(b)	$1.50	$8.50
(c)	$3.04	$4.56

7. Fill in the missing entries.

	Original Price	% Markdown	Sale Price
(a)	$75	$20	___
(b)	$1700	___	$1275
(c)	___	$20	$64

	Orig. Price	% Markdown	Sale Price
(a)	$75	20	$60
(b)	$1700	25	$1275
(c)	$76.80	20	$64

8. Henry's baseball team has played 42 games and won 28 games.

   (a) What percent of the games played did they win?
   (b) What percent of the games played did they lose?

   $66\frac{2}{3}\%$

	Won	Played
Games	28	42
Percent	x	100

   $\dfrac{28}{42} = \dfrac{x}{100}$
   $42x = 280$
   $x = 66\frac{2}{3}$

   (b) $33\frac{1}{3}\%$

9. A person's optimal exercise heart rate is found by following these steps:
   1. Subtract person's age from 220
   2. Find 70% of this difference
   3. Find 80% of this difference
   4. The optimal rate is between the latter two numbers

(a) What is your optimal heart rate range?
(b) What is the optimal heart rate range for a 40 year
old person?

(a) Suppose your age is 20
220 − 20 = 200
0.70 × 200 = 140
0.80 × 200 = 160
Range 140 - 160
126 − 144
220 − 40 = 180
0.70 × 180 = 126
0.80 × 180 = 144
(b) Range 126 - 144

10. The average person uses 70 liters of water per day for showers. 40% of total household water per person is used for showers. What is the average total amount of water used per day per person?

175 liters

	Showers	Total
Water Used	70	x
Percent	40	100

$\dfrac{70}{x} = \dfrac{40}{100}$
$40x = 7000$
$x = 175$

11. In 1938 one quart of motor oil cost 30¢.
In 1992 motor oil costs 99¢.
What is the percent increase in the price from 1938 to 1992?

230%
increase in price = 0.99 − 0.30
= 0.69
% increase = 0.69 ÷ 0.30 = 2.3
= 230%

12. Following is one tax table from a recent form.

If Your Taxable Income Is: **Your Tax Is:**
Not over $1000   5.1% of taxable income
Over $1000 but not over $1500 $51 + 6.2% of excess over $1000
Over $1500 but not over $2000 $82 + 7.3% of excess over $1500
Over $2000 but not over $3000 $118.50 + 8.4% of excess over $2000
Over $3000 $160.50 + 9.5% of excess over $3000

(a) What will the tax be for a taxable income of $2470?

154 CHAPTER 6 From Rational Numbers to Real Numbers

(b) If your tax is $287.10, what was your taxable income?

> (a) $157.98   Since 2000 < 2470 < 3000 we have tax = $118.50 + 8.4% of excess over 200
> = $118.50 + (0.084) (470)
> = $118.50 + 39.48 = $157.98
> (b) $4,332.63

13. At the end of the day, Doni has a total of $11.91.84 in the cash register. This represents sales for the day plus a 4% sales tax. What were her sales for the day?

> $1,146
>
> | | Sales | Total |
> |---|---|---|
> | Register | x | $1191.84 |
> | Percent | 100 | 104 |
>
> $$\frac{x}{1191.84} = \frac{100}{104}$$
> $104x = (100)(1191.84)$
> $104x = 119148$
> $x = 1146$

14. The Mustangs have played 20 games and won 15 for a winning percentage of 75%. How many games will they need to win in a row in order to have a winning percentage of 80%?

> 5
>
> | | Won | Played |
> |---|---|---|
> | Mustangs | 15 + x | 20 + x |
> | Percent | 80 | 100 |
>
> $$\frac{15+x}{20+x} = \frac{80}{100}$$
> $(15 + x)(100) = (20 + x)(80)$
> $1500 + 100x = 1600 + 80x$
> $20x = 100$
> $x = 5$

15. Which of the following will give the greatest amount of discount?

(a) A 15% markdown.
(b) A 10% markdown first and then a 5% discount.

> 15% markdown

Section 3 The System of Rational Numbers

Cover the right side of the page and work on the left, then check your work

1. Do each each of the following using mental computation.

   (a) $\dfrac{1}{4} \cdot (15 \cdot 8)$

   (b) $6\dfrac{1}{5} \cdot 15$

   (c) $\left(\dfrac{3}{4} + \dfrac{2}{3}\right) + \dfrac{1}{3}$

   (d) $(18 + 15) \cdot \dfrac{1}{3}$

   (a) 30 $\quad \dfrac{1}{4} \cdot (15 \cdot 8) = \left(\dfrac{1}{4} \cdot 8\right) \cdot 15$
   $= 2 \cdot 15 = 30$

   (b) 93

   (c) $1\dfrac{3}{4}$

   $\left(\dfrac{3}{4} + \dfrac{2}{3}\right) + \dfrac{1}{3} = \dfrac{3}{4} + \left(\dfrac{2}{3} + \dfrac{1}{3}\right)$

   $= \dfrac{3}{4} + 1 = 1\dfrac{3}{4}$

   (d) 11

2. Write each of the following as repeating decimals.

   (a) $\dfrac{4}{7}$

   (b) $\dfrac{2}{15}$

   (c) $\dfrac{5}{18}$

   (a) $\dfrac{4}{7} = 0.\overline{571428}$

   (b) $\dfrac{2}{15} = 0.1\overline{3}$

   (c) $\dfrac{5}{18} = 0.2\overline{7}$

3. Write each decimal as a reduced fraction.

   (a) $2.4\overline{8}$

   (b) $2.\overline{48}$

   (c) $2.4\overline{84}$

   (d) $0.\overline{248}$

(a) $\dfrac{246}{99}$

Let $N = 2.\overline{48}$
Then $100N = 248.\overline{48}$
$- N = 2.\overline{48}$
$99N = 246$
$N = \dfrac{246}{99}$

(b) $\dfrac{224}{90}$

(c) $\dfrac{2460}{990}$

$1000N = 2484.\overline{84}$
$- 10N = 24.\overline{84}$
$990N = 2460$
$N = \dfrac{2460}{990}$

(d) $\dfrac{248}{999}$

4. Find three numbers, equally spaced, between $\dfrac{1}{3}$ and $\dfrac{3}{5}$.

$\dfrac{6}{15}, \dfrac{7}{15}, \dfrac{8}{15}$

5. Order the set of decimals from smallest to largest.
 {3.4, 3.$\overline{4}$, 3.$\overline{43}$, 3.4$\overline{3}$, 3.43}

3.4, 3.43, 3.4$\overline{3}$, 3.$\overline{43}$, 3.$\overline{4}$

6. Find a fraction to represent each of the following.

 (a) $0.\overline{2a}$
 (b) $0.\overline{2ab}$

(a) $\dfrac{2a}{99}$

Let $N = 0.\overline{2a}$
Then $100N = 2a.\overline{2a}$
$- N = 0.\overline{2a}$
$99N = 2a$
$N = \dfrac{2a}{99}$

(b) $\dfrac{2ab}{999}$

7. Place the following numbers in the correct region of the diagram. $\dfrac{3}{8}, -\dfrac{3}{4}, -5, 4, 0$

Section 3   The System of Rational Numbers   157

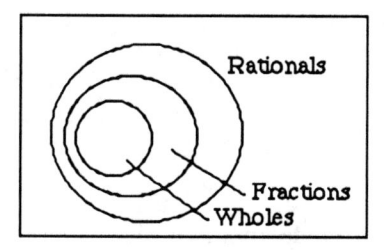

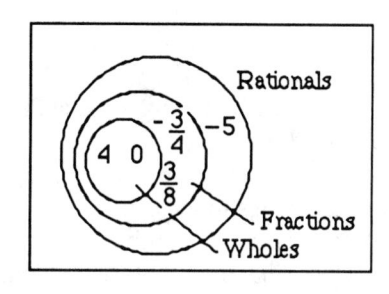

8. Answer the following questions about the operation table given below.

   (a) Which element is the identity?
   (b) Which element is the inverse for **a** ?

+	a	b	c
a	b	c	a
b	c	a	b
c	a	b	c

   (a) c
   (b) b
   $a + b = c$ (the identity)

9. Classify as true or false.

   (a) Every integer is a rational number.
   (b) Some rational numbers are not whole numbers.
   (c) The integers are a subset of the whole numbers.
   (d) The integers are closed under subtraction.

   (a) True
   (b) True
   (c) False
   (d) True

10. Is the following system a field? Why or why not?

+	0	1	2
0	0	1	2
1	1	2	3
2	2	3	0
3	3	0	1

×	0	1	2	3
0	0	0	0	0
1	0	1	2	3
2	0	2	0	2
3	0	3	0	3

No. Under multiplication 2 does not have an inverse.

11. Which of the following are true for the set of whole numbers?

   (a) −1 belongs to the set.
   (b) The set has an additive identity.
   (c) The set has a multiplicative identity.
   (d) Each member of the set has an additive inverse.
   (e) The set is closed under addition.
   (f) The set is closed under division.

   (b), (c), (e)

12. Which of the following are true for the set of integers?

   (a) −1 belongs to the set.
   (b) The set has an additive identity.
   (c) The set has a multiplicative identity.
   (d) Each member of the set has an additive inverse.
   (e) The set is closed under addition.
   (f) The set is closed under division.

   (a), (b), (c), (d), (e)

13. Complete the table below so that the operation is commutative.

*	a	b	c	d
a	b	d	a	c
b		c	b	
c	a	b	c	d
d			d	b

*	a	b	c	d
a	b	d	a	c
b	d	c	b	a
c	a	b	c	d
d	c	a	d	b

14. It can be observed that $\frac{2}{3} < \frac{3}{4} < \frac{4}{5}$. Show that this is true in general. Specifically, show that $\frac{n}{n+1} < \frac{n+1}{n+2}$ when $n \geq 0$.

$\frac{n}{n+1} < \frac{n+1}{n+2}$ will be true if

$n(n+2) < (n+1)(n+1)$

Then, $n^2 + 2n < n^2 + 2n + 1$

Section 4 The Real Number System

Cover the right side of the page and work on the left, then check your work

1. Classify each of the following as rational or irrational.

   (a) $4 + \pi$
   (b) $\dfrac{5\sqrt{3}}{2\sqrt{3}}$
   (c) $0.45674567\ldots$
   (d) $2.010011000111\ldots$

   (a) irrational
   (b) rational $\dfrac{5\sqrt{3}}{2\sqrt{3}} = \dfrac{5}{2}$
   (c) rational
   (d) irrational   does not repeat.

2. Solve each of the following inequalities on the domain of real numbers.

   (a) $x - 3 > 4$
   (b) $-3x + 5 < 10$
   (c) $7x + 3 \geq 15$
   (d) $4x - 3 \geq x + 3$

   (a) $x > 7$
   (b) $x > \dfrac{-5}{3}$
   $-3x + 5 < 10$
   $-3x < 5$
   $x > \dfrac{-5}{3}$
   (c) $x \geq \dfrac{12}{7}$
   (d) $x \geq 2$
   $4x - 3 \geq x + 3$
   $4x \geq x + 6$
   $3x \geq 6$
   $x \geq 2$

3. Evaluate each of the following.

   (a) $\sqrt{121}$
   (b) $-\sqrt{225}$
   (c) $\sqrt[3]{216}$
   (d) $\sqrt{16 \cdot 9}$

(a) 11
(b) −154
(c) 6
(d) 12
$\sqrt{16 \cdot 9} = \sqrt{16} \cdot \sqrt{9}$
$= 4 \cdot 3 = 12$

4. Simplify each of the following radicals.

   (a) $\sqrt{60}$
   (b) $\sqrt{120}$
   (c) $\sqrt{72}$
   (d) $\sqrt{882}$

(a) $2\sqrt{15}$
$\sqrt{60} = \sqrt{4 \cdot 15} = \sqrt{4} \cdot \sqrt{15} = 2\sqrt{15}$
(b) $2\sqrt{30}$
(c) $6\sqrt{2}$
$\sqrt{72} = \sqrt{9 \cdot 4 \cdot 2} = \sqrt{9} \cdot \sqrt{4} \cdot \sqrt{2} = 3 \cdot 2\sqrt{2} = 6\sqrt{2}$
(d) $21\sqrt{2}$
$\sqrt{882} = \sqrt{2 \cdot 9 \cdot 49} = 3 \cdot 7\sqrt{2} = 21\sqrt{2}$

5. Perform the indicated operation.

   (a) $3\sqrt{7} - 4\sqrt{7}$
   (b) $7\sqrt{3} + 9\sqrt{3}$
   (c) $6\sqrt{7} + 4\sqrt{63}$
   (d) $5\sqrt{40} - 3\sqrt{10}$

(a) $-\sqrt{7}$
$3\sqrt{7} - 4\sqrt{7} = (3 - 4)\sqrt{7} = -1\sqrt{7}$
(b) $16\sqrt{3}$
(c) $19\sqrt{7}$
$7\sqrt{3} + 4\sqrt{63} = 7\sqrt{3} + 4\sqrt{9 \cdot 7} = 7\sqrt{3} + 12 = 19\sqrt{7}$
(d) $7\sqrt{10}$

6. Evaluate each of the following.

   (a) $49^{\frac{-1}{2}}$
   (b) $(-64)^{\frac{1}{3}}$

162 CHAPTER 6 From Rational Numbers to Real Numbers

(c) $(7^{\frac{1}{2}})^4$

(d) $\dfrac{9^{-\frac{1}{2}}}{3^{-2}}$

(a) $\dfrac{1}{7}$

$49^{-\frac{1}{2}} = \dfrac{1}{49^{\frac{1}{2}}} = \dfrac{1}{\sqrt{49}} = \dfrac{1}{7}$

(b) -4

(c) 49

$(7^{\frac{1}{2}}) = 7^{\frac{1}{2} \cdot 4} = 7^2 = 49$

(d) 3

7. Write each of the following in simplest form.

(a) $\sqrt{40}$

(b) $\sqrt{\dfrac{180}{162}}$

(c) $\sqrt[3]{729}$

(d) $\sqrt[5]{343}$

(a) $2\sqrt{10}$

(b) $\dfrac{\sqrt{10}}{3}$

$\sqrt{\dfrac{180}{162}} = \dfrac{\sqrt{36 \cdot 5}}{\sqrt{81 \cdot 2}} = \dfrac{6\sqrt{5}}{9\sqrt{2}} =$

$\dfrac{6\sqrt{5}}{9\sqrt{2}} \cdot \dfrac{\sqrt{2}}{\sqrt{2}} = \dfrac{6\sqrt{10}}{9 \cdot 2} = \dfrac{6\sqrt{10}}{18} =$

$\dfrac{\sqrt{10}}{3}$

(c) 9

(d) 3

8. Rationalize the denominator of each of the following.

(a) $\dfrac{4}{\sqrt{3}}$

(b) $\dfrac{\sqrt{7}}{\sqrt{12}}$

(c) $\dfrac{\sqrt{7}}{\sqrt{14}}$

(a) $\dfrac{4\sqrt{3}}{3}$

$\dfrac{4}{\sqrt{3}} = \dfrac{4}{\sqrt{3}} \cdot \dfrac{\sqrt{3}}{\sqrt{3}} = \dfrac{4\sqrt{3}}{3}$

(b) $\dfrac{\sqrt{84}}{12}$

(c) $\dfrac{\sqrt{2}}{2}$

$\dfrac{\sqrt{7}}{\sqrt{14}} = \dfrac{\sqrt{7} \cdot \sqrt{14}}{\sqrt{14} \cdot \sqrt{14}} = \dfrac{\sqrt{98}}{14}$

$= \dfrac{\sqrt{49 \cdot 2}}{14} = \dfrac{7\sqrt{2}}{14} = \dfrac{\sqrt{2}}{2}$

9. Arrange the following in order from largest to smallest.

$\dfrac{3}{5}$, $0.\overline{66}$, $0.6\overline{8}$, 0.678678867888

$\dfrac{3}{5}$, $0.\overline{66}$, 0.678678867888, $0.6\overline{8}$

10. Show each of the following on a number line.

(a) $|x| \leq 5$
(b) $|x - 1| \geq 3$

(a)

[number line with solid segment from -5 to 5]

(b)

[number line with solid rays extending left from -2 and right from 4]

$|x - 1| \geq 3$
$x - 1 \geq 3$ or $x - 1 \leq {}^-3$
$x \geq 4$ or $x \leq {}^-2$

11. Prove that $2\sqrt{3}$ is an irrational number.

Assume that $\dfrac{a}{b} = 2\sqrt{3}$ is a rational number. Then

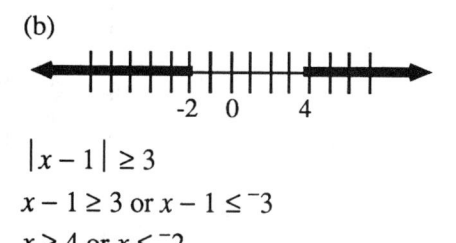

164 CHAPTER 6 From Rational Numbers to Real Numbers

$\left(\dfrac{a}{b}\right)^2 = 4 \cdot 3 = 12$ and $a^2 = 12b^2$

Factor a into prime factors. One of the factors will be 12 since $a^2 = 12b^2$. a^2 contains an even number of 12's as factors while $2b^2$ contains an odd number of 12's as factors. Thus, we have a contradiction since The Fundamental Theorem of Arithmetic says prime factorization is unique.

12. Find the value(s) of x that make each of the following statements true.

   (a) $\sqrt{x} = 16$
   (b) $\sqrt{x} > 0$
   (c) $\sqrt{-x} = 16$

(a) $x = 256 \quad 16 \cdot 16 = 256$
(b) $x > 0$
(c) $x = -256$
$\sqrt{-x} = \sqrt{-(-256)} = \sqrt{256} = 16$

13. Use the Pythagorean Theorem to construct $\sqrt{3}$ on a number line.

First, construct $\sqrt{2}$

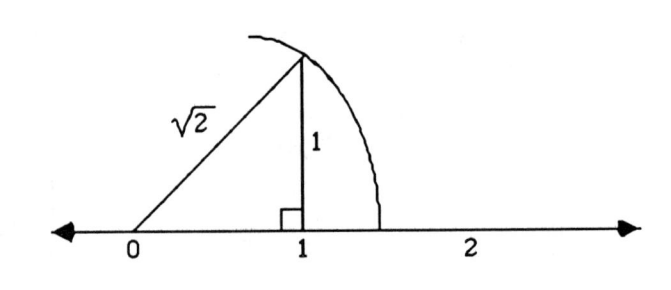

then,

Section 4   The Real Number System   165

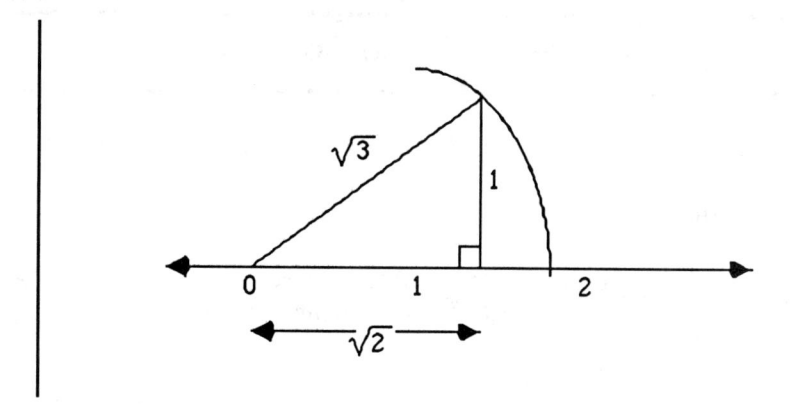

[Figure: Number line showing construction of $\sqrt{3}$ on number line. A right triangle with legs $\sqrt{2}$ (along number line from 0) and 1 (vertical), hypotenuse $\sqrt{3}$, with an arc from the top of the triangle down to the number line.]

14. Determine whether the following statement is true or false.

$$\sqrt{x+y} = \sqrt{x} + \sqrt{y}$$

False.
If $x = 4$ and $y = 9$, then

$$\sqrt{x+y} = \sqrt{4+9} = \sqrt{13} \text{ and}$$
$$\sqrt{x} + \sqrt{y} = \sqrt{4} + \sqrt{9} = 2 + 3 = 5$$
$$\sqrt{13} \neq 5$$

15. Let R be the set of real numbers, Q the set of rational numbers, I the set of integers, and W the set of natural numbers. For each equation, determine for which sets of numbers the equation will have a solution.

	R	Q	I	W
$x^2 + 2 = 7$				
$x^2 = 7$				
$4x + 3 = 9$				
$\sqrt{x} = -3$				
$x^2 = 16$				

	R	Q	I	W
$x^2 + 2 = 7$	yes	no	no	no
$x^2 = 7$	yes	no	no	no
$4x + 3 = 9$	yes	yes	no	no
$\sqrt{x} = -3$	no	no	no	no
$x^2 = 16$	yes	yes	yes	yes

Chapter 7
Consumer Mathematics
Section 1 Some Comparisons of Interest Rates

Cover the right side of the page and work on the left, then check your work

1. Estimate each of the following.

   (a) Fred Hudson invests $3150 at a simple interest rate of 5.7%. How much interest will he earn the first year?

   (b) $P = \$500, r = 0.09, t = 3$ years $I = ?$

       (a) $180 $3150 × 0.057 ≈
                    3000 × 0.06 = $180
       (b) $150 $I = 500 × 0.09 × 3 ≈
               (500 × 0.10) × 3 ≈ 50 × 3 = $150

2. Find the simple interest for each of the following loans.

   (a) $100 for 2 years at 7%
   (b) $500 for 3 years at 11%
   (c) $600 for 1 year at $5\frac{1}{4}$%
   (d) $800 for $2\frac{1}{2}$ years at 9%

       (a) $I = \$14$ $I = 100 × 0.07 × 2$
                           $= 7 × 2 = \$14$
       (b) $I = \$165$
       (c) $I = \$31.50$ $I = 600 × 5\frac{1}{4}$% × 1
                                 $= 600 × 0.0525 × 1$
                                 $= \$31.50$

3. What will be the amount to be repaid for each loan in Exercise 2?

       (a) $114   Amount repaid =
       Amount borrowed + Interest
                         = 100 + 14
                         = $114
       (b) $665
       (c) $631.50
       (d) $980

4. Find the compound amount and compount interest in each of the following.

(a) $600 at 4%, compounded quarterly for 1 year.
(b) $2500 at 11%, compounded monthly for 5 years.
(c) $5000 at 9%, compounded semiannually for 3 years.

(a) Amount = $624.36

$$A = 600\left(1 + \frac{0.04}{4}\right)^{4(1)}$$
$$= 600(1 + 0.01)^4$$
$$= 600(1.01)^4$$
$$= 600(1.0406)$$
$$= \$624.36$$

Interest = $24.36

(b) $A = \$4279.67 \quad I = \$1779.67$
(c) $A = \$6511.30 \quad I = \$1511.30$

$$A = 5000\left(1 + \frac{0.09}{2}\right)^{2(3)}$$
$$= 5000(1 + 0.045)^6$$
$$= 5000(1.045)^6$$
$$= 5000(1.3023)$$
$$= \$6511.30$$

5. Complete the following table. $1000 at 8% interest.

Compounded	Annually	Quarterly	Monthly
Balance After 1 Year			
Balance After 10 Years			

Compounded	Annually	Quarterly	Monthly
Balance After 1 Year	$1080	$1082.43	$1083.43
Balance After 10 Years	$2158.93	$2280.04	$2309.60

6. Find the simple interest rate on the following loan. Assume a 360 day year.

168　CHAPTER 7　Consumer Mathematics

$P = \$4500$
Amount repaid in 90 days $\approx \$4572$

Rate = 6.4%

$I = Prt$ and $4572 - 4500 = 72$, so
$72 = (4500)\,(r)\left(\dfrac{1}{4}\right) \leftarrow \dfrac{90}{360}$
$72 = (4500)\left(\dfrac{1}{4}\right)(r)$
$72 = 1125\,r$
$0.064 = r$

7. How much should be deposited in an account paying 8% compounded annually in order to have a balance of $8000 nine years from now.

$4002
$A = 8000 \quad r = 0.08 \quad t = 9$
$8000 = P\left(1 + \dfrac{0.08}{1}\right)^9$
$8000 = P\,(1.08)^9$
$8000 = P\,(1.999)$
$\$4002 = P$

8. Find the present value of the money for $9,000 due in $5\dfrac{1}{2}$ years if money is worth 10% compounded semiannually.

$5262.23
$A = 900 \quad r = 0.10 \quad t = 5\dfrac{1}{2}$
$9000 = P\left(1 + \dfrac{0.10}{2}\right)^{2\left(5\frac{1}{2}\right)}$
$9000 = P\,(1 + 0.05)^{2\left(\frac{11}{2}\right)}$
$9000 = P\,(1.05)^{11}$
$9000 = P\,(1.7103)$
$\$5262.23 = P$

9. How long will it take for the following amounts to double at 9% interest compounded semi-annually.

(a) $1000
(b) $2000
(c) $3000
(d) N dollars

(a) 8 years

$$2000 = 1000 \left(1 + \frac{0.09}{2}\right)^{2t}$$

$$2000 = 1000 (1 + 0.045)^{2t}$$

$$2000 = 1000 (1.045)^{2t}$$

$$2 = (1.045)^{2t}$$

Try some values for t.

t	$2t$	(1.045)	
1	2	1.0920	Too small
10	20	2.4117	Too large
7	14	1.8519	Too small
8	16	2.0223	

(b) 8 years
(c) 8 years
(d) 8 years $\quad 2N = N \left(1 + \frac{0.09}{2}\right)^{2t}$

$$2N = N (1 + .045)^{2t}$$

$$2 = (1.045)^{2t}$$

10. Scott Johnson purchases some land for $50,000. If the annual inflation rate is 4% per year, what will the land be worth in 20 years?

$109,556.16

$$A = 50000 \left(1 + \frac{0.04}{1}\right)^{20}$$

$$= 50000 (1 + .04)^{20}$$

$$= 50000 (1.04)^{20}$$

$$= 50000 (2.1911)$$

$$= \$109,556.16$$

11. A teacher whose salary is $25,000 has salary increases of 4.0%, 5.0%, 4.9%, and 5.6% in 4 consecutive years. If the annual rate of inflation during this period was 5.3%, did the teacher's salary keep up with inflation?

170 CHAPTER 7 Consumer Mathematics

No.

End of Year	1	2	3	4
Increase	1000	1300	1337.70	1603.71
New Salary	26000	27300	28637.70	30241.41

For inflation, we have $A = 25000\left(1 + \dfrac{0.053}{1}\right)^4$

$= 25000\,(1.053)^4 = \$30736.44$

12. Which of the following would pay the most interest for one year?

    6% compounded quarterly
    5.5% compounded monthly

    6% compounded quarterly
    $A = 1000\left(1 + \dfrac{0.06}{4}\right)^4$ 6% rate

    > 1000 is chosen as an arbitrary starting amount.

    $A = 1000\,(1.015)^4$
    $A = \$1061.36$
    Interest paid + 1061.36 − 1000 = $61.36

    $A = 1000\left(1 + \dfrac{0.055}{12}\right)^{12}$
    $A = 1000\,(1.0046)^{12}$
    $A = \$1056.62$
    Interest = 1056.62 − 1000 = $56.62

13. The population of Warrensburg in 1990 was 15,500. The population is expected to increase at a rate of 5% annually for the next 6 years. What will the population be at the end of six years?

    20,771 $A = 15500\left(1 + \dfrac{0.05}{1}\right)^6$
    $= 15500\,(1.05)^6$
    $= 20771.48$

Section 1 Some Comparisons of Interest Rates 171

14. Shelly has a balance due of $400 on her credit card. The interest rate is $1\frac{1}{2}\%$ simple interest per month on the unpaid balance. If Shelly pays $220 toward her balance this month, and does not charge anything else next month, what will be the balance due on her charge account next month?

$182.70
400 − 220 = 180 unpaid balance
180 × 0.015 = interest
$2.70 = interest
Balance = unpaid balance + interest
= 180 + 2.70
= $182.70

15. Compute the amount that Mr. Staat would have after 3 months if he had deposited $100 each month in a savings account paying 5% compounded monthly.

$302.53

Month	1	2	3
Amount Deposited	$100	$100	$100
Total Amount	$100.42	$201.26	$302.53

month 1 $A = 100\left(1 + \frac{.05}{12}\right)^{12(\frac{1}{12})} \leftarrow$ one month
$= 100(1.0042)^1$
$= \$100.42$

month 2 $A = 200.42\left(1 + \frac{0.05}{12}\right)^{12(\frac{1}{12})}$
$= 200.42(1.0042)^1$
$= \$201.26$

month 3 $A = 301.26\left(1 + \frac{0.05}{12}\right)^{12(\frac{1}{12})}$
$= 301.26(1.0042)$
$= \$302.53$

Section 2 A Calculator Approach to Annuities

Cover the right side of the page and work on the left, then check your work

1. Find the amount for each of the following annual annuities.

    (a) $R = 500$, $i = 7\%$, $n = 8$
    (b) $R = 1000$, $i = 9\%$, $n = 20$

    Find the amount of each of the following annuities.

	Periodic Payment	Number of Deposits per Year	Number of Years	Interest Rate	Compounding Period
2.	$100	1	15	12%	Annually
3.	$50	4	3	6.5%	Quarterly
4.	$200	52	6	6%	Weekly
5.	$80	12	10	8%	Monthly

1. (a) $S = \$5129.90$
$R = 500$, $i = 0.07$, $n = 8$

$$S = 500\left[\frac{(1 + 0.07)^8 - 1}{0.07}\right]$$

$$= 500\left[\frac{(1.07)^8 - 1}{0.07}\right] = \$5{,}129.90$$

(b) $S = \$51{,}160.12$

2. $S = \$3727.97$
$R = 100 \quad i = 0.12 \quad n = 15$

$$S = 100\left[\frac{(1 + 0.12)^{15} - 1}{0.12}\right]$$

$S = \$3727.97$

3. $S = \$656.64$
$R = 100 \quad i = \dfrac{0.065}{4} = 0.01625 \quad n = 12$

$$S = 50\left[\frac{(1 + 0.01625)^{12} - 1}{0.01625}\right]$$

4. $S = \$75{,}011.37$
$R = 200 \quad i = \dfrac{0.06}{52} = 0.00115$
$n = 52.6 = 312$

$$S = 200\left[\frac{(1 + 0.00115)^{312} - 1}{0.00115}\right]$$
5. $S = \$14638.95$

6. Tom and Susan decide to deposit $50 a month into a saving's account for their son. The account earns 6%, compounded monthly. What is the balance after 10 years?

$S = \$8193.97$
$R = 50 \quad i = \dfrac{0.06}{12} = 0.005$
$n = 12 \cdot 10 = 120$
$$S = 50\left[\frac{(1 + 0.005)^{120} - 1}{0.005}\right]$$
$S = \$8193.97$

7. Eileen wants to save enough money to make a down payment on a house. She decides to save $200 a month. The account she chooses pays $7\frac{1}{2}\%$, compounded monthly. After 3 years, how large a down payment can she make?

$S = \$8046.28$
$R = 200$
$i = \dfrac{0.075}{12} = 0.00625$
$n = 12 \cdot 3 = 36$
$$S = 200\left[\frac{(1 + 0.00625)^{36} - 1}{0.00625}\right]$$
$S = \$8046.28$

8. Mr. Hoffman deposits $500 a quarter for 5 years into an account that pays 11% compounded quarterly. At the end of the five years how much is the balance of his account?

$S = \$13{,}098.70$
$R = 500 \quad i = \dfrac{0.11}{4} = 0.0275$
$n = 4 \cdot 5 = 20$
$$S = 500\left[\frac{(1 + 0.0275)^{20} - 1}{0.0275}\right]$$
$S = \$13{,}098.70$

9. Ed Parker deposited $3500 per year into a retirement account. The account pays 12%

174 CHAPTER 7 Consumer Mathematics

compounded annually. Make a table showing how his money accumulates for the first 6 years.

Period	Interest	Deposit	Increase In Fund	Amount In Fund
1	0	$3500	$3500.00	$3500.00
2	$420.00	$3500	$3920.00	$7420.00
3	$890.40	$3500	$4390.40	$11810.40
4	$1417.25	$3500	$4917.25	$16727.65
5	$2007.32	$3500	$5507.32	$22234.97
6	$2668.20	$3500	$6168.20	$28403.17

10. Ivana B. Alone deposited $300 a month for 5 years into an account that pays 7%, compounded monthly. At the end of 5 years, she leaves the account untouched for 5 more years. What is her balance at the end of the 10 years?

Balance = $30,401.90

$$S = 300 \left[\frac{(1 + 0.0058)^{60} - 1}{0.0058}\right]$$

$S = \$21445.65$

$$A = 21{,}445.65 \left(1 + \frac{0.07}{12}\right)^{60}$$

$= 21{,}445.65 \ (1.4176)$
$= \$30{,}401.90$

11. Jack Fox deposits $50 a quarter for 10 years into an account that pays 6.5%, compounded quarterly. At the end of the 10 years he transfers the money to a savings account that pays 8%, compounded monthly. If he leaves the money in the second account for 5 years, what will his balance be at the end of the 15 year period?

Balance = $4155.67

$$S = 50 \left[\frac{(1 + 0.01625)^{40} - 1}{0.01625}\right]$$

$S = \$2786.33$

$$A = 2786.33 \left(1 + \frac{0.08}{12}\right)^{6}$$

$= 2786.33 \ (1.4898)$
$= \$4155.67$

12. Ragu wants to save money to purchase a car that costs $12,000. How much money must be put

each month into an annuity that pays $7\frac{1}{2}\%$, compounded monthly, in order to have the $12,000 in 5 years?

$R = \$165.46$

$$\$12,000 = R\left[\frac{(1 + 0.00625)^{60} - 1}{0.00625}\right]$$

$12,000 = R \ (72.5271)$

$\$165.46 = R$

13. How much must be deposited each week into an annuity that pays 6%, compounded weekly, to have a balance of $50,000 after 10 years?

$R = \$69.99$

$$\$50,000 = R\left[\frac{(1 + 0.00115)^{520} - 1}{0.00115}\right]$$

$50,000 = R \ (714.3378)$

$\$69.99 = R$

14. Tina is planning a retirement program. She is 43 years old and will retire at age 62. How much should she save each month in an annuity paying 6.5%, compounded monthly, in order to have a balance of $100,000 at retirement?

$R = \$223.09$

$$100,000 = R\left[\frac{(1 + 0.00542)^{228} - 1}{0.00542}\right]$$

$100,000 = R \ (448.2527)$

$\$223.09 = R$

15. Professor Ragavan is planning a retirement program. He is 30 years old and will retire at age 62. How much should he save each month in an annuity paying 7.5%, compounded monthly, in order to withdraw (starting at retirement) $1000 per month for 20 years.

$R = \$150.89$

To withdraw 1000/month for 20 years will require:
$1000 \times 12 \times 20 = \$240,000$.
Then:

$$240,000 = R\left[\frac{(1 + 0.00625)^{304} - 1}{0.00625}\right]$$

$240,000 = R \ (1590.5843)$

$\$150.89 = R$

Section 3 Present Value of An Annuity

Cover the right side of the page and work on the left, then check your work

Find the present value for each of the following annual annuities.

1. $R = \$2000$, $i = 7\%$, $n = 10$

$P = \$14,047.16$

$P = 2000 \left[\dfrac{1 - (1 + 0.07)^{-10}}{0.07} \right]$

$= 2000 \, (7.0236)$

$= \$14,047.16$

2. $R = \$10,000$, $i = 9\dfrac{1}{4}\%$, $n = 8$

$P = \$54837.62$

Find the present value for each of the following annuities.

	Periodic Payment	Number of Payments Per Year	Number of Years	Interest Rate	Comp. Period
3.	$500	12	2	5.5%	Monthly
4.	$5000	4	20	10%	Quarterly
5.	$300	2	9	8%	Semi-annu.
6.	$250	52	7	9%	Weekly

3. $P = \$11,336.68$

$R = \$100 \qquad i = \dfrac{0.055}{12} = 0.0046$

$n = 2 \cdot 12 = 24$

$P = 500 \left[\dfrac{1 - (1 + 0.0046)^{-24}}{0.0046} \right]$

$= 500 \, (22.6734) = \$11,336.68$

4. $P = \$172,259.09$

$R = \$5000 \quad i = \dfrac{0.10}{4} = 0.025$

$n = 20 \cdot 4 = 80$

$P = 5000 \left[\dfrac{1 - (1 + 0.025)^{-80}}{0.025} \right]$

$= 5000 \, (34.4518) = \$172,259.09$

5. $P = \$3797.79$

6. $P = \$67,812.74$

$R = \$250 \quad i = \dfrac{0.09}{52} = 0.0017$

$n = 7 \cdot 52 = 364$

$$P = 250\left[\frac{1-(1+0.0017)^{-364}}{0.0017}\right]$$
$$= 250\,(271.2509) = \$67{,}812.74$$

For each of the following, find the monthly payment and total payment.

7. A loan of $1500 at 15% interest compounded monthly for 12 months.

   $R = \$135.39$
   $P = 1500 \quad i = \dfrac{0.15}{12} = 0.0125$
   $n = 1 \cdot 12 = 12$
   $$1500 = R\left[\frac{1-(1+0.0125)^{-12}}{0.0125}\right]$$
   $1500 = R\,(11.0793)$
   $\$135.39 = R$
   Total payment $= 135.39 \times 12 = \$1624.68$

8. A debt of $20,000 at 9% interest compounded monthly for 15 years.

   $R = \$202.85$
   Total payment $= 202.85 \times 12 \times 15 = \$36513$

9. A debt of $400 at 18% interest compounded monthly for 6 months.

   $R = \$70.21$
   $P = \$400 \quad i = \dfrac{0.18}{12} = 0.015$
   $n = \dfrac{1}{2} \cdot 12 = 6$
   $$400 = R\left[\frac{1-(1+0.015)^{-6}}{0.015}\right]$$
   $400 = R\,(5.6972)$
   $\$70.21 = R$
   Total payment $= 70.21 \times 6 = \$421.60$

10. Cynthia purchases a new stereo system for $500 with a loan from the credit union. The loan is to be repaid in 12 monthly installments at an interest rate of $11\frac{1}{2}\%$.
    (a) What will her monthly payment be?
    (b) How much interest does she pay the first month?
    (c) What is her balance after 1 payment?

    (a) $R = \$44.31$

178   CHAPTER 7   Consumer Mathematics

$P = 500 \quad i = \dfrac{0.115}{12} = 0.0096$

$n = 1 \cdot 12 = 12$

$500 = R \left[\dfrac{1 - (1 + 0.0096)^{-12}}{0.0096} \right]$

$500 = R \,(11.2836)$

$\$44.31 = R$

(b) interest = $4.80

Interest = $(500) \times (0.0096)$

$ = \$4.80 \qquad \uparrow$

 interest rate
 per period

(c) balance = $460.49

Balance =

Principal + interest − payment

$= 500 + 4.80 - 44.31 = \$460.49$

11. Dean must repay a loan of $900 at 16% interest compounded monthly. He has a choice of 12 monthly payments or 18 monthly payments.
   (a) What is the monthly payment for each plan?
   (b) What is the total payment for each plan?

(a) 12 payments − $81.64

$900 = R \left[\dfrac{1 - (1 + 0.0133)^{-12}}{0.0133} \right]$

$900 = R \,(11.0240)$

$\$81.64 = R$

18 payments − $56.55

$900 = R \left[\dfrac{1 - (1.0133)^{-18}}{0.0133} \right]$

$900 = R \,(15.9141)$

$\$56.55 = R$

(b) Total payment (12 months) =
$12 \times 81.64 = \$979.68$
Total payment (18 months) =
$18 \times 56.55 = \$1017.90$

12. Nathalie takes out a loan for $1000 at an interest rate of 14%. She can afford monthly payments of no more than $50. Which of the payment plans should she choose?

   (a) 12 months
   (b) 18 months
   (c) 24 months
   (d) 30 months

Payment for 12 months = $89.81
Total payment = $1078.92
Payment for 18 months = $61.93
Total payment = $1114.74
Payment for 24 months = $\boxed{\$48.03}$
Total payment = $1152.72
Payment for 30 months = $39.72
Total payment = $1191.16
∴ Teresa should choose the 24 month plan.

13. For the loan in Exercise 12, assume that Nathalie could afford payments of up to $65 a month. Which plan should she choose then?

Nathalie should choose the 18 month plan since she could afford the payments and pay less total payment than with the 24 month plan.

14. Compute the payment necessary to finance a refrigerator for $1200 at 18% interest compounded monthly for 12 months.

$R = \$110.02$

$$1200 = R\left[\frac{1 - (1 + 0.015)^{-12}}{0.015}\right]$$

$1200 = R\,(10.9075)$
$\$110.02 = R$

15. Make an amortization table for Exercise 14.

Months	Outstanding Principal	Interest Due	Payment	Principal Repaid Each Period
1	$1200.00	$18.00	$110.02	$92.02
2	1107.98	16.62	110.02	93.40
3	1014.58	15.22	110.02	94.80
4	919.78	13.80	110.02	96.22
5	823.56	12.35	110.02	97.67
6	725.89	10.89	110.02	99.13
7	626.76	9.40	110.02	100.62
8	526.14	7.89	110.02	102.13
9	424.01	6.36	110.02	103.66
10	320.35	4.81	110.02	105.21
11	215.14	3.23	110.02	106.79
12	108.35	1.63	109.98	108.35

Section 4 Annual Percentage Rate

Cover the right side of the page and work on the left, then check your work

1. Calculate the interest paid for the following loans at the given add-on interest rate.

   (a) $2000 at 9% add-on interest for 3 years.
   (b) $4500 at 11.5% add-on interest for 5 years.

   (a) Interest = $540
   Interest = (0.09) (2000) (3) = $540
   (b) Interest = $2587.50
   Interest = (0.115) (4500) (5) = $2587.50

   Compute the finance charges for each of the following loans.

2. A $800 loan to be repaid in 36 payments of $25.50 each.

   Finance charges = $118
   Finance charges = (36) (25.50) − 800 = $118

3. A $7000 loan with 7% add-on interest for 4 years with a $25 credit check and a $50 maintenance contract fee.

   Finance charges = $2035

   Interest = (0.07)(7000)(4) = $1960
   Finance charges
     = $1960 interest
         25 credit check
      + 50 maintenance fee
      $2035

4. A $1200 loan for 3 years with $160 interest and $45 insurance fee.

   Finance charges = $205

5. A $3700 loan to be repaid in 48 payments of $82 each.

   Finance charges = $236

6. Compute the monthly payments for Exercise 3.

   Monthly payment = $188.23
   Monthly payment = $\frac{(7000 + 2035)}{48}$
   = $\frac{9035}{48}$ = $188.23

7. Compute the monthly payments for exercise 4.

Section 4   Annual Percentage Rate   181

Monthly payment = $39.03

8. The Hoffmans purchased a new couch for $500. They agree to pay for the couch with 24 payments of $25 each. What is the finance charge?

Finance charge = $100
Finance charge = (24) (25) − 500
= 600 − 500 = $100

9. A stereo is advertised for $550. It may be purchased by making a down payment of $100 with monthly payments for 6 months of $78.50. What is the finance charge?

Finance charge = $21
Finance charge = (6) (78.50) − 450
= 471 − 450 = $21

10. Mike Dew needs $450 to buy a bicycle. The loan company indicates that he will have finance charges of $78. If he is to repay the loan in 12 monthly payments, what will be the amount of each payment?

Monthly payment = $44

Monthly payment = $\dfrac{450 + 78}{12}$

= $\dfrac{528}{12}$ = $44

11. Which of the following options would result in a lower monthly payment for a loan of $2500?

(a) 8% add-on for 4 years.
(b) $7\dfrac{1}{4}$% add-on for 6 years.

Option (b)
(a) Interest = (0.08) (2500) (4) = 800

Monthly payment = $\dfrac{2500 + 800}{48}$

= $\dfrac{3300}{48}$ = $68.75

(b) Interest = (0.0725) (2500) (6) = $1087.50

Monthly payment = $\dfrac{2500 + 1087.50}{72}$

= $\dfrac{3587.50}{72}$ = $49.83

12. Which of the options in exercise 11 will result in the largest dollar amount in finance charges?

option (a)

182 CHAPTER 7   Consumer Mathematics

13. Beimers borrows $900, to be repaid in 24 equal monthly payments of $41.50.
    (a) What are the finance charges?
    (b) What is the APR (rounded to nearest $\frac{1}{4}$%)?

   (a) $96   (24)(41.50) − 900 = 96
   (b) 1. $\frac{(\$96)(100)}{900} = \$10.67$
   2. Find the row in Table A-1 for 24 payments. Find the entry in that row nearest 10.67. The nearest entry is 10.75.
   3. The percentage rate at the top of the column is 10.00%. This is the APR.

14. Granite bank will loan $3000 for 36 months at $8\frac{1}{2}$% add-on interest. A loan company will loan $3000 to be repaid in 36 monthly payments of $92.50 each. Which option has
    (a) the smaller finance charges?
    (b) the smaller APR?

   (a) Loan company
   Granite Bank finance charge = $(0.085)(3000)(\frac{36}{12}) = \$765$
   Loan company finance charge = $(36)((92.50) − 3000 = \$330$
   (b) Loan company
   Granite Bank APR = 13.75%
   Loan company APR = 10.00%

15. Busnach can choose one of three options to pay for a lawn mower that sells for $898. Which of the three options give the best APR?
    (a) $898 loan from his grandfather to be repaid in 24 equal monthly payments of $45.
    (b) Pay the lawn mower company 36 equal payments of $38.50.
    (c) $898 loan from a bank with 6% add-on interest for 30 months.

   (a) APR = 13.75%
   Finance charge = (24)(45) − 898 = $182
   $\frac{(182)(100)}{898} = \$20.27$
   APR = 13.75%
   (b) APR = 13.75%
   (c) APR = 11.00%   Brian should choose the bank loan.

Chapter 8
An Introduction to Probability Theory

Section 1 The Language of Probability

Cover the right side of the page and work on the left, then check your work

1. Write the sample space for each of the following.

   (a) Spinner 1 is spun once.
   (b) Spinner 2 is spun once.
   (c) Spinner 2 is spun once and then spinner 1 is spun once.
   (d) Spinner 1 is spun twice.

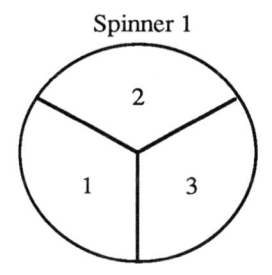

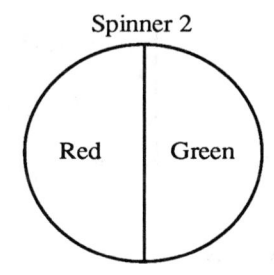

   Spinner 1: divided into three equal sectors labeled 1, 2, 3.
   Spinner 2: divided into two equal sectors labeled Red, Green.

   (a) $S = \{1, 2, 3\}$
   (b) $S = \{\text{red, green}\}$
   (c) $S = \{(\text{red}, 1), (\text{red}, 2), (\text{red}, 3), (\text{green}, 1), (\text{green}, 2), (\text{green}, 3)\}$
   (d) $S = \{(1, 1), (1, 2), (1, 3), (2, 1), (2, 2), (2, 3), (3, 1), (3, 2), (3, 3)\}$

2. Assign a probability to each outcome in the sample spaces of exercise 1.

   (a) $P(1) = P(2) = P(3) = \frac{1}{3}$
   (b) $P(\text{red}) = P(\text{green}) = \frac{1}{2}$
   (c) $P = \frac{1}{6}$ for each outcome
   (d) $P = \frac{1}{9}$ for each outcome.

3. An experiment consists of selecting one of ten digits (0 - 9) from a container. List each of the following:

(a) The sample space.
(b) The event consisting of outcomes that the digit is less than 6.
(c) The event consisiting of outcomes that the digit is even.
(d) The event consisiting of outcomes that the digit is prime.

> (a) $S = \{0, 1, 3, 4, 5, 6, 7, 8, 9\}$
> (b) $E_1 = \{0, 1, 2, 3, 4, 5\}$
> (c) $E_2 = \{0, 2, 4, 6, 8\}$
> (d) $E_3 = \{2, 3, 5, 7\}$

4. Find the probability of each of the events described in Exercise 3.

> (b) $P(E_1) = \dfrac{6}{10}$
> (c) $P(E_2) = \dfrac{5}{10}$
> (d) $P(E_3) = \dfrac{4}{10}$

5. A card is drawn from an ordinary deck. What is the probability of getting:

(a) a face card.
(b) a club.
(c) not a face card.

> (a) $P \text{ (face card)} = \dfrac{12}{52}$
> Jack, Queen, King for each of 4 suits.
> (b) $P \text{ (club)} = \dfrac{13}{52}$
> (c) $P \text{ (not a face cared)} = \dfrac{40}{52}$
> $40 = 52 - 12$
> ↑ face cards

Answer the questions relative to the following spinner:

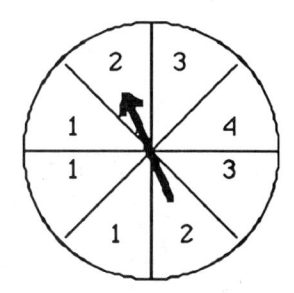

6. What is the probability of the pointer stopping on an even number?

$P \text{ (even number)} = \dfrac{3}{8}$

7. What is the probability of it stopping on a 1?

$P(1) = \dfrac{3}{8}$

8. What is the probability of it stopping on an odd number?

$P \text{ (odd number)} = \dfrac{5}{8}$

9. What is the probability of it stopping on a multiple of 2?

$P \text{ (multiple of 2)} = \dfrac{3}{8}$ ← 2 and 4 are multiples of 2.

10. A drawer contains 5 black socks, 4 blue socks, and 2 red socks. One sock is drawn from the drawer. Find the probability for each of the following:

    (a) The sock is black.
    (b) The sock is red.
    (c) The sock is black or red.

(a) $P \text{ (black)} = \dfrac{5 \leftarrow \text{number of black socks}}{11 \leftarrow \text{total number of socks}}$

(b) $P \text{ (red)} = \dfrac{2}{11}$

(c) $P \text{ (black or red)} = \dfrac{7}{11}$

11. What, if anything, is wrong with each of the following?

(a) The probability that the Royals win the World Series is $\frac{3}{5}$.

(b) There are 12 months. Thus, the probability that a person is born in February is $\frac{1}{12}$.

(a) $\frac{3}{5} + \frac{1}{2} = \frac{6}{10} + \frac{5}{10} = \frac{11}{10}$ ← total
probability > 1

(b) There are fewer days in February than in other months. P (born in February) may be less than for other (longer) months.

12. A box contains balls lettered a, b, c, and d. Tabulate a sample space for each of the following.

   (a) A ball is drawn, its letter noted, and a second ball is drawn.
   (b) A ball is drawn, its letter noted, and it is returned to the box. A second ball is then drawn and its letter is noted.

(a) $S = \{(a, a), (a, b), (a, c), (a, d),$
$(b, a), (b, b), (b, c), (b, d), (c, a),$
$(c, b), (c, c), (c, d) (d, a), (d, b),$
$(d, c), (d, d)\}$.

(b) S = same as in part (a).

13. Answer each of the following questions about the experiment in Exercise 12.

   (a) What is the probability the first ball drawn is the letter a?
   (b) What is the probability the first ball drawn is the letter c?
   (c) After the first ball is put back in the box, what is the probability the second ball drawn is c?

(a) $\frac{1}{4}$
(b) $\frac{1}{4}$
(c) $\frac{1}{4}$

Section 1 The Language of Probability 187

14. A roulette wheel has 38 slots. The first 36 slots are numbered 1 to 36. Half of these 36 slots are black and the other half are red. The other 2 slots are numbered 0 and 00 and are green in color. As the wheel spins, an ivory ball falls into one the slots.

   (a) What is the probability the ball lands in a red slot?
   (b) What is the probability the ball lands on an odd number?
   (c) What is the probability the ball lands on a red slot or in 00?

   (a) $\dfrac{18}{38}$ ← number of red slots / ← total number of slots
   (b) $\dfrac{18}{38}$
   (c) $\dfrac{19}{38}$

15. The weather report predicts that tomorrow there is a 70% chance for rain. What is the probability that it will not rain tomorrow?

   $\dfrac{30}{100}$ or 30%

Section 2 Empirical Probability & the Fundamental Principle of Counting

Cover the right side of the page and work on the left, then check your work

1. How many different three-digit numbers can be made from the digits 1, 2, 3, 4, 5, and 6? Each digit can be used only once.

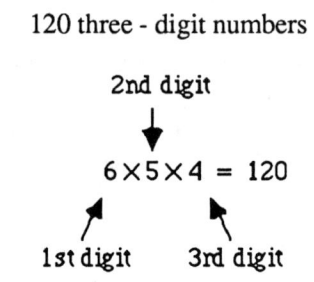

   120 three-digit numbers

   2nd digit
   ↓
   6 × 5 × 4 = 120
   ↑   ↖
   1st digit   3rd digit

2. How many different three-digit numbers can be made in Exercise 1 if a digit can be used more than once?

   216 three-digit numbers
   6 · 6 · 6 = 216

3. Use a tree diagram to illustrate the possible outcomes when a coin is tossed four times.

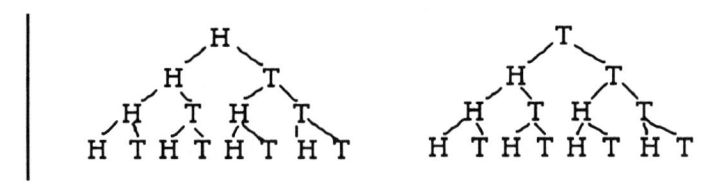

   ```
           H                           T
         /   \                       /   \
        H     T                     H     T
       / \   / \                   / \   / \
      H   T H   T                 H   T H   T
     /\  /\ /\  /\               /\  /\ /\  /\
    H T H T H T H T             H T H T H T H T
   ```

4. The telephone prefix for a university is 543. The prefix is followed by four digits. How many telephone numbers are possible before a new prefix will be needed?

   10,000
   10 · 10 · 10 · 10 = 10,000 different phone numbers

5. The results of a college mathematics course are summarized in the table below.

Section 2   Empirical Probability and the Fundamental Principle of Counting   **189**

	Female	Male
Passed The Course	22	30
Failed The Course	5	8

(a) What is the probability that a student passed the course?
(b) What is the probability that a female failed the course?
(c) What is the probability that a person who passed is female?

$$(a)\ P(\text{passed}) = \frac{27}{65}$$
number passed = 22 + 5 = 27
total number = 22 + 5 + 30 + 8 = 65
$$(b)\ P(\text{female failed}) = \frac{5}{27}$$
$$(c)\ P = \frac{22}{52}$$

6. If a person takes a five-question true / false test, what is the probability that he/she will answer all five correctly by quessing.

$$P(5\ \text{correct}) = \frac{1 \cdot 1 \cdot 1 \cdot 1 \cdot 1}{2 \cdot 2 \cdot 2 \cdot 2 \cdot 2} =$$
$$\frac{1}{32} \longleftarrow \text{only one correct for each question}$$

7. The following is a record of the senior History exam at Jones College.

Score	Number of Students
91-100	25
81-90	127
71-80	223
61-70	106
51-60	74
41-50	20

(a) Assign a probability to the event of making a score of 91-100.
(b) Assign a probability to the event of making a score of 71-90.
(c) Assign a probability of making a score below 71.

(a) $P(91-100) = \dfrac{25}{575}$

(b) $P(71-90) = \dfrac{223+127}{515} = \dfrac{350}{575}$

(c) $P(<71) = \dfrac{106+74+20}{575} = \dfrac{200}{575}$

8. Peter's Pan Pizza Palace has 3 kinds of salads, 15 kinds of pizza, and 4 kinds of desserts. How many different three-course meals could be ordered?

number of meals = $3 \cdot 15 \cdot 4 = 180$

9. Television stations have call letters that must begin with a 'W' or a 'K'. They must have a total of 3 or 4 letters.

   (a) How many sets of three-letter call letters are possible?
   (b) How many sets of four-letter call letters are possible?

(a) number of 3-letter = $2 \cdot 26 \cdot 26 = 1352$

   ↑
   W or K

(b) number of 4-letter call letters = $2 \cdot 26 \cdot 26 \cdot 26 = 35{,}152$

10. In a family of exactly three children, what is the probability that:

    (a) all three are boys?
    (b) the first 2 children are girls?
    (c) at least one is a boy?

(a) $\dfrac{1}{8}$

(b) $\dfrac{2}{8}$

(c) $\frac{7}{8}$

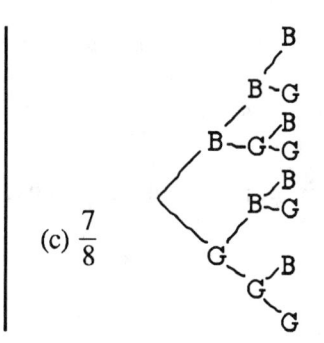

```
            B
          B-G
            B
      B-G-G
            B
        B-G
    G
          B
      G
          G
```

11. In a drawer, there are 9 black socks and 11 blue socks. Two socks are drawn from the drawer. What is the probability of getting:

   (a) one blue and one black sock?
   (b) a matching pair?

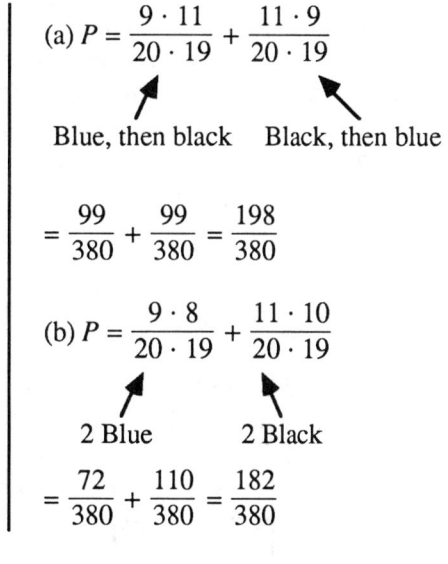

(a) $P = \dfrac{9 \cdot 11}{20 \cdot 19} + \dfrac{11 \cdot 9}{20 \cdot 19}$

　　　　↑　　　　　　↖

Blue, then black　　Black, then blue

$= \dfrac{99}{380} + \dfrac{99}{380} = \dfrac{198}{380}$

(b) $P = \dfrac{9 \cdot 8}{20 \cdot 19} + \dfrac{11 \cdot 10}{20 \cdot 19}$

　　　　↑　　　　　　↘

　　　2 Blue　　　　　2 Black

$= \dfrac{72}{380} + \dfrac{110}{380} = \dfrac{182}{380}$

12. The license plate pattern for a particular state consists of two letters followed by four digits. How many different license plates can be made if:

   (a) the letter O and digit 0 cannot be used?
   (b) no letter can be repeated?
   (c) no letter and no digit can be repeated?

(a) number = $25 \cdot 25 \cdot 9 \cdot 9 \cdot 9$ = 455,625
(b) number = $26 \cdot 25 \cdot 10 \cdot 10 \cdot 10$ = 650,000
(c) number = $26 \cdot 25 \cdot 10 \cdot 9 \cdot 8$ = 468,000

13. Cathy Curious tosses a thumbtack 500 times and records the number of times it landed point up or

point down. The results are summarized below.

# Times Point Up	#Times Point Down
178	322

(a) What is the probability that such a thumbtack will land point down on a given toss?
(b) If Cathy tosses the tack 10,000 more times, how many times would she expect it to land point up?

$$\text{(a) } \frac{178}{500} = 0.356$$

$$\text{(b) } \frac{N}{10000} = \frac{178}{500}$$
$$500N = 1{,}780{,}000$$
$$N = 3{,}576$$

14. A combination lock has numbers on it ranging from 0 to 39. A given combination consists of three numbers. If a combination is chosen at random, what is the probability that the first two numbers are multiples of five and the third number is a multiple of 7?

$$P = \frac{8 \cdot 8 \cdot 6}{40 \cdot 40 \cdot 40} \quad \begin{array}{l}\leftarrow \text{2 multiples of 5 and 1 multiple of 7} \\ \leftarrow \text{total combinations}\end{array}$$
$$= \frac{384}{64{,}000}$$

15. The number of symbols on each of the three dials of a slot machine are shown in the table below.

Symbol	Dial 1	Dial 2	Dial 3
Bar	1	3	1
Bell	1	3	3
Orange	3	6	7
Cherry	7	7	0
Lemon	3	0	4
Plum	5	1	5
Total	20	20	20

(a) What is the probability of getting three oranges?
(b) What is the probability of getting three bells?

$$\text{(a) } P = \frac{3 \cdot 6 \cdot 7}{20 \cdot 20 \cdot 20} = \frac{126}{8{,}000}$$

$$\text{(b) } P = \frac{1 \cdot 3 \cdot 3}{20 \cdot 20 \cdot 20} = \frac{9}{8{,}000}$$

Section 3 Counting Techniques Using Permutations and Combinations

Cover the right side of the page and work on the left, then check your work

1. Evaluate each of the following.

   (a) $P(7, 5)$
   (b) $P(12, 12)$
   (c) $P(6, 2)$
   (d) $P(53, 3)$

   (a) $P(7, 5) = \dfrac{7!}{(7-5)!} = \dfrac{7 \cdot 6 \cdot 5 \cdot 4 \cdot 3 \cdot 2 \cdot 1}{2!} = \dfrac{7 \cdot 6 \cdot 5 \cdot 4 \cdot 3 \cdot 2 \cdot 1}{2 \cdot 1} = 2520$

   (b) $P(19, 9) = 9! = 362{,}880$
   (c) $P(6, 2) = 30$
   (d) $P(53, 3) = \dfrac{53!}{(53-3)!} = \dfrac{53!}{50!} = 140{,}556$

2. Evaluate each of the following.

   (a) $C(7, 5)$
   (b) $C(6, 4)$
   (c) $C(40, 37)$
   (d) $C(r, 3)$

   (a) $C(7, 5) = \dfrac{7!}{5!\,(7-5)!} = \dfrac{7 \cdot 6 \cdot 5 \cdot 4 \cdot 3 \cdot 2 \cdot 1}{(5 \cdot 4 \cdot 3 \cdot 2 \cdot 1)(2!)} =$

   $\dfrac{7 \cdot 6 \cdot \cancel{5} \cdot \cancel{4} \cdot \cancel{3} \cdot \cancel{2} \cdot \cancel{1}}{(\cancel{5} \cdot \cancel{4} \cdot \cancel{3} \cdot \cancel{2} \cdot \cancel{1})(2!)}$

   $= \dfrac{42}{2} = 21$

   (b) $C(6, 4) = \dfrac{6!}{4!\,2!} = 15$

   (c) $C(40, 37) = \dfrac{40!}{37!\,3!} = \dfrac{40 \cdot 39 \cdot 38}{3 \cdot 2 \cdot 1} = 9880$

   (d) $C(r, 3) = \dfrac{r!}{3!\,(r-3)!} = \dfrac{(r)(r-1)(r-2)}{6}$

194   **CHAPTER 8**   An Introduction to Probability Theory

3. Which of the following are true?

   (a) $5! = 5 \cdot 4!$
   (b) $2! \cdot 3! = 6!$
   (c) $4! + 4! = 8!$
   (d) $\dfrac{6!}{5!} = 6$
   (e) $\dfrac{6!}{3!} = 2!$

   (a) True
   $$5 \cdot 4! = 5 \cdot 4 \cdot 3 \cdot 2 \cdot 1 = 5!$$
   (b) False
   $$2!\, 3! = (2 \cdot 1)(3 \cdot 2 \cdot 1)$$
   $$\neq 6 \cdot 5 \cdot 4 \cdot 3 \cdot 2 \cdot 1$$
   (c) False
   $$4! + 4! = (4 \cdot 3 \cdot 2 \cdot 1) + (4 \cdot 3 \cdot 2 \cdot 1)$$
   $$\neq 8 \cdot 7 \cdot 6 \cdot 5 \cdot 4 \cdot 3 \cdot 2 \cdot 1$$
   (d) True
   $$\frac{6!}{5!} = \frac{6 \cdot 5 \cdot 4 \cdot 3 \cdot 2 \cdot 1}{5 \cdot 4 \cdot 3 \cdot 2 \cdot 1} = 6$$
   (e) False
   $$\frac{6!}{3!} = \frac{6 \cdot 5 \cdot 4 \cdot 3 \cdot 2 \cdot 1}{3 \cdot 2 \cdot 1} \neq 2!$$

4. In how many ways can the letters in the word SCRABBLE be rearranged?

   20, 160
   $$P(8, 8) = \frac{8!}{2!} = 8 \cdot 7 \cdot 6 \cdot 5 \cdot 4 \cdot 3$$
   $$= 20{,}160$$

5. How many three-person committees can be formed from a group of eight people?

   56
   $$C(8, 3) = \frac{8!}{3!\, 5!} = \frac{8 \cdot 7 \cdot 6}{3 \cdot 2 \cdot 1} = 56$$

6. A class has 25 members.

   (a) In how many ways can a president, vice president, secretary, and treasurer be chosen?
   (b) How many committees of five persons can be chosen?

   (a) 69,000
   $$P(25, 3) = \frac{25!}{22!} = 25 \cdot 24 \cdot 23 =$$

Section 3 Counting Techniques using Permutations and Combinations

> 69,000
> (b) 53,130
> $C(25, 5) = \dfrac{25!}{5!\,20!} =$
> $\dfrac{25 \cdot 24 \cdot 23 \cdot 22 \cdot 21}{5 \cdot 4 \cdot 3 \cdot 2 \cdot 1}$
> $= 53,130$

7. Four juniors and four seniors must be appointed to the prom committee. There are 14 juniors and 9 seniors to choose from. How many different committees could be selected.

> 84,084
> $C(14, 4) \cdot C(9, 3)$
> $= \dfrac{14 \cdot 13 \cdot 12 \cdot 11}{4 \cdot 3 \cdot 2 \cdot 1} \cdot \dfrac{9 \cdot 8 \cdot 7}{3 \cdot 2 \cdot 1}$
> $= 1001 \cdot 84$
> $= 84,084$

8. A basketball coach wants to try every possible 5-player combination. If there are 12 players on the team, how many such 5-player combinations are there?

> 792
> $C(12, 5) = \dfrac{12!}{5!\,7!} =$
> $\dfrac{12 \cdot 11 \cdot 10 \cdot 9 \cdot 8}{5 \cdot 4 \cdot 3 \cdot 2 \cdot 1}$
> $= 792$

9. A committee of three people is to be selected from a group made up of six Americans, five Canadians, and three Italians.

   (a) What is the probability that the committee consists of all Americans?
   (b) What is the probability the committee has no Americans?

> (a) 0.33
> $P(E) = \dfrac{C(6,3)}{C(14,3)} =$
> $\dfrac{\dfrac{6 \cdot 5 \cdot 4}{3 \cdot 2 \cdot 1}}{\dfrac{14 \cdot 13 \cdot 12}{3 \cdot 2 \cdot 1}}$
> $= \dfrac{120}{364} \approx 0.33$

196   **CHAPTER 8**   An Introduction to Probability Theory

(b) 0.15

$$P(E) = \frac{C(8,3)}{C(14,3)} = \frac{\frac{8 \cdot 7 \cdot 6}{3 \cdot 2 \cdot 1}}{\frac{14 \cdot 13 \cdot 12}{3 \cdot 2 \cdot 1}}$$

$$= \frac{56}{364} \approx 0.15$$

10. What is Lea's probability of getting three kings and 2 jacks in a 5-card hand?

0.000009

$$P(E) = \frac{\underset{\text{Kings}}{C(4,3)} \cdot \underset{\text{Jacks}}{C(4,2)}}{C(52,5)} = \frac{4 \cdot 6}{2{,}598{,}960}$$

$= 0.000009$

11. A four-volume set of books is placed on a shelf. What is the probability that books will be in order (volumes 1-4) from left to right?

0.04

$$P(E) = \frac{1}{P(4,4)} \quad \longleftarrow \text{only one correct order}$$

$$= \frac{1}{24} \approx 0.04$$

12. Consider the following figure. How many lines can be drawn if each line must contain a pair of points?

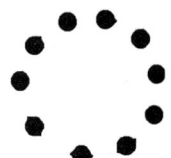

45

$$C(10,2) = \frac{10!}{2! \, 8!} = \frac{10 \cdot 9}{2 \cdot 1} = 45$$

↑

choose 2 points at a time from 10

Section 3 Counting Techniques using Permutations and Combinations 197

13. In how many ways can four couples be seated in a row of eight chairs if no couple is separated?

    24
    $P(4, 4) = 24$
    ↑
    arrangements of 4 couples

14. The probability that a given light bulb is defective is $\frac{6}{100}$. Suppose 5 bulbs are selected from a lot of 100. What is the probability that three of the 5 bulbs defective?

    0.002
    $$P(E) = \frac{C(94,2) \cdot C(6,3)}{C(100,5)}$$
    $$= \frac{\frac{94 \cdot 93}{2} \cdot \frac{6 \cdot 5 \cdot 4}{3 \cdot 2}}{\frac{100 \cdot 99 \cdot 98 \cdot 97 \cdot 96}{5 \cdot 4 \cdot 3 \cdot 2 \cdot 1}}$$
    $$= \frac{8742 \cdot 20}{75{,}287{,}520} \approx 0.002$$

15. Jim Freel knows that there are ten horses in the race. Help him find the probability that horse number 4 finishes first and horse number 8 finishes second.

    The probability is 0.01.
    $P(E) =$
    $\frac{1 \cdot 1 \cdot P(8, 8)}{P(10, 10)}$ ← arrangements of remaining horses
    $= \frac{8!}{10!} = \frac{1}{9 \cdot 10} \approx 0.01$

Section 4 Properties of Probability

Cover the right side of the page and work on the left, then check your work

1. Suppose $P(A) = 0.7$, $P(B) = 0.4$, and $P(A \cap B) = 0.3$ Compute each of the following.

   (a) $P(\overline{B})$
   (b) $P(A \cup B)$
   (c) $P(\overline{A \cup B})$
   (d) $P(\overline{A \cap B})$

   (a) $P(\overline{B}) = 0.6$
   $P(B) = 1 - P(\overline{B}) = 1 - 0.4 = 0.6$
   (b) $P(A \cup B) = 0.8$
   $P(A \cup B) = P(A) + P(13) - P(A \cap B)$
   $= 0.7 + 0.4 - 0.3 = 0.8$
   (c) $P(\overline{A \cup B}) = 0.2$
   $P(\overline{A \cup B}) = 1 - P(A \cup B)$
   $= 1 - 0.8 = 0.2$
   (d) $P(\overline{A \cap B}) = 0.7$
   $P(\overline{A \cap B}) = 1 - P(A \cap B)$
   $= 1 - 0.3 = 0.7$

2. Suppose $P(A) = 0.6$, $P(B) = 0.4$, and $P(A \cup B) = 0.7$. Compute each of the following.

   (a) $P(A \cap B)$
   (b) $P(A \cup \overline{B})$
   (c) $P(\overline{A \cap B})$

   (a) $P(A \cap B) = 0.3$
   $P(A \cup B) = P(A) + P(B) - P(A \cap B)$
   $0.7 = 0.6 + 0.4 - P(A \cap B)$
   $0.7 = 1.0 - P(A \cap B)$
   $0.7 - 1.0 = - P(A \cap B)$
   $- 0.3 = - P(A \cap B)$
   $0.3 = P(A \cap B)$

   (b) $P(A \cup \overline{B}) = 0.9$
   $P(A \cup \overline{B}) = P(A) + P(\overline{B}) - P(A \cap \overline{B})$
   $= 0.6 + 0.6 - P(A \cap B)$
   $= 0.6 + 0.6 - 0.3$
   $= 0.9$
   $P(A) = 0.6$
   $P(B) = 0.4$
   $P(A \cup B) = 0.7$

Section 4 Properties of Probability   199

```
┌─────────────────────────┐
│ A                     B │
│    ╭────╮╭────╮         │
│   ╱     ╳     ╲         │
│  │ 0.3 │0.3│ 0.1│       │
│   ╲     ╳     ╱         │
│    ╰────╯╰────╯         │
└─────────────────────────┘
```

(c) $P(\overline{A \cap B}) = 0.7$

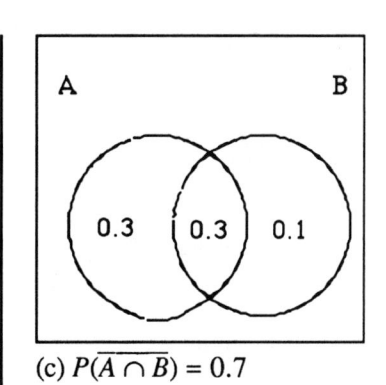

3. In a survey, 40% of the people drink Brand X, 70% of the people drink Brand Y, and 20% of the people drink both. What is the probability that a person drinks Brand X or Brand Y?

$P(\text{Brand X or Brand Y}) = 0.90$

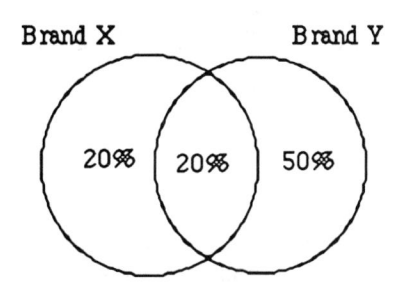

Brand X   Brand Y
```
    ╭────╮╭────╮
   ╱     ╳     ╲
  │ 20% │20%│ 50%│
   ╲     ╳     ╱
    ╰────╯╰────╯
```

4. A card is drawn from a 52-card deck. What is the probability that it is:

   (a) either a red card or a ten
   (b) not a queen
   (c) either a face card or a club
   (d) not a face card or not a club

(a) $P(\text{Red or Ten}) = \dfrac{28}{52}$

$P(\text{Red or Ten}) =$
$P(\text{Red}) + P(\text{Ten}) - P(\text{Red and Ten})$
$= \dfrac{26}{52} + \dfrac{4}{52} - \dfrac{2}{52}$
$= \dfrac{28}{52}$

(b) $P(\text{not a queen}) = \dfrac{48}{52}$

(c) $P(\text{face card or club}) = \dfrac{22}{52}$

$P(\text{face or club}) =$

200 CHAPTER 8 An Introduction to Probability Theory

$P(\text{face}) + P(\text{club}) - P(\text{face and club})$

(d) $P(\text{not face or not club}) = \dfrac{49}{52}$

$P(F \cup C) = P(F) + P(C) - P(F \cap C)$
$= \dfrac{40}{52} + \dfrac{39}{52} - \dfrac{30}{52} = \dfrac{49}{52}$

5. In a family of exactly three children what is the probability that:

   (a) at least two are girls?
   (b) none of them are girls?
   (c) either two or three are boys?

   (a) $\dfrac{4}{8}$
   (b) $\dfrac{1}{8}$
   (c) $\dfrac{4}{8}$

6. Write in words the complement for each part of Exercise 5.

   (a) at most one is a girl
   (b) at least one is a girl
   (c) at most one is a boy

7. In a survey of 100 students, it was found that 35 students were taking algebra, 32 were taking biology, 40 were taking history, 18 were taking algebra and biology, 21 were taking algebra and history, 24 were taking biology and history, and 15 were taking all three. If a student is chosen at random, what is the probability for each of the following?

   (a) the student is taking only algebra.
   (b) the student is not taking any of the three subjects.
   (c) the student is taking algebra and history, but not biology.

   (a) $\dfrac{11}{100}$
   (b) $\dfrac{41}{100}$
   (c) $\dfrac{6}{100}$

8. What are the odds in favor of a family having three children, all of whom are boys?

Section 4   Properties of Probability   201

9. The odds against Bruiser winning his next prize fight are 3 to 7. What is the probability that he will win the fight?

1 to 7

$P(3 \text{ boys}) = \dfrac{1}{8}$

odds in favor $= P(E) : P(\overline{E})$

$= \dfrac{1}{8} : \dfrac{7}{8} = 1 \text{ to } 7$

$\dfrac{7}{10}$

odds against = 3 to 7
odds in favor = 7 to 3

$P(E) = \dfrac{7}{7+3} = \dfrac{7}{10}$

10. On a single roll of a pair of dice, what are the odds in favor of rolling a sum of 6?

6 to 30

$P(\text{seven}) = \dfrac{6}{36}$

odds in favor $= \dfrac{6}{36} : \dfrac{30}{36} = 6 : 30$

11. A coin is tossed four times. Which one of the following is the complement of the event that at least two heads appear?

(a) at least two tails appear.
(b) one head appears.
(c) at most one head appears.
(d) at most two tails appear.

(c)

12. The table below gives the results of a recent survey.

	C	$\overline{C}$	Total
M	68	253	321
$\overline{M}$	25	154	179
Total	93	407	500

M - students taking a math course
$\overline{M}$ - students not taking a math course
C - students having a personal computer
$\overline{C}$ - students not having a personal computer

What is the probability that a student selected at random:

(a) has a computer?
(b) is taking a math course and has a computer?
(c) is taking a math course or has a computer?
(d) is not taking a math course or does not have a computer?

(a) $\dfrac{93}{500}$

(b) $\dfrac{68}{500}$

(c) $\dfrac{346}{500}$

$P(M \cup E) = P(M) + P(C) - P(M \cap C)$
$= \dfrac{321}{500} + \dfrac{93}{500} - \dfrac{68}{500} = \dfrac{346}{500}$

(d) $\dfrac{432}{500}$

$P(\overline{M} \cup \overline{C}) = P(\overline{M}) + P(\overline{C}) - P(\overline{M} \cap \overline{C})$
$= \dfrac{179}{500} + \dfrac{407}{500} - \dfrac{154}{500} = \dfrac{432}{500}$

13. For a single roll of two dice, tabulate each of the following.

    (a) Event A = {sum of 7}
    (b) Event B = {6 < sum < 10}
    (c) Event A and B
    (d) Event A or B

(a) A = {(1, 6), (6, 1), (2, 5), (5, 2), (3, 4), (4, 3)}

(b) B = {(1, 6), (6, 1), (2, 5), (5, 2), (3, 4), (4, 3), (2, 6), (6, 2), (3, 5), (5, 3), (4, 4), (3, 6), (6, 3), (4, 5), (5, 4)}

14. For the events from Exercise 13 calculate each of the following.

    (a) $P(A)$
    (b) $P(\overline{A})$
    (c) $P(A \cup B)$

(a) $P(A) = \dfrac{6}{36}$

(b) $P(\overline{A}) = \dfrac{30}{36}$

(c) $P(A \cup B) = \dfrac{15}{36}$

Section 5 Expected Value and the Probability of Compound Events

Cover the right side of the page and work on the left, then check your work

1. If $P(A) = 0.6$, $P(B) = 0.4$ and $P(A \cap B) = 0.3$.
   Compute each of the following.

   (a) $P(A|B)$
   (b) $P(B|A)$
   (c) $P(B|\overline{A})$
   (d) $P(A \cup \overline{B})$

   (a) 0.75
   $$P(A|B) = \frac{P(A \cap B)}{P(B)} = \frac{0.3}{0.4} = 0.75$$
   (b) 0.6
   $$P(B|A) = \frac{P(A \cap B)}{P(A)} = \frac{0.3}{0.5} = 0.6$$
   (c) $$P(B|\overline{A}) = \frac{P(B \cap \overline{A})}{P(\overline{A})} = \frac{0.1}{0.4} = 0.25$$

	A	$\overline{A}$	Total
B	0.3	0.1	0.4
$\overline{B}$	0.3	0.3	0.6
Total	0.6	0.4	1.0

   (d) $P(A \cup \overline{B}) =$
   $P(A) + P(\overline{B}) - P(A \cap \overline{B})$
   $= 0.6 + 0.6 - 0.3$
   $= 0.9$

2. Given the following table, compute the following probabilities.

   (a) $P(A)$
   (b) $P(D)$
   (c) $P(B \cap C)$
   (d) $P(A|C)$
   (e) $P(B)$

204 CHAPTER 8 An Introduction to Probability Theory

	C	D	Total
A	1/3	1/5	8/15
B	4/15	1/5	7/15
Total	3/5	2/5	1

(a) $\dfrac{8}{15}$

(b) $\dfrac{2}{5}$

(c) $\dfrac{4}{15}$

(d) $\dfrac{5}{9}$

$$P(A|C) = \dfrac{P(A \cap C)}{P(C)} = \dfrac{\frac{1}{3}}{\frac{3}{5}} = \dfrac{5}{9}$$

(e) $\dfrac{8}{15}$

3. Two standard dice are thrown. What is the probability:

   (a) of a sum of 7, given that the sum is odd?
   (b) of a sum of 7, given that one die is a 3?

(a) $\dfrac{1}{3}$

$$P(7|\text{odd}) = \dfrac{P(7 \cap \text{odd})}{P(\text{odd})}$$

$$= \dfrac{\frac{6}{36}}{\frac{1}{2}} = \dfrac{6}{18} = \dfrac{1}{3}$$

(b) $\dfrac{2}{11}$

$$P(7/\text{one die} = 3) = \dfrac{P(7 \cap \text{one die} = 3)}{P(\text{one die} = 3)}$$

$$= \dfrac{\frac{2}{36}}{\frac{11}{36}} = \dfrac{2}{11}$$

4. A penny, a nickel, and a quarter are tossed. What is the probability of getting 3 tails, given that the quarter is tails?

$$P(3 \text{ tails/Quarter} = \text{tails}) = \frac{P(3 \text{ tails} \cap \text{Quarter} = \text{tails})}{P(\text{Quarter} = \text{tails})} = \frac{\frac{1}{8}}{\frac{1}{2}} = \frac{1}{4}$$

5. A box contains seven white balls and five black balls. Two balls are drawn from the box.

    (a) What is the probability of getting two white balls.
    (b) What is the probability that the first ball is black and the second is white.

    (a) $\frac{42}{121}$
    W_1 = 1st ball white
    W_2 = 2nd ball white
    $P(W_1 \cap W_2) = P(W_1) \cdot P(W_2 \mid W_1)$
    $= \frac{7}{12} \cdot \frac{6}{11} = \frac{42}{121}$

    (b) $\frac{35}{121}$
    B_1 = 1st ball black
    W_2 = 2nd ball white

    $P(B_1 \cap W_2) = P(B_1) \cdot P(W_2 \mid B_1)$
    $= \frac{5}{12} \cdot \frac{7}{11} = \frac{35}{121}$

6. Late Larry sets two alarm clocks so that he will be sure to wake up on time. One clock works 80% of the time and the other works 60% of the time. What is the probability that Larry will get up on time (one or both of the clocks work)?

    $\frac{92}{100}$
    C_1 = first clock works
    C_2 = second clock works
    $P(C_1 \cap C_2) = P(C_1) \cdot P(C_2 \mid C_1)$
    ↑
    $= \frac{8}{10} \cdot \frac{6}{10}$ independent events

$$= \frac{48}{100}$$

$$P(C_1 \cup C_2) = P(C_1) + P(C_2) - P(C_1 \cap C_2)$$

$$= \frac{8}{10} + \frac{6}{10} - P(C_1 \cap C_2)$$

$$= \frac{8}{10} + \frac{6}{10} - \frac{48}{100}$$

$$= \frac{14}{10} - \frac{48}{100} = \frac{92}{100}$$

7. In a certain region of the western United States the probability of a deer being healthy is $\frac{19}{20}$. If a deer is healthy the probability of it being killed by a predator is $\frac{1}{200}$. If it is not healthy, the probability of it being killed is $\frac{2}{5}$. If a deer is chosen at random from the region, what is the probability that it is killed by a predator.

$$\frac{99}{4000}$$

K = Killed by predator
H = healthy

$$P(K \cap H) = P(H) \cdot P(K|H)$$

$$= \frac{19}{20} \cdot \frac{1}{200} = \frac{19}{4000}$$

$$P(K \cap \overline{H}) = P(\overline{H}) \cdot P(K|\overline{H})$$

$$= \frac{1}{20} \cdot \frac{2}{5} = \frac{2}{100}$$

$$P((K \cap H) \cup (K \cap \overline{H})) = P(K \cap H) + P(K \cap \overline{H})$$

$$= \frac{99}{4000}$$

8. Refering to exercise 5, suppose the following game is played. Two balls of the same color drawn results in winning $10. Two balls of different color results in losing $3. What is the expected value of this game?

$6.86
Exp. value =
P(same color) $\cdot$ 10 + P(diff. color) $\cdot$ 3

Section 5   Expected Value and the Probability of Compound Events

$$= \frac{62}{121} \cdot 10 + \frac{70}{121} \cdot 3$$
$$= \frac{620}{121} + \frac{210}{121} = \frac{830}{121} = \$6.86$$

9. A box contains 5 one-dollar bills, 3 ten-dollar bills and 2 one hundred dollar bills. A player draws one bill from the box at random. What is the expected value?

$24.00
$23.50
E.V. =
$$\frac{5}{10} \cdot 1 + \frac{3}{10} \cdot 10 + \frac{2}{10} \cdot 10$$
$$= \frac{5}{10} + 3 + 20 = \$23.50$$

10. A school sells 5000 lottery tickets. There is one $500 prize, five $100 prizes, and ten $10 prizes. What is the expected value for one ticket?

$0.62
E.V. =
$$\frac{1}{5000} \cdot 500 + \frac{5}{5000} \cdot 100 + \frac{10}{5000} \cdot 10$$
$$= \frac{1}{10} + \frac{1}{2} + \frac{1}{50} = \$0.62$$

11. A box contains three blue balls, five white balls, and four yellow balls. A person draws balls from the box until a blue ball is drawn. What is the probability that a blue ball is drawn for the first time on:

(a) the first draw.
(b) the second draw.
(c) the third draw.

(a) $\frac{3}{12}$

(b) $\frac{27}{132}$

B_1 = blue ball on first draw
B_2 = blue ball second draw
B_3 = blue ball on third draw
$P(\overline{B}_1 \cap \overline{B}_2) = P(\overline{B}_1) \cdot P(B_2|\overline{B}_1)$

$$= \frac{9}{12} \cdot \frac{3}{11} = \frac{27}{132}$$

(c) $\frac{216}{1320}$

$P((\bar{B}_1 \cap \bar{B}_2) \cap B_3)) = P(\bar{B}_1 \cap \bar{B}_2) \cdot P(B_3|\bar{B}_1 \cap \bar{B}_2) = \frac{72}{132} \cdot \frac{3}{10} = \frac{216}{1320}$

Alternate solution:

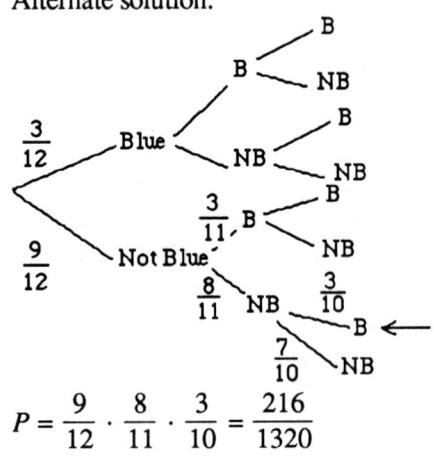

$$P = \frac{9}{12} \cdot \frac{8}{11} \cdot \frac{3}{10} = \frac{216}{1320}$$

12. 65% of a company's employees are female and 50% of the female employees are part-time. What is the probability that an employee chosen at random is:

(a) female and part-time?
(b) part-time, given that the employee is female?

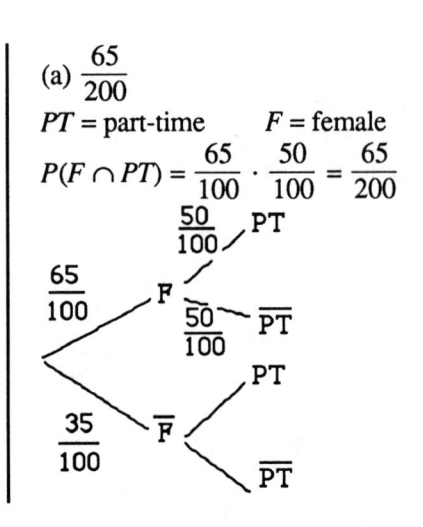

(a) $\frac{65}{200}$

PT = part-time F = female

$P(F \cap PT) = \frac{65}{100} \cdot \frac{50}{100} = \frac{65}{200}$

Section 5 Expected Value and the Probability of Compound Events   209

(b) $\dfrac{50}{100} = \dfrac{1}{2}$

$P(PT|F) = \dfrac{P(PT \cap F)}{P(F)}$

$= \dfrac{\frac{65}{200}}{\frac{65}{100}} = \dfrac{1}{2}$

13. A cereal company has 10 different action characters distributed randomly in each box of cereal (one character per box). Use Table 9 to estimate the average number of boxes that would have to be purchased to obtain all 10 characters.

29.5 boxes
Let the digits 0 - 9 in the table represent the 10 action characters. 0 will represent the tenth character.

Use rows of Table 9 and count number of digits (boxes) until all of 0 - 9 are selected.

Trial Number	Number of Boxes
1	12
2	42
3	34
4	17
5	24
6	28
7	29
8	26
9	41
10	42

Average number of boxes = 29.5

14. If it rains today, the probability it will rain tomorrow is $\dfrac{7}{10}$. If it does not rain today, the probability it will rain tomorrow is $\dfrac{2}{10}$. It is raining today. Use Table 9 to predict (simulate) the weather for the next five days.

Use digits 0 - 6 to represent rain tomorrow, digits 7 - 9 to represent no rain tomorrow when it is raining today. When it is not raining today, use digits 0 - 1 to represent rain tomorrow and digits 2 - 9 to represent no rain tomorrow.

Day	Digit from Table 9	Weather
Today	------	Rain
1	8	No Rain
2	7	No Rain
3	0	Rain
4	1	Rain
5	6	Rain

15. A single card is drawn from a standard deck.

(a) What is the probability that it is a face card and a club?
(b) What is the probability that it is a jack, given that it is a face card?

(a) $\dfrac{3}{52}$
$F = $ face card $C = $ club
$P(F \cap C) = P(F) \cdot P(C|F)$
$= \dfrac{12}{52} \cdot \dfrac{3}{12} = \dfrac{3}{52}$

(b) $\dfrac{4}{12}$
$J = $ jack $F = $ face card
$P(J|F) = \dfrac{P(J \cap F)}{P(F)}$
$= \dfrac{\frac{4}{52}}{\frac{12}{52}} = \dfrac{4}{12}$

Chapter 9
The Uses and Misuses of Statistics

Section 1 Organizing Data

Cover the right side of the page and work on the left, then check your work

1. Answer each of the questions about the graph shown below.

   (a) Which month had the highest average temperature and what was it?
   (b) Between which two months was there the greatest increase in average temperature?

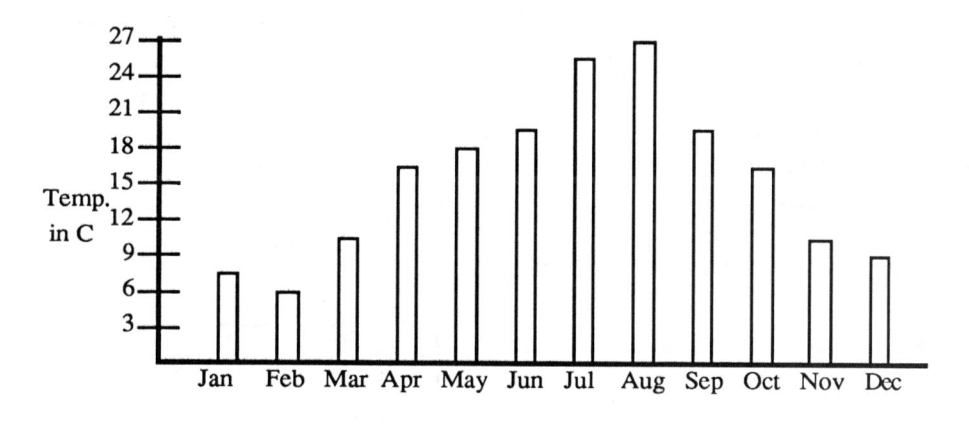

   [Bar graph: Temp. in C vs. months Jan–Dec]

   (a) August - 27^0
   (b) March to April - 6^0 increase

2. Make a frequency distribution to summarize the following data.

   Score on Math Test 1

   95 60 55 60 65 90 85 85 80 75
   70 75 80 65 80 80 70 55 95 90
   75 80 85 90 70 75 65 86 75 90

Score	Tallies	Frequency
95	II	2
90	IIII	4
85	IIII	4
80	IIII	5
75	IIII	5
70	III	3
65	III	3
60	II	2
55	II	2

3. For the data:

18 43 29 27 22 26 34 25 28 11
30 38 17 28 24 29 33 26 31 13

(a) Construct a frequency distribution of minimal integral length with 5 classifications, with the first class starting at 10.
(b) What are the class marks?
(c) What are the class boundaries?

(a)

Score	Frequency
11-17	3
18-24	3
25-31	10
32-38	3
39-45	1

(b) 14, 21, 28, 35, 42
(c) $10.5 - 17.5$, $17.5 - 24.5$, $24.5 - 31.5$, $31.5 - 38.5$, $38.5 - 45.5$

4. Make a histogram for the data in Exercise 3.

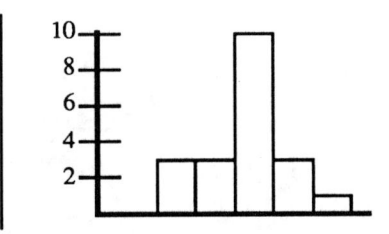

Section 1   Organizing Data   213

5. Draw a circle graph to represent the following data.

Crop	Percent of Farm Land
Corn	18%
Wheat	43%
Barley	8%
Cotton	12%
Soybeans	19%

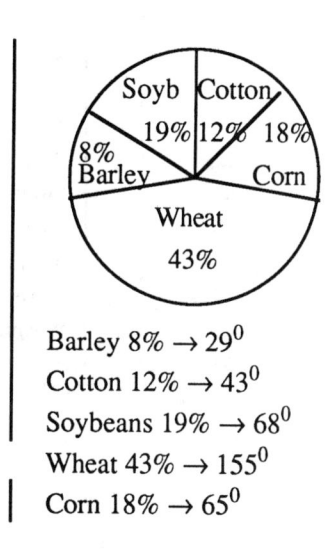

Barley 8% → 29^0
Cotton 12% → 43^0
Soybeans 19% → 68^0
Wheat 43% → 155^0
Corn 18% → 65^0

6. Draw a line graph for the data in Exercise 5.

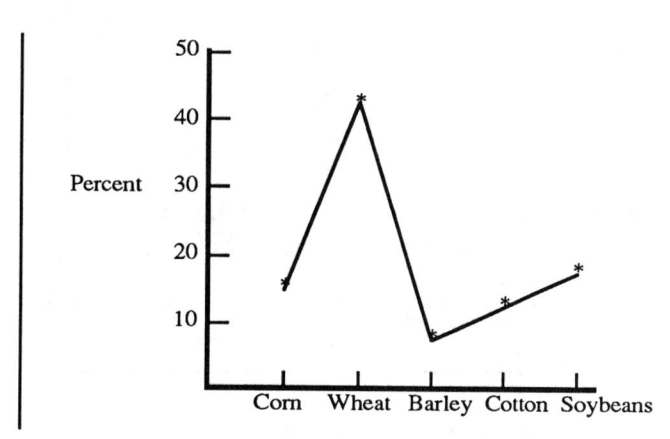

7. Make a stem and leaf plot for the following data.

214 CHAPTER 9 The Uses and Misuses of Statistics

President	Age at Death	President	Age at Death
L. Johnson	64	A. Johnson	66
Kennedy	46	Lincoln	56
Eisenhower	78	Buchanan	77
Truman	88	Pierce	64
F. Roosevelt	63	Fillmore	74
Hoover	90	Taylor	65
Coolidge	60	Polk	53
Harding	57	Tyler	71
Wilson	67	W. Harrison	68
Taft	72	Van Buren	79
T. Roosevelt	60	Jackson	78
McKinley	58	J. Q. Adams	80
B. Harrison	67	Monroe	73
Cleveland	71	Madison	85
Arthur	57	Jefferson	83
Garfield	49	J. Adams	90
Hayes	70	Washington	67
Grant	63		

Ages of Presidents at Death

```
9 | 0 0
8 | 0 3 5 8
7 | 0 1 1 2 3 4 7 8 8 9
6 | 0 0 3 3 4 4 5 6 7 7 8
5 | 3 6 7 7 8
4 | 6 9
```

8. Five coins are tossed 75 times. The following table shows the distribution for the number of tails.

Number of Tails	0	1	2	3	4	5
Frequency	3	12	23	22	11	4

(a) Draw a bar graph for this data.
(b) Draw a line graph for this data.

(a)

(b)

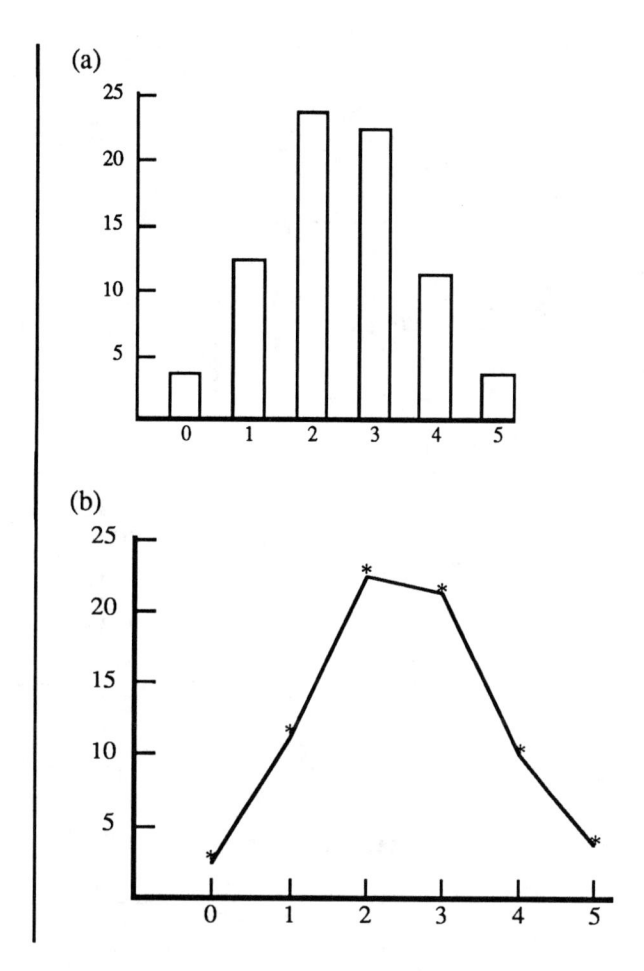

9. A list of the heights (in centimeters) of 20 boys and 20 girls is given below.

Boys	Girls
127 135 141 135 131	136 127 139 128 147
120 134 140 143 126	144 129 134 139 140
137 131 142 139 130	133 121 119 146 133
124 143 126 129 155	122 138 137 141 142

(a) Make a frequency distribution, starting the first interval at 115 and using 5 intervals of minimum integral length.

(b) Make a histogram for the data for each group.

Answers follow:

(a)

Height	Boys Frequency	Girls Frequency
115-123	1	3
124-132	9	3
133-141	7	10
142-150	2	4
151-159	1	0

(b)

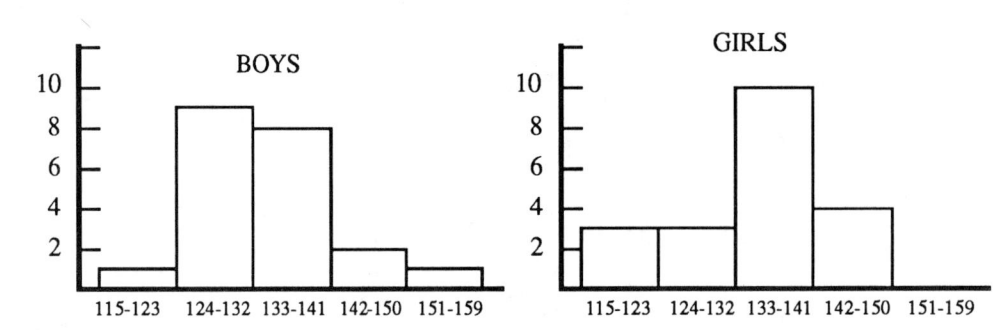

10. Make a stem and leaf plot for the data in Exercise 9.

Boys Heights Leaves	Stem	Girls Heights Leaves
	11	9
9 7 6 6 4 0	12	1 2 7 8 9
9 7 5 5 4 1 1 0	13	3 3 4 6 7 8 9 9
3 3 2 1 0	14	0 1 2 4 6 7
5	15	

11. The grades for a mathematics for elementary teachers course are given in the table.

Grade	Frequency
A	5
B	10
C	12
D	6
F	3

(a) Draw a bar graph for this data.
(b) Draw a circle graph for this data.

(a)

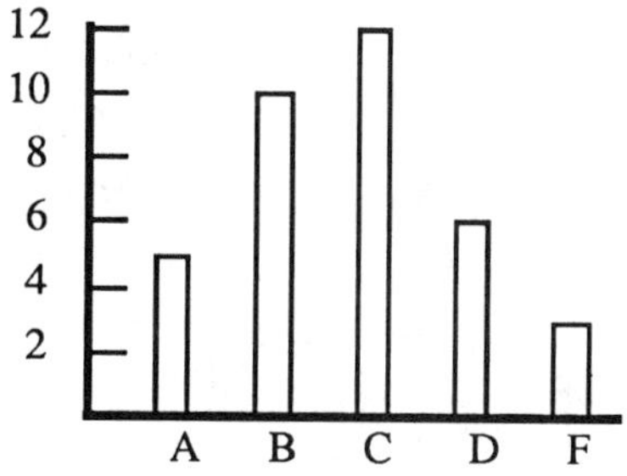
12
10
8
6
4
2
A B C D F

(b)

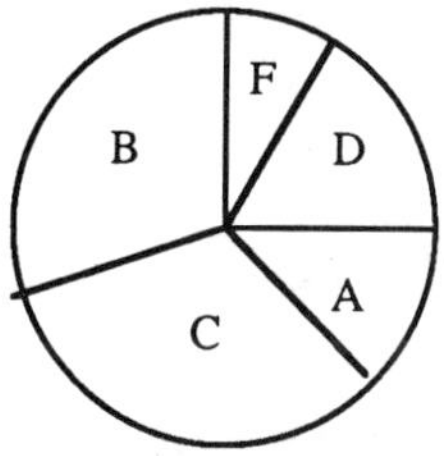
F
B
D
A
C

$A = 5 \rightarrow 50^0 \left(\frac{5}{36} \times 360^0\right)$

$B = 10 \rightarrow 100^0$

$C = 12 \rightarrow 120^0$

$D = 6 \rightarrow 60^0$

$F = 3 \rightarrow 30^0$

12. What is wrong with the following bar graph?

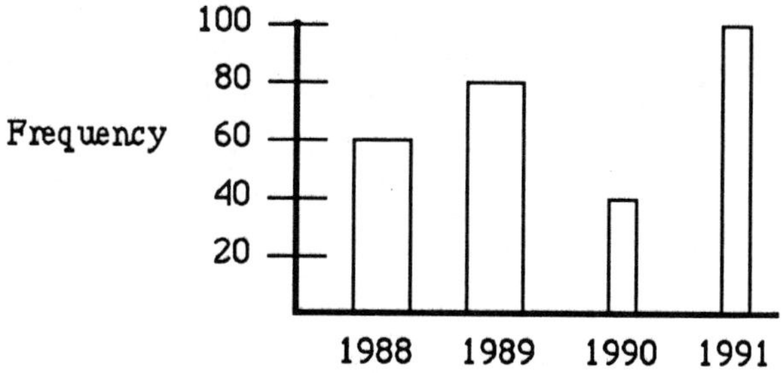
100
80
Frequency
60
40
20
1988 1989 1990 1991

Bars for 1990 and 1991 are too narrow compared to the bars for 1988 and 1989.

218 CHAPTER 9 The Uses and Misuses of Statistics

13. Use a frequency polygon to represent the following data.

Salary	Number of Employees
15,000-24,999	200
25,000-34,999	110
35,000-44,999	50
45,000-54,999	15
55,000-64,999	5

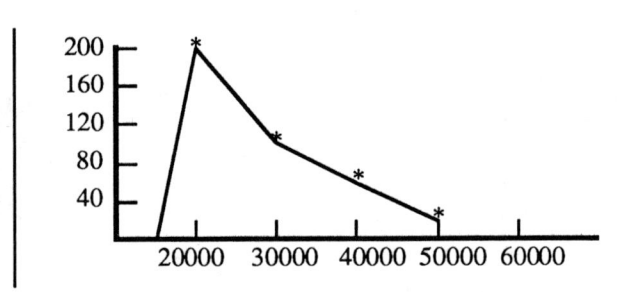

14. The golf scores for 40 golfers are given below. Make a line graph to represent this data.

Score	Frequency
66	2
67	3
68	2
69	3
70	6
71	7
72	9
73	4
74	2
75	2

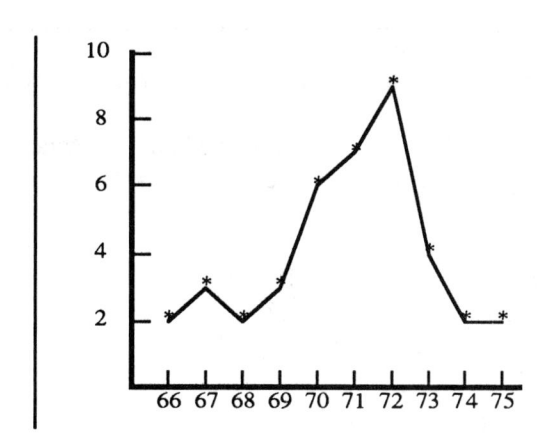

15. No-Pain claims that its product has more pain relieving ingredient than the other leading brands. The graph below is used to support this claim. What, if anything, is misleading about the graph?

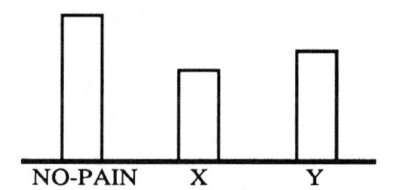

There is no vertical scale. We have no way of knowing what the bars represent. Suppose the actual values are:

No-Pain = 500 mg
Brand X = 480 mg
Brand Y = 485 mg

We can make the differences appear larger by constructing the vertical scale differently. Consider the two graphs below.

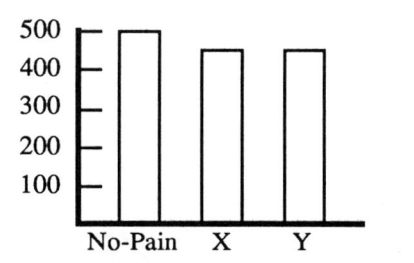

Differences appear small

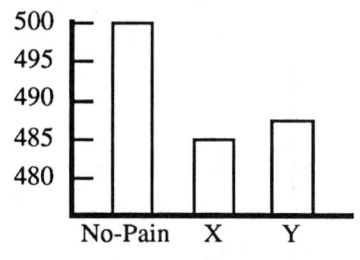

Differences appear large

Section 2 What is Average?

Cover the right side of the page and work on the left, then check your work

1. Compute the arithmetic mean, median, and mode for each of the following sets of data.

   (a) 5, 5, 6, 7, 9, 10
   (b) 2, 5, 5, 7, 8, 8, 8, 10
   (c) 3, 7, 11, 15, 19, 23, 27
   (d) 2, 5, 8, 6, 3, 4, 8, 11, 12, 1

   (a) mean = 7
   $$\text{mean} = \frac{5+5+6+7+9+10}{6}$$
   $$= \frac{42}{6} = 7$$
   median = 6.5
   mode = 5

   middle scores
   ↓ ↓
   Median - 5 5 6 7 9 10
   $$\text{Median} = \frac{6+7}{2} = 6.5$$

   (b) mean = 6.625
       median = 7.5
       mode = 8
   (c) mean = 15
       median = 15
       mode = no mode
   (d) mean = 6
       median = 5.5
       mode = 8

2. Find the mean and median annual salaries from the table given below.

Salary	Number of Workers
$17,000	2
20,000	4
26,000	4
35,000	3
38,000	12
44,000	8
52,000	4
81,000	2
150,000	1

mean = $41,275

x	f	xf
17,000	2	34,000
20,000	4	80,000
26,000	4	104,000
35,000	3	105,000
38,000	12	456,000
44,000	8	352,000
52,000	4	208,000
81,000	2	162,000
150,000	1	150,000
Total:	40	1,651,000

$$\bar{x} = \frac{1,651,000}{40} = \$41,275$$

median = $38,000
the median will be between the 20th and 21st pieces of data.
20th = 38,000
21st = 38,000 $\Big\}$ median = $38,000

3. In each part, give a set of data with the given characteristics.

   (a) The mean equals the median.
   (b) The mean is greater than the median.
   (c) The mean is less than the median.

(a) 1, 2, 3, 4, 5
$$\text{mean} = \frac{1+2+3+4+5}{5} = \frac{15}{5} = 3$$
median = 3

(b) 1, 2, 3, 5, 9
$$\text{mean} = \frac{1+2+3+5+9}{5} = \frac{20}{5} = 4$$
median = 3

(c) 1, 2, 3, 4, 4
$$\text{mean} = \frac{1+2+3+4+4}{5} = \frac{14}{5} = 2.8$$
median = 3

222   CHAPTER 9   The Uses and Misuses of Statistics

4. A student has scores of 72, 80, 65, 78, and 60 on her first five 100-point tests. What must she get on the next 100-point test to have a mean of 70?

   Will need a score of 65

   $$70 = \frac{\text{sum of 6 tests}}{6}$$
   420 = sum of 6 tests

   sum of 5 tests =
   72 + 80 + 65 + 78 + 60 = 355
   sixth text = 420 − 355 = 65

5. The mean age of the 175 students at Ridgeview Elementary is 8. The mean age of the 12 adults at the school is 40. What is the mean age of all persons at Ridgeview?

   mean age ≈ 10.05 years

   Total of students ages = 175 · 8
                             ↑   ↑
                         number mean
                  of students = 1400
   Total of adults ages = 12 · 40 = 480
   Mean of all persons = $\frac{1400 + 480}{175 + 12}$
   = $\frac{1880}{187}$ ≈ 10.05 years

6. Teresa's spring semester grades are given below. Find her grade point average for the semester.
   (A = 4, B = 3, C = 2, D = 1, F = 0)

Course	Grade	Credits
Math	B	4
History	A	3
Biology	C	3
Golf	D	1
Art	A	4

   Grade Point Average = 3.4

   Grade Point Average =
   $\frac{4 \cdot 3 + 3 \cdot 4 + 3 \cdot 2 + 1 \cdot 1 + 4 \cdot 5}{4 + 3 + 3 + 1 + 4}$

Section 2   What is Average   223

$$= \frac{51}{15} = 3.4$$

7. The heights of two basketball teams are given below.

   (a) Find the mean height for each team.
   (b) Does the mean accurately reflect the heights of the teams?

Cougars	Mules
70 in.	72 in.
71	73
72	74
72	78
90	78

   (a)   Mean Height
Cougars	Mules
75 in.	75 in.

   (b) No. Does not reflect the one extremely tall player of 90 inches.

8. Mel owns a shoe store and decides to order only one size shoe for the coming year. He looks at last year's sales figures, which are itemized by size. Should Mel order the mean, median, or mode shoe size?

   The mode shoe size. This would be the size sold most often.

9. A magazine reports that the average family in the United States has 2.35 children. What average is being used? Explain your answer.

   The mean is the average used here. The mode would be the number of children that the most families have (1, 2, 3, 4, etc.). The median would be the middle number or halfway between the two middle numbers (1.5, 2.5, etc.).

10. Find the 10th and 75th percentiles for the following set of data.

    35 40 37 53 31 26 62 39 42 48
    52 36 21 84 42 57 53 41 39 60

224 CHAPTER 9 The Uses and Misuses of Statistics

$P_{10} = 28.5$

Arrange data:
21 26 31 35 36 37 39 39 40 41
42 42 48 52 53 53 57 60 62 84

$\left(\frac{1}{10}\right)(20) = 2$

$P_{10} = \frac{26 + 31}{2} = 28.5$

$P_{75} = 53$

$\left(\frac{75}{100}\right)(20) = 15$

$P_{75} = \frac{53 + 53}{2} = 53$

11. Brian's last math test score was 7th from the top score. There were 70 scores on the test. What was his percentile score?

$P = \frac{63}{70}$ ← Scored higher than 63 of the students

$= 0.9$

12. Consider the following set of data.

Name	Age
Terry	43
Teresa	44
Brian	18
Kristin	15
LB	8

(a) What is the mean age?
(b) What will the mean of the ages be 5 years from now?
(c) What will the mean of the ages be 20 years from now?
(d) What conclusion can you make from (a), (b), and (c)?

(a) Mean = 25.6

Mean = $\frac{43 + 44 + 18 + 15 + 8}{5}$

= 25.6

(b) Mean = 30.6

$$\text{Mean} = \frac{48 + 49 + 23 + 20 + 13}{5}$$
$$= 30.6$$
(c) Mean = 45.6
$$\text{Mean} = \frac{63 + 64 + 38 + 35 + 28}{5}$$
$$= 45.6$$
(d) If you add **n** to each piece of data, the mean will increase by **n**.

13. The mean salary of 20 people is $30,000. How much is the mean increased by the addition of one person with a salary of $75,000?

Increase = $2,142.86
Total of 20 people = 20 · 30,000
$$= 600,000$$
Total of 21 people
$$= 600,000 + 75,000 = 675,000$$
$$\text{Mean of 21 people} = \frac{675,000}{21}$$
$$\approx \$32,142.86$$
Increase = 32,142.86 − 30,000
$$= 2,142.86$$

14. Find the mean percent price increase for 100 consumer items.

% Increase	Frequency
0 - 2	5
3 - 5	14
6 - 8	37
9 - 11	26
12 - 14	17
15 - 17	1

Mean = 8.17%
Mean =
$$\frac{5 \cdot 1 + 14 \cdot 4 + 37 \cdot 7 + 26 \cdot 10 + 17 \cdot 13 + 1 \cdot 16}{5 + 14 + 37 + 26 + 17 + 1}$$
$$= \frac{817}{100} = 8.17$$

15. Professor Hampton found the following results on a recent test from his Math for Elementary Teachers classes.

226 **CHAPTER 9** The Uses and Misuses of Statistics

8:00 Class   11:00 Class

Mean for Math Minors: 86   Mean for Math Minors: 82
Mean for others: 72   Mean for others: 61
Dr. Hampton shared these results with the 28 students in each of his classes. He further revealed the following.

8:00 Class   11:00 Class
Overall Mean: 74   Overall Mean: 76

Assuming that Dr. Hampton's computations are correct, explain how this surprising result could occur.

What do we know?
1) Each class has 28 students.
2) More scores of 11:00 class are close to 82. More scores of 8:00 class are close to 72.

Constructing a plan
Let m = number of math minors in 8:00 class
e = all other elementary education majors in 8:00 class
Then, we know $m + e = 28$ and

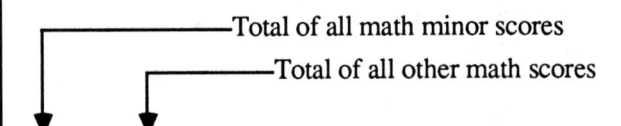

$$\frac{86m + 72e}{m + e} = 74$$

⎯Total of all math minor scores
⎯Total of all other math scores
⎯Total number of scores

But we also know that $e = 28 - m$

Making this substitution we get,

$$\frac{86m + 72(28 - m)}{28} = 74$$

$$\frac{86m + 2016 - 72m}{28} = 74$$

$14m + 2016 = 2072$
$14m = 56$
$m = 4$ and $e = 24$
In a similar way, for the 11:00 class we find,
$m = 20$ and $e = 8$

Section 3 How to Measure Scattering

Cover the right side of the page and work on the left, then check your work

1. For each of the following sets of data, find the range, the variance, and standard deviation.

   (a) 3, 7, 11, 15, 19, 23
   (b) 2, 5, 8, 6, 3, 4, 8, 11, 12, 1, 17, 23

   (a) Range = 20
   Variance = 56
   Standard deviation ≈ 7.48

x	x^2
3	9
7	49
11	121
15	225
19	361
23	529
78	1294

   $$\bar{x} = \frac{78}{6} = 13$$

   $$s_x^2 = \frac{1294 - 6(13)^2}{6-1} = \frac{280}{5} = 56$$

   $$s_x = \sqrt{56} \approx 7.48$$

   (b) Range = 22
   Variance ≈ 42.61
   Standard deviation ≈ 6.53

2. For Exercise 1(b) construct a box and whisker plot.

   1 2 3 4 5 6 8 8 11 12 17
         ↑   ↑   ↑
      $Q_1 = 3.5$ Median = 7 $Q_3 = 11.5$

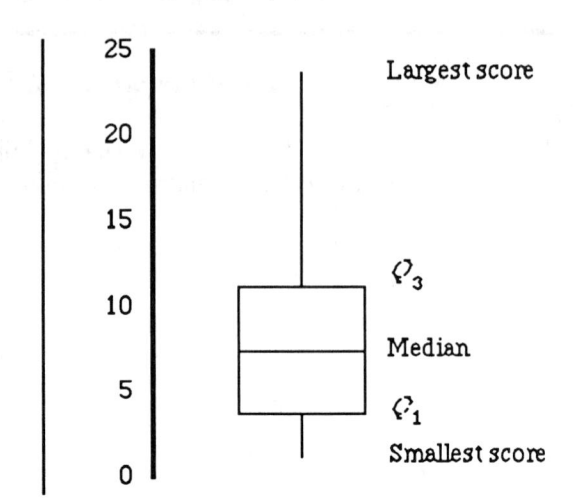

228　CHAPTER 9　The Uses and Misuses of Statistics

```
25 ┤
   │                    ├── Largest score
20 ┤                    │
   │                    │
15 ┤                    │
   │                 ┌──┤
10 ┤                 │  │── $Q_3$
   │                 ├──┤
 5 ┤                 │  │── Median
   │                 └──┤── $Q_1$
 0 ┤                    ├── Smallest score
```

3. For Exercise 1(b) construct a stem and leaf chart. From this chart find $Q_3 + 1.5\ IQR$ and $Q_1 - 1.5\ IQR$. Are there any outliers?

Stems	Leaves
0	1 2 3 4 5 6 8 8
1	1 2 7
2	3

$Q_1 = 3.5$
Med = 7
$Q_3 = 11.5$

$IQR = 11.5 - 3.5 = 8$

$Q_3 + 1.5\ IQR = 11.5 + 1.5(8) = 23.5$
$Q_1 - 1.5\ IQR = 3.5 - 1.5(8) = {}^-8.5$
There are no outliers.

4. The table below gives the salaries for the employees of a small company. Find the standard deviation of the salaries.

Salary	Number of Employees
$18,000	3
$22,000	5
$30,000	6
$35,000	10
$40,000	3
$50,000	2
$80,000	1

Standard deviation = $12,187

x	f	xf	x^2	x^2f
18	3	54	324	972
22	5	110	484	2420
30	6	180	900	5400
35	10	350	1225	12250
40	3	120	1600	4800
50	2	100	2500	5000
80	1	80	6400	6400
Total	30	994		37242

$$\bar{x} = \frac{994}{30} = 33.13 \text{ (convert to thousands)}$$
$$\bar{x} = \$33,133.3$$

$$s_x^2 = \frac{37242 - 30(33.13)^2}{29} = 148.533$$

$$s_x = \sqrt{148.533} = 12.187$$
(convert to thousands) = $12,187

5. A certain standardized test has a mean of 500 and a standard deviation of 100. Convert each of the following test scores to z-scores.

   (a) 600
   (b) 400
   (c) 650
   (d) 250

   (a) $z = 1$ $z = \frac{600 - 500}{100} = 1$
   (b) $z = ^-1$
   (c) $z = 1.5$ $z = \frac{650 - 500}{100} = 1.5$
   (d) $z = ^-2.5$

6. Find the standard deviation of the following data.

Class	Frequency
0 - 4	5
5 - 9	10
10 - 14	20
15 - 19	25
20 - 24	15

230 CHAPTER 9 The Uses and Misuses of Statistics

Standard deviation = 5.77

x	f	xf	x^2	$x^2 f$
2	5	10	4	20
7	10	70	49	490
12	20	240	144	2880
17	25	425	289	7225
22	15	330	484	7260
Total	75	1075		17875

$$\bar{x} = \frac{1075}{75} = 14.33$$

$$s_x^2 = \frac{12875 - 75(14.33)^2}{74} = 33.33$$

$$s_x = \sqrt{33.33} = 5.77$$

7. Find the median of the data given in Exercise 6.

> Median = 17

8. In 1985, the students from Wilson School scored an average of 240 on the state mathematics exam. In 1990, students from Wilson scored an average of 300 on the state exam. The overall mean on the 1985 test was 220 with a standard deviation of 20, while the mean on the 1990 test was 290 with a standard deviation of 20. The local newspaper reported that the students at Wilson had improved significantly on the math exam from 1985 to 1990. How would you respond to this claim?

> Wilson School did better, compared to the state average in 1985 (1 standard deviation above mean compared to 0.5 standard deviation above the mean in 1990).
>
> For 1985: $z = \dfrac{240 - 220}{20} = 1$
>
> For 1990: $z = \dfrac{300 - 290}{20} = 0.5$

9. Consider the data from Exercise 4.

   (a) How many of the employees have salaries within one standard deviation of the mean?

(b) How many of the employees have salaries within two standard deviations of the mean?

(a) 24 employees

$\bar{x} \approx \$33,133$
$s_x \approx \$12,187$
$\bar{x} + s_x = 33,133 + 12,187 = 45,320$
$\bar{x} - s_x = 33,133 - 12,187 = 20,946$

(b) 29 employees

$\bar{x} + 2s_x = 33,133 + 2(12,187)$
$\quad = 57,507$
$\bar{x} - 2s_x = 33,133 - 2(12,187) = 8,759$

10. Recall Exercise 7, Section 2 of Chapter 9.

    (a) Find the standard deviation for each team.
    (b) How could you use the information from part (a) to compare the teams?

(a) Cougars - 8.40   Mules - 2.83

For Cougars:
$\bar{x} = \dfrac{375}{5} = 75$

$s_x^2 = \dfrac{28409 - 5(75)^2}{4}$
$\quad = 71$

$s_x = \sqrt{71} = 8.40$

For Mules:
$\bar{x} = \dfrac{375}{5} = 75$

$s_x^2 = \dfrac{28157 - 5(75)^2}{4}$
$\quad = 8$

$s_x = \sqrt{8} = 2.83$

x	x^2
70	4900
71	5041
72	5184
72	5184
90	8100
375	28409

x	x^2
72	5184
73	5329
74	5476
78	6084
78	6084
375	28157

(b) Since the standard deviation for the Mules is less than that for the Cougars, there must be more variation in the heights of the Cougars.

11. Make a box and whisker plot for the data found in Exercise 7 from Section 1 of Chapter 9.

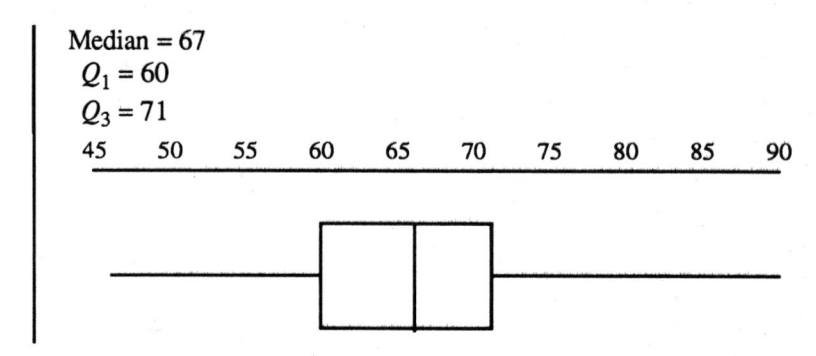

Median = 67
$Q_1 = 60$
$Q_3 = 71$

45 50 55 60 65 70 75 80 85 90

12. Which of the following golfers had the most consistent scores.

Golfer	Scores
Jones	75 65 74 74 67
Smith	69 70 72 71 73
Ortiz	77 75 67 65 71
White	79 63 78 64 71

Golfer	Mean	Standard Deviation
Jones	71	4.64
Smith	71	1.58 ←—Smallest
Ortiz	71	5.10
White	71	7.52

Smith was the most consistent.

13. A college entrance test has three parts: verbal, quantative, and reasoning. The table below gives the means and standard deviations for each part.

	Mean	Standard Deviation
Verbal	110	10
Quantitative	120	16
Reasoning	60	5

(a) Jerry scored 118 on verbal, 132 on quantitative, and 65 on reasoning. Determine her z-score for each part of the test.

(b) On which part did she score the relative highest?
(c) On which part did she score the relative lowest?

(a) Verbal z-score = 0.8
Quantitative z-score = 0.75
Reasoning z-score = 1.0
Verbal z-score = $\dfrac{118 - 110}{10} = 0.8$
(b) Reasoning
(c) Quantitative

14. For the following set of data, find the mean and standard deviation.

    $k, k + 2, k + 3, k + 4, k + 6$

Mean = $k + 3$

Mean = $\dfrac{k + k + 2 + k + 3 + k + 4 + k + 6}{5}$
= $\dfrac{5k + 15}{5} = k + 3$

x	$x - \bar{x}$	$(x - \bar{x})^2$
k	-3	9
$k + 2$	-1	1
$k + 3$	0	0
$k + 4$	1	1
$k + 6$	3	9
$5k + 15$		20

Standard deviation = 2
$s_x^2 = \dfrac{20}{5} = 4$
$s_x = \sqrt{4} = 2$

15. The Wright Brothers decide to test their airplane at Kitty Hawk because they find that the average wind speed there is 15 m.p.h. (not too low or too high)

    (a) What other information do they need to know about the winds at Kitty Hawk?
    (b) Suppose a sample of the wind speeds for twenty days at Kitty Hawk looks like the following. Find the standard deviation for this data.

20 25 1 0 19 30 2 3 28 29
10 11 20 21 10 1 0 2 3 24

(a) Information on the variation in the wind speed from day to day.

(b) $\bar{x} = \dfrac{259}{20} = 12.95$

$s_x^2 = \dfrac{5677 - 20(12.95)^2}{19} = 122.26$

$s_x = \sqrt{122.26} = 11.06$

(See the table below)

x	f	xf	x^2	$x^2 f$
0	2	0	0	0
1	2	2	1	2
2	2	4	4	8
3	2	6	9	18
10	2	20	100	200
11	1	11	121	121
19	1	19	361	361
20	2	40	400	800
21	1	21	441	441
24	1	24	576	576
25	1	25	625	625
28	1	28	784	784
29	1	29	841	841
30	1	30	900	900
	20	259		5677

Section 4 The Normal Distribution

Cover the right side of the page and work on the left, then check your work

1. Find the area under the standard normal curve that lies between the following pairs of values of z.

   (a) $z = 0$ to $z = 1.5$
   (b) $z = 0$ to $z = 0.36$
   (c) $z = {}^-0.45$ to $z = 0$
   (d) $z = {}^-1.4$ to $z = 1.4$

   (a) area = 0.4332
   $P(0 \le z \le 1.5) = 0.4332$
   (b) area = 0.1406
   (c) area = 0.1736
   $P({}^-0.45 \le z \le 0) = P(0 \le z\ 0.45)$
   $= 0.1736$
   (d) area = 0.8384
   $P({}^-1.4 \le z \le 1.4) = 2 \cdot P(0 \le z \le 1.5)$
   $= 2(0.4192) = 0.8384$

2. Find the following probabilities from Table 19.

   (a) $P(z < {}^-1.47)$
   (b) $P(z < 1.51)$
   (c) $P(z > 1.8)$
   (d) $P({}^-1.4 \le z \le 1.6)$

   (a) $P = 0.4292$
   $P(z < {}^-1.47) = P(z > 1.47) = 0.4292$
   (b) $P = 0.9345$
   $P(z < 1.51) = 0.4345 + 0.5000$
   $= 0.9345$
   (c) $P = 0.0359$
   $P(z > 1.8) = 1 - P(z < 1.8)$
   $= 1 - (0.5000 + P(0 < z < 1.8)$
   $= 1 - (0.5000 + 0.4641)$
   $= 1 - 0.9641 = 0.0359$
   (d) $P = 0.8644$

3. Assume that x is normally distributed with mean of 20 and standard deviation of 6. Find the following probabilities.

   (a) $P(x \ge 20)$
   (b) $P(x \ge 30)$
   (c) $P(x \ge 35)$

(d) $P(8 \leq x \leq 25)$

(a) $P = 0.5$
$$z = \frac{20 - 20}{6} = 0$$
$P(x \geq 20) = P(z \geq 0) = 0.5$

(b) $P = 0.5475$
$$z = \frac{30 - 20}{6} \approx 1.67$$
$P(x \geq 30) = P(z \geq 1.67)$
$= 1 - P(z \leq 1.67) = 1 - 0.4525$
$= 0.5475$

(c) $P = 0.5062$

(d) $P = 0.7739$
$$z_1 = \frac{8 - 20}{6} = {}^-2$$
$$z_2 = \frac{25 - 20}{6} = 0.83$$
$P(8 \leq x \leq 25) = P({}^-2 \leq z \leq 0.83)$
$= 0.7739$

4. In a certain normal distribution, the mean is 30 and the standard deviation is 8. What is the probability that a given piece of data will lie within 5 units of 30?

$P = 0.4714$
We want $P(25 \leq x \leq 35)$
$$z_1 = \frac{25 - 30}{8} = {}^-0.625$$
$$z_2 = \frac{35 - 30}{8} = 0.625$$
$P(25 \leq x \leq 35) =$
$P({}^-0.625 \leq z \leq 0.625)$
$= 2 \cdot P(z \leq 0.63) = 2(0.2357)$
$= 0.4714$

5. For Exercise 4, calculate the percent of the data that will be in the following intervals.

(a) 28.5 to 36.5
(b) 20.5 to 40.5
(c) 32.5 to 38.5

(a) 36.63%
(b) 78.79%
$$z_1 = \frac{20.5 - 30}{8} \approx {}^-1.19$$

$$z_2 = \frac{40.5 - 30}{8} \approx 1.31$$
%(20.5 ≤ x ≤ 40.5) =
%(⁻1.19 ≤ z ≤ 1.31) = 0.7879
(c) 23.37%

6. On a given college entrance exam the mean score is 550 and the standard deviation is 100. (Assume a normal distribution)

   (a) Out of 1,000 students, how many would be expected to score between 450 and 750?
   (b) Suppose Disco Tech College requires a score of 650 or higher for admission. What is the probability that any given student will be accepted?

   (a) 820
   $$z_1 = \frac{450 - 550}{100} = {}^{-}1.0$$
   $$z_2 = \frac{750 - 550}{100} = 2$$
   $P(450 \le x \le 750) = P(^-1 \le z \le 2)$
   $= P(0 \le z \le 1) + P(0 \le z \le 2)$
   $= 0.3413 + 0.4772 = 0.8185$
   82% of 1,000 = 820 students

   (b) $P = 0.1587$

7. Boxes of a certain cereal say they contain 14 ounces. In reality, the manufacturer fills the boxes so that the mean weight is 14.2 ounces with a standard deviation of 0.1 ounce. If the weights are normally distributed, what percent of the boxes contain 14 ounces or more?

   98%
   $$z = \frac{14 - 14.2}{0.1} = {}^-2$$
   $P(x \ge 14) = P(z \ge {}^-2) = 0.9772$

8. Videotron tests the picture tubes for its television sets and finds the mean time to failure to be 3 years with a standard deviation of 0.5 years. If the company sells the televisions with a warranty of 2 years, what percent of the the failures will the company have to pay for?

238 CHAPTER 9 The Uses and Misuses of Statistics

2%
Will have to pay for TVs that fail before 2 years
$$z = \frac{2-3}{0.5} = {}^-2$$
$P(x \leq 2) = P(z \leq {}^-2)$
$= P(z \geq 2) = 1 - P(z \leq 2)$
$= 1 - 0.9722 = 0.0228$

9. The mean IQ score for 2,000 students is 100 with a standard deviation of 12. If the scores are normally distributed find each of the following.

   (a) How many students have IQ scors between 88 and 112?
   (b) How many students have IQ scores between 76 and 124?
   (c) How many students have IQ scores greater than 130?

(a) 1,360
$P(88 \leq x \leq 112) = P({}^-1 \leq z \leq 1)$
$= 2 \cdot P(z \leq 1) = 2(0.3413) = 0.6826$
68% of 2,000 = 1,360
(b) 1900
(c) 12
$P(x \geq 130) = P(z \geq 2.5) =$
$1 - P(z \leq 2.5) = 1 - 0.9938 = 0.0062$
6% of 2,000 = 12

10. For Exercise 9, find P_{30} and P_{90}.

$P_{30} = 93.76$
$P_{30} \Rightarrow z = {}^-0.52$
$${}^-0.52 = \frac{P_{30} - 100}{12}$$
$P_{30} = 93.76$

11. Weights for employees at a certain company are normally distributed with a mean of 70 kg and standard deviation of 5 kg. What percent of the employees weigh between 60 kg and 80 kg?

95.44%
$P(60 \leq x \leq 80) = P({}^-2 \leq z \leq 2)$
$= 2 \cdot P(z \leq 2) = 2(0.4772) = 0.9544$

12. State University uses the college entrance exam described in Exercise 6. State U. will accept students who score in the top 10% of the scores on the test. What is the minimum score needed to be accepted?

678
$P(z \geq n) = 0.10$
$P(z \leq n) = 1 - P(z \geq n)$
$= 1 - 0.10 = 0.90$
$P(z \leq n) = 0.90 = 0.5000 + 0.4000$
We look in Table 19 for the entry closest to 0.4000.
This is 0.3997 which corresponds to $z = 1.28$. Then,
$1.28 = \dfrac{x - 150}{100}$
$128 = x - 550$
$678 = x$

13. A manufacturer has been given a contract to manufacture parts which must be between 1.25 and 1.35 centimeters in length. Testing indicates that the manufactruer's machine produces parts whose lengths are normally distributed with a mean length of 1.3 cm and a standard deviation of 0.06 cm. What percentage of the parts meet the contract specifications?

59%
$P(1.25 \leq x \leq 1.35) =$
$P(^-0.83 \leq z \leq 0.83)$
$= 2 \cdot P(z \leq 0.83) = 2(0.2967)$
$= 0.5934$

14. The Mongoose automobile is tested for mileage. The mean miles per gallon is 25 with a standard deviation of 4 miles. For a car chosen at random, what is the probability that it can travel 480 miles on a 20 gallon tank of gas?

$P = 0.5987$
m.p.g. $= \dfrac{480}{20} = 24$
$P(x \geq 24) = P(z \geq {}^-0.25)$
$= 0.5000 + P(z \leq 0.25)$
$= 0.5000 + 0.0987 = 0.5987$

Chapter 10

Informal Geometry

Section 1 Some Basic Ideas of Geometry

Cover the right side of the page and work on the left, then check your work

1. Label each statement as true or false. If false, explain why.

   (a) If there are two points common to a line and a plane, then the entire line is in the plane.
   (b) If two distinct lines do not intersect, then they are parallel.
   (c) If three points are coplanar, then they must be collinear.
   (d) $\overrightarrow{AB} = \overrightarrow{BA}$

   (a) True
   (b) False. The lines could be skew lines.
   (c) False. Consider the figure.

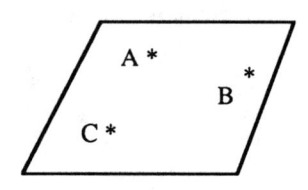

   (d) False. Consider the figure.

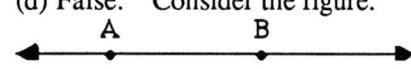

2. From the following figure find each of the following.

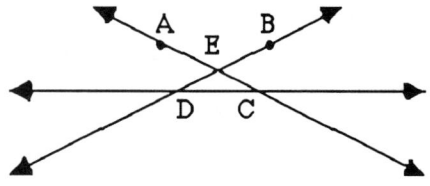

   (a) A segment containing point D
   (b) A ray containing point B
   (c) A segment containing $\overline{AE}$
   (d) Two rays containing point A

   (a) $\overline{DC}, \overline{DE},$ or $\overline{DB}$
   (b) $\overline{BE}, \overline{DB},$ or $\overline{EB}$
   (c) $\overline{AE}$ or $\overline{AC}$
   (d) $\overrightarrow{AE}, \overrightarrow{EA},$ or $\overrightarrow{CA}$

3. Label points A, B, C, D, and E on a line so that each of the following is true.
   A is on the ray $\overrightarrow{BC}$
   C is on the ray $\overrightarrow{AB}$
   D is on the ray $\overrightarrow{CE}$
   D is on the ray $\overrightarrow{EA}$
   B is on the ray $\overrightarrow{DC}$

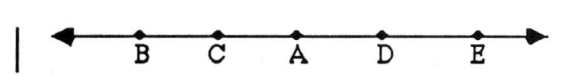

4. Using the figure, determine each of the following.

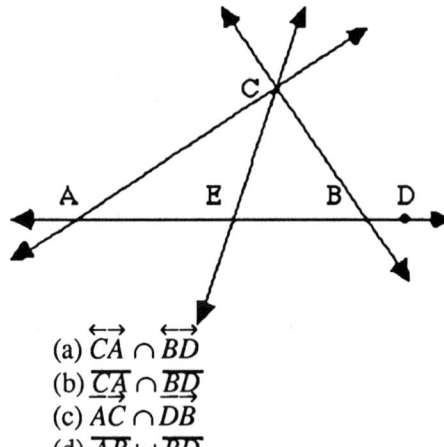

   (a) $\overleftrightarrow{CA} \cap \overleftrightarrow{BD}$
   (b) $\overrightarrow{CA} \cap \overrightarrow{BD}$
   (c) $\overrightarrow{AC} \cap \overrightarrow{DB}$
   (d) $\overline{AB} \cup \overline{BD}$

   (a) A
   (b) ∅
   (c) A
   (d) AD

5. Using the following figure name 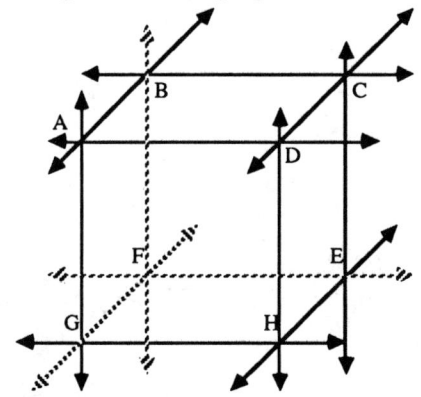
   (a) two parallel lines
   (b) two skew lines
   (c) two intersecting lines

242   CHAPTER 10   Informal Geometry

(a) $\overleftrightarrow{AB}$ and $\overleftrightarrow{CD}$, $\overleftrightarrow{BC}$ and $\overleftrightarrow{AD}$, $\overleftrightarrow{AC}$ and $\overleftrightarrow{DH}$ are some examples
(b) $\overleftrightarrow{AB}$ and $\overleftrightarrow{CE}$, $\overleftrightarrow{BC}$ and $\overleftrightarrow{DH}$, $\overleftrightarrow{GH}$ and $\overleftrightarrow{BF}$ are some examples
(c) $\overleftrightarrow{AB}$ and $\overleftrightarrow{BC}$, $\overleftrightarrow{CD}$ and $\overleftrightarrow{DH}$, $\overleftrightarrow{CE}$ and $\overleftrightarrow{HE}$ are some examples

6. Suppose l is a line and P is a point not on l.

   (a) How many lines intersecting l may be drawn through P?
   (b) How many planes contain l and P?

(a) infinite number
(b) one

7. Draw a figure to illustrate each of the following.

   (a) Two line segments whose intersection is a point
   (b) Two line segments whose intersection is a segment.

(a) $\overline{AB} \cap \overline{CD} = E$

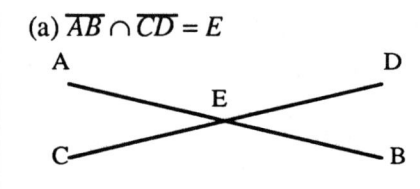

(b) $\overline{AC} \cap \overline{AB} = \overline{AB}$

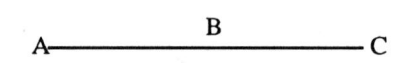

8. (a) Give three different names for line l.
   (b) Name two different rays on l containing point T
   (c) $\overrightarrow{RT} \cap \overrightarrow{TR} = ?$

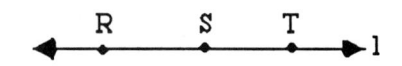

(a) $\overleftrightarrow{RS}$, $\overleftrightarrow{RT}$, $\overleftrightarrow{ST}$
(b) $\overrightarrow{RT}$, $\overrightarrow{ST}$, $\overrightarrow{TS}$, $\overrightarrow{TR}$ Any two of these
(c) $\overline{RT}$

9. What geometric concepts are suggested by these physical situations?
   (a) A flashlight
   (b) The ceiling in a room
   (c) Railroad tracks

(a) a ray
(b) a plane
(c) parallel lines

10. Using the line given, find each of the following.

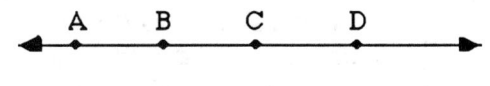

    A   B   C   D

(a) $\overline{AC} \cup \overline{BD}$
(b) $\overrightarrow{AC} \cap \overline{BD}$
(c) $\overrightarrow{BC} \cap \overleftarrow{CD}$
(d) $\overrightarrow{AB} \cup \overrightarrow{BC}$

(a) $\overline{AD}$
(b) $\overline{BC}$
(c) $\overleftarrow{CD}$
(d) $\overrightarrow{AC}$

11. How many different rays can you name on this line?

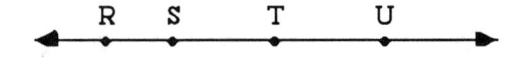

    R   S   T   U

$\overrightarrow{RS}, \overrightarrow{RT}, \overrightarrow{TU}, \overrightarrow{UT}, \overrightarrow{TS}, \overrightarrow{SR}$

12. Give a geometric explanation of why a three-legged stool is always stable and a four-legged stool will sometimes wobble.

The end points of the tree legs must lie in a single plane. The endpoints of the four legs could lie in more than one plane; hence, it would wobble.

13. Use a drawing to show whether the following statement is true or false.

"If two distinct lines are parallel to a third line, then the two lines are parallel to each other."

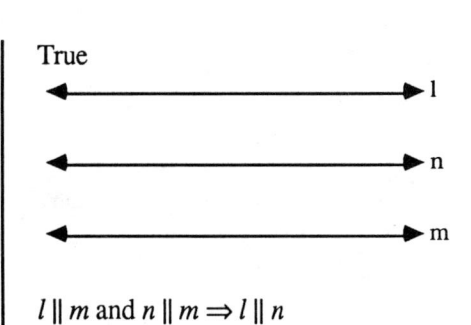

True

⟵————————⟶ l

⟵————————⟶ n

⟵————————⟶ m

$l \parallel m$ and $n \parallel m \Rightarrow l \parallel n$

244 CHAPTER 10 Informal Geometry

14. Use the fact that there is one and only one plane containing three distinct noncollinear points to prove that a line and a point not on the line determine a plane.

> Consider a line l and point P not on l. Choose points A and B on l. Thus, we know points A, B, and P are not collinear, since $P \notin l$. Then, there is only one plane containing point A, B, and P.

15. Classify the following pairs of segments as parallel or not parallel.

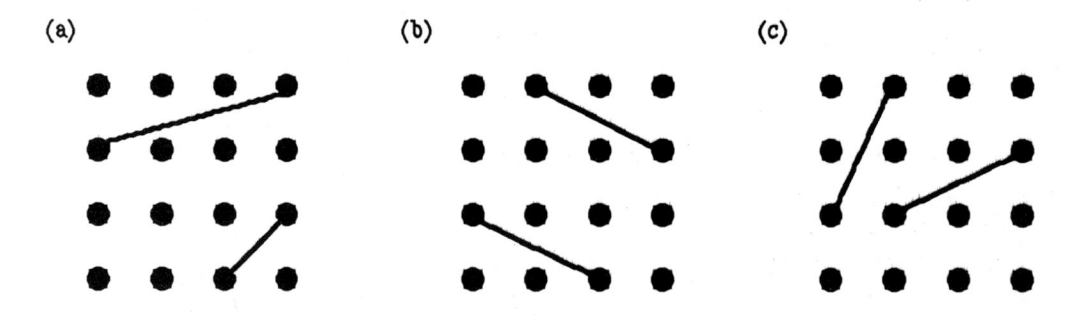

(a)

(b)

(c)

> (a) Not parallel
> (b) Parallel
> (c) Not parallel

Section 2 Lines, Planes, and Angles

Cover the right side of the page and work on the left, then check your work

1. Classify each of the following as true or false.

   (a) Two distinct planes either intersect in a line or are parallel.
   (b) If two distinct lines intersect, there is one and only one plane containing the lines.
   (c) If a plane α contains one line l, but not another line m, and l is parallel to m, then α is parallel to m.
   (d) A line parallel to each of two intersecting planes is parallel to the line of intersection of these planes.

   | (a) True
   | (b) True
   | (c) True
   | (d) True

2. Illustrate the following with a drawing.

   "If line l is perpendicular to line m and line n is perpendicular to line m, then l is not necessarily perpendicular to n."

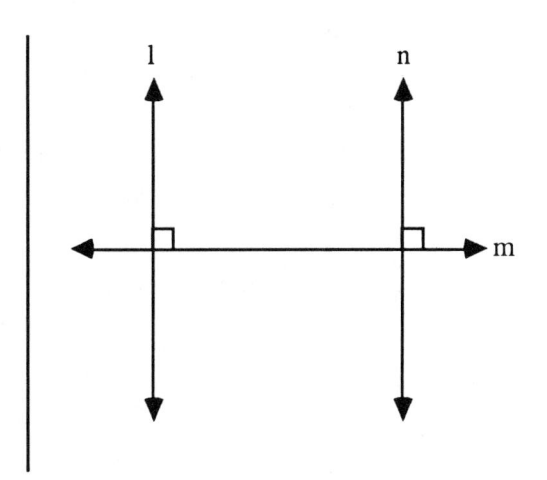

3. Perform each of the following operations.

   (a) 19^0 34′ 32″
       $+23^0$ 53′ 47″

   (b) 87^0 29′ 15″
       -17^0 48′ 26″

246   **CHAPTER 10**   Informal Geometry

Answers follow.

(a) 43^0 28' 19"

$$\begin{array}{r} 19^0 \ 34' \ 32" \\ +23^0 \ 53' \ 47" \\ \hline 79" \end{array} \rightarrow \begin{array}{r} 19^0 \ 34' \ 32" \\ 23^0 \ 53' \ 47" \\ \hline 19 \end{array} \rightarrow \begin{array}{r} 19^0 \ 34' \ 32" \\ 23^0 \ 53' \ 47" \\ \hline 88' \ 19" \end{array} \rightarrow \begin{array}{r} 19^0 \ 34' \ 32" \\ 23^0 \ 53' \ 47" \end{array}$$

$$\rightarrow \begin{array}{r} 19^0 \ 34' \ 32" \\ 23^0 \ 53' \ 47" \\ \hline 28' \ 19" \end{array} \rightarrow \begin{array}{r} 19^0 \ 34' \ 32" \\ 23^0 \ 53' \ 47" \\ \hline 43^0 \ 28' \ 19" \end{array}$$

(b) 69^0 40' 49"

4.   If $m(\angle 2) = 60^0$, find each of the following.

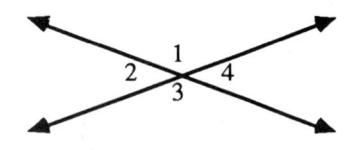

(a) $m(\angle 1)$
(b) $m(\angle 3)$
(c) $m(\angle 4)$

(a) 120^0
$\angle 1$ and $\angle 2$ are supplementary angles
(b) 120^0
(c) 60^0
$\angle 2$ and $\angle 4$ are vertical angles

5.   For each of the following sets of angles, tell which pairs of angles are adjacent and which are vertical.

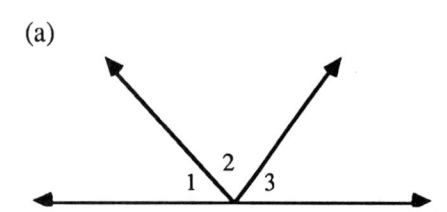

(a)

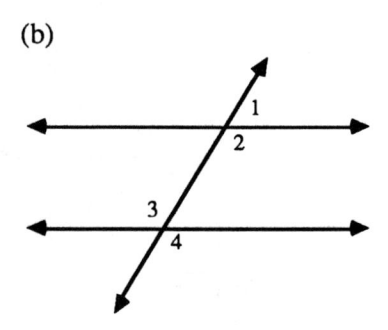

(b)

(a) ∠1 and ∠2 are adjacent
∠2 and ∠3 are adjacent
(b) ∠1 and ∠2 are adjacent
∠3 and ∠4 are vertical

6. For the following figure find each each of the following, given that **m(∠2) = 110⁰**

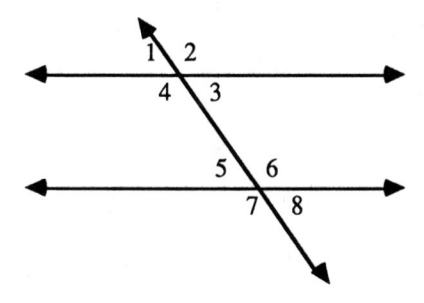

(a) $m(\angle 1)$
(b) $m(\angle 3)$
(c) $m(\angle 4)$
(d) $m(\angle 5)$
(e) $m(\angle 6)$
(f) $m(\angle 7)$

(a) 70^0
(b) 70^0
(c) 110^0
(d) 70^0
(e) 110^0
(f) 70^0

7. Make sketches that show the intersection of two angles as:

(a) exactly three points
(b) more than four points

248 CHAPTER 10   Informal Geometry

(a)

(b)

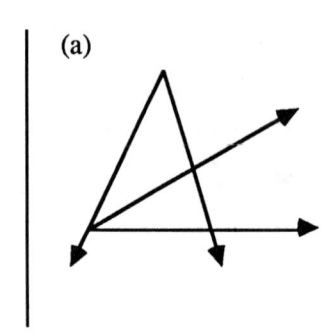

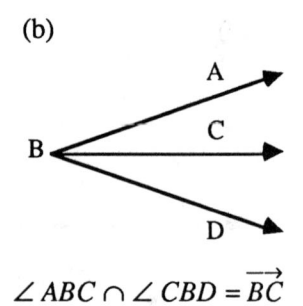

$\angle ABC \cap \angle CBD = \overrightarrow{BC}$

8. If three lines meet in a single point, how many pairs of vertical angles are formed?

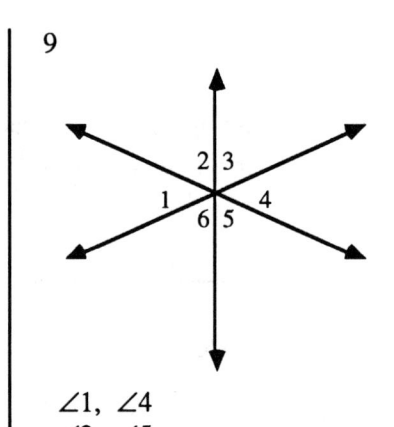

9.

∠1, ∠4
∠2, ∠5
∠3, ∠6
∠1 + ∠2, ∠4 + ∠5
∠2 + ∠3, ∠5 + ∠6
∠3 + ∠4, ∠1 + ∠6
∠1 + ∠2 + ∠3, ∠4 + ∠5 + ∠6
∠2 + ∠3 + ∠4, ∠1 + ∠6 + ∠5
∠3 + ∠4 + ∠5, ∠1 + ∠2 + ∠6

9. In each figure below, are *m* and *n* parallel lines? Explain your answer.

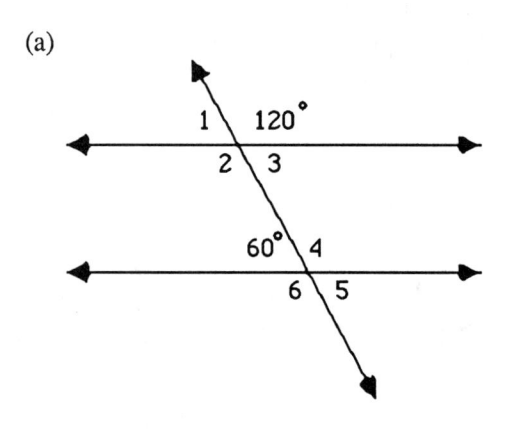

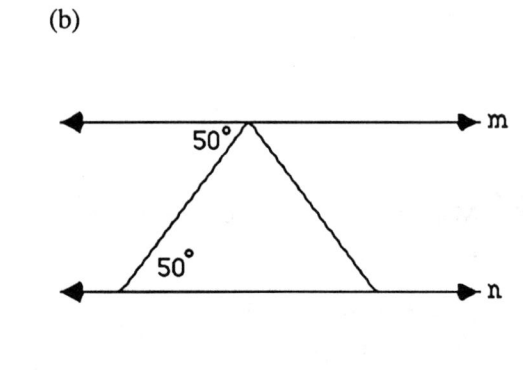

(a)

(b)

Section 2   Lines Planes, and Angles   249

(a) yes
∠3 is equal to the 60⁰ angle.
Alternate interior angles are equal.

(b) yes

10. If $m(\angle 1) = 4x°$ and $m(\angle 2) = (6x - 20)°$ and $\angle 1$ and $\angle 2$ are supplementary, what is the value of x?

$x = 20°$ $4x + (6x - 20) = 180$
$4x + 6x - 20 = 180$
$10x = 200°$
$x = 20°$

11. In the figure below, find one adjacent angle and one vertical angle for each of the following.

(a) ∠ABD
(b) ∠JDH
(c) ∠BCD

(a) ∠DBC is adjacent
∠CBE is vertical

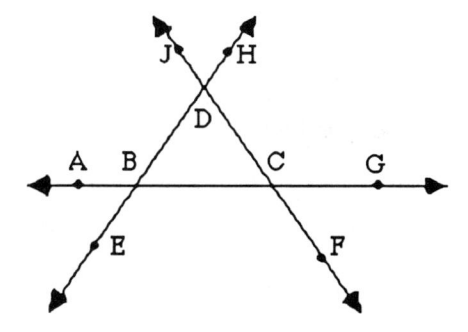

(b) ∠JDB is adjacent
∠BDC is vertical

(c) ∠DCG is adjacent
∠FCG is vertical

12. Express each of the following in degrees, minutes, and seconds without decimals.

(a) $3.9°$
(b) $26.17°$

250 CHAPTER 10   Informal Geometry

(a) $3° 54'$
$3.9° = 3° + (.9)(60')$
$= 3° + 54'$
(b) $26° 10' 12''$
$26.17° = 26° + (.17)(60')$
$= 26° + 10.2' = 26° + 10' + (.2)(60'')$
$= 26° + 10' + 12''$

13. (a) If $m(\angle 1) = 70°$ and $m(\angle 2) = 20°$ then $\angle 1$ and $\angle 2$ are _____ angles.
    (b) Any two right angles are _____ angles.
    (c) $\angle ABC$ and $\angle DEF$ are supplementary and $m(\angle ABC) = 57°$.
    $m(\angle DEF) =$ _____.

(a) complementary
(b) supplementary
(c) $123°$

14. Is it possible for a line to be perpendicular to two distinct lines in a plane and not be perpendicular to the plane?

not possible

Section 3 Simple Closed Curves

Cover the right side of the page and work on the left, then check your work

1. Identify each of the figures as simple or not simple.

   (a)　　　　　　　　　(b) 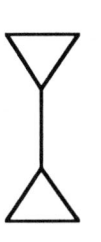　　　　　　　　　(c)

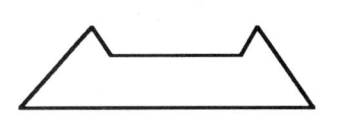

   (d) 　　　　　　　　　(e) 　　　　　　　　　(f)

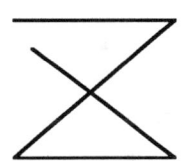

	(a) simple
	(b) not simple
	(c) not simple
	(d) simple
	(e) simple
	(f) not simple

2. Classify each of the figures in Exercise 1 as closed or not closed.

	(a) closed
	(b) closed
	(c) closed
	(d) closed
	(e) not closed
	(f) not closed

3. Which of the following regions are convex?

   (a)　　　　　　　　　(b) 　　　　　　　　　(c)

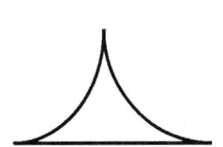

(d)

| (a), (d)

4. Consider the following set of capital letters.

   A B C D E F G H I J K L M
   N O P Q R S T U V X Y Z

   (a) Which are simple curves?
   (b) Which are closed curves?

   | (a) C, D, G, L, M, N, O, S, V, W, Z
   | (b) D, O

5. Draw each of the following curves.

   (a) closed but not simple
   (b) not closed and not simple
   (c) a non-convex hexagon
   (d) a convex octagon

   Answers follow.
   (a)   (b)    (c) 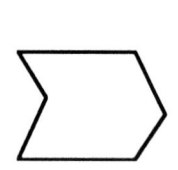   (d)

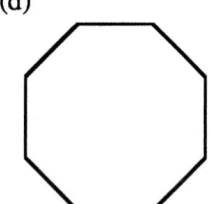

6. For each of the following, draw two parallelograms that satisfy the conditions.

   (a) Their intersection is a single point
   (b) Their intersection is exactly three points
   (c) Their intersection is exactly one line segment

(a)

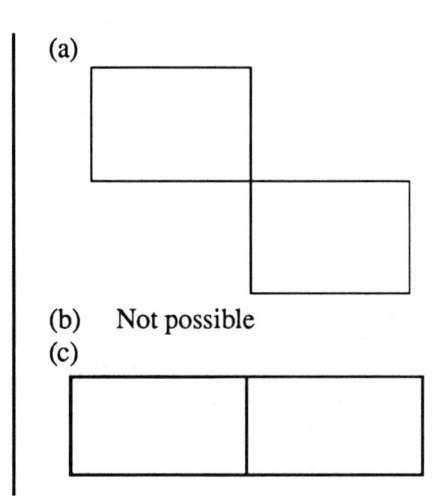

(b) Not possible
(c)

7. Can a triangle have two obtuse angles? Justify your answer.

No. Sum of the angles of the triangle must be 180^0. Two obtuse angles would have measure greater than 180^0

8. Can a parallelogram have four acute angles? Justify your answer.

No. Sum of four acute angles would be less than 360^0.

9. In the following figure, determine whether point X is inside or outside the curve.

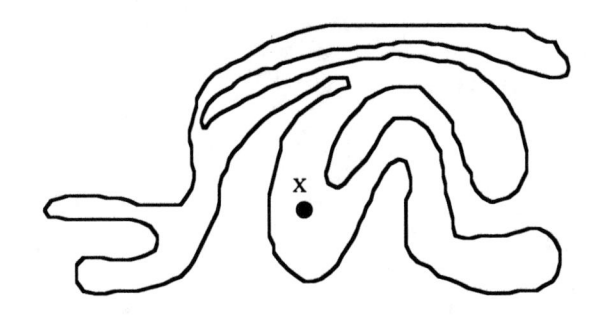

Inside

10. How many diagonals does each of the following have?

   (a) 30-gon
   (b) 100-gon

254 CHAPTER 10 Informal Geometry

(a) 420 Thirty vertices with 28 diagonals from each. We must divide by 2 so we count each diagonal only once.
Diagonals = $\dfrac{30 \cdot 28}{2} = 420$
(b) 4,900

11. How many triangles are in the following figure?

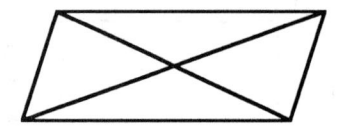

8

12. Classify each of the following as true or false.

(a) Every isosceles triangle is equilateral.
(b) Every equilateral triangle is isosceles.
(c) Some rectangles are rhombi.
(d) Some right triangles are isosceles.
(e) Every rhombus is a regular quadrilateral.

(a) False
(b) True
(c) True
(d) True
(e) False

13. What regular *n*-gons satisfy each of the following?
(a) vertex angle of $162°$
(b) vertex angle of $178.2°$
(c) central angle of $2°$
(d) central angle of $5°$

(a) 20-gon
(b) 200-gon
(c) 180-gon
(d) 72-gon

14. Why is the following figure a trapezoid?

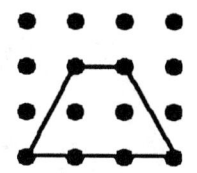

| It has exactly one pair of parallel sides.

15. Use a Venn diagram to illustrate the relationships among the sets of quadrilaterals (Q), trapezoids (T), parallelograms (P), rhombi (H), and squares (S).

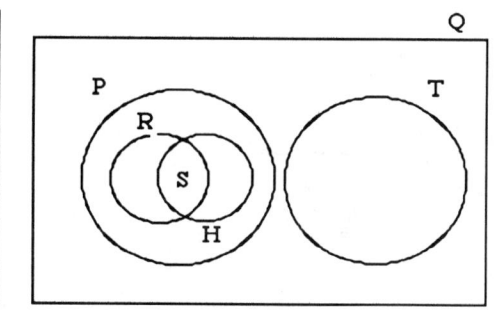

Section 4 Patterns in Nature and Art

Cover the right side of the page and work on the left, then check your work

1. In which of the following drawings is there a line of symmetry?

   (a)　　　　　(b) 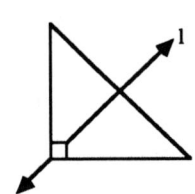　　　　　(c)

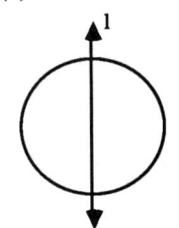

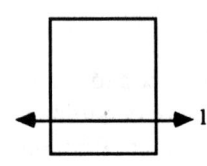

   (d)

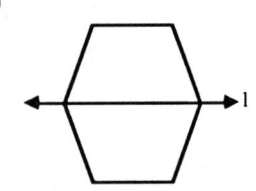

   (a) line of symmetry
   (b) no line of symmetry
   (c) no line of symmetry
   (d) line of symmetry

2. Draw all lines of symmetry for each of the following figures.

   (a)　　　　　(b) 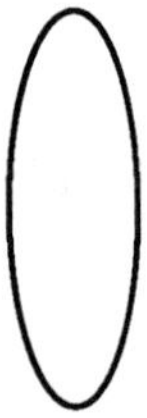　　　　　(c) 　　　　　(d)

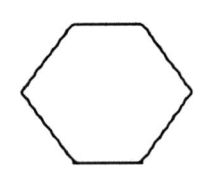

Answers follow:
(a) 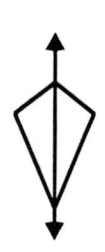
(b) 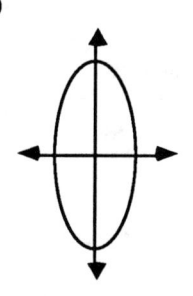
(c) 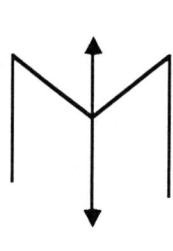
(d)

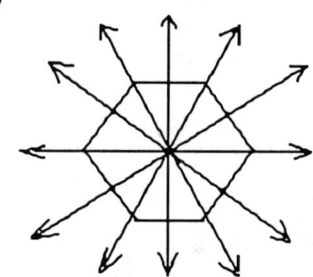

3. How many lines of symmetry are there for each of the following figures?

(a) 
(b) 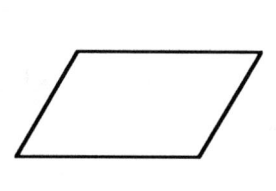
(c)

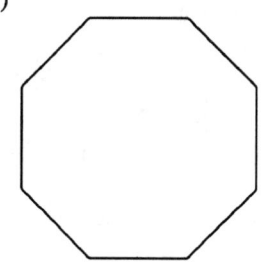

(a) 4
(b) 2
(c) 16

4. Which of the letters below have:

(a) one line of symmetry?
(b) two lines of symmetry?
(c) rotational symmetry?

C B D Y

(a) C, B, D, Y
(b) none
(c) B, D

5. With which of the following polygons can one tesselate the plane? Draw a portion of the plane to show the tesselation.

(a)

(b)

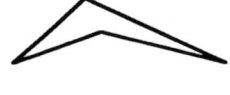

(c)

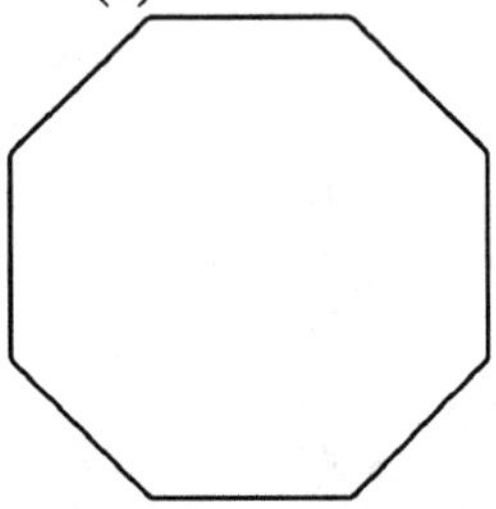

(a)

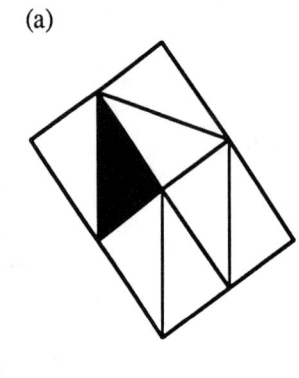

(b)

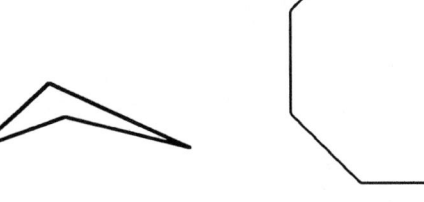

(c) Not possible

6. On a lattice draw a tesselation using the figures shown.

(a)

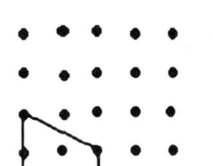

(b)

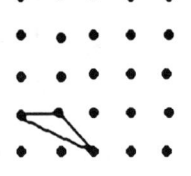

Section 4 Patterns in Nature and Art   259

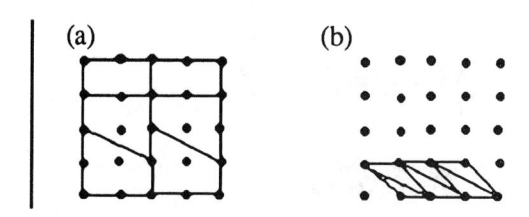

(a) (b)

7. A portion of a tesselation is shown below. Some angles have been labeled. Are lines *m* and *n* parallel? Why or why not?

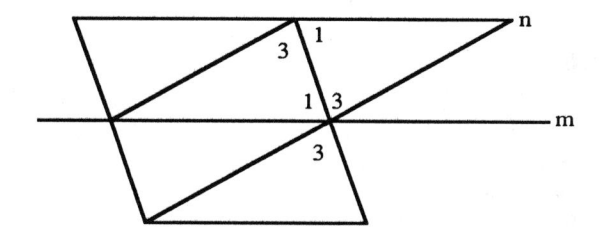

Yes. Alternate interior angles (1 or 3) are equal.

8. (a) Draw the vertex figure for the following tesselation.

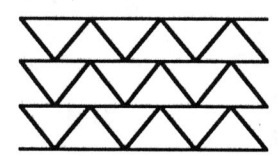

(b) Is this a regular tesselation?

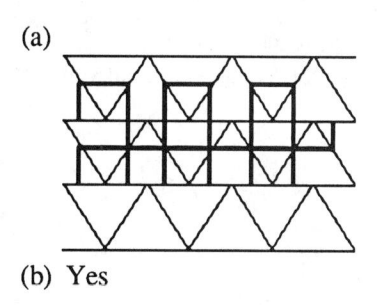

(a)

(b) Yes

9. Which of the flags below have:

(a) reflectional symmetry? How many lines?
(b) rotational symmetry? What angles?

260 CHAPTER 10   Informal Geometry

(i)

(ii)

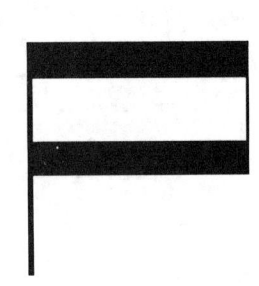

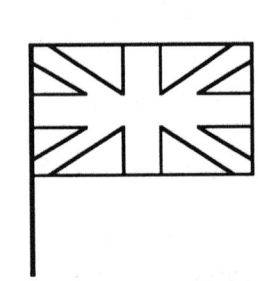

         (a) i) - 2 lines   ii) - 4 lines
         (b) i) - 180^0 iii) - 180^0

10. Fill in the blanks relative to a regular *n*-gon.

     (a) If *n* is even, half the lines of symmetry connect a _____ to the opposite _____
     (b) If n is odd, each line of symmetry goes through a _____ and the _____ of the opposite side.

         (a) vertex; vertex
         (b) vertex; midpoint

11. For each of the following:

(i)

(ii)

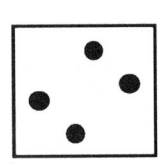

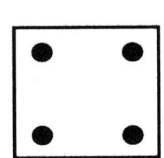

     (a) draw the lines of reflection.
     (b) describe the rotational symmetries.

         (a)   (b)

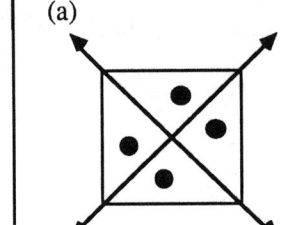

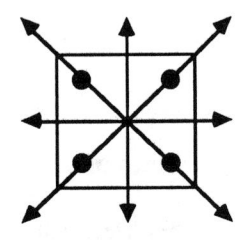

12. A portion of a triangular lattice is given.

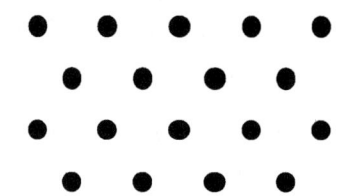

    (a) Can an equilateral triangle be drawn on this lattice?
    (b) Can a square by drawn on this lattice?
    (c) Can an isosceles triangle be drawn on this lattice?
    (d) Can a scalene, non-equilateral triangle be drawn on this lattice?

    | (a) yes
    | (b) no
    | (c) no
    | (d) yes

13. Answer each of the following yes or no. If no, draw a counterexample.

    (a) If a figure has rotational symmetry, it must have reflection symmetry.
    (b) If a figure has reflection symmetry, it must have rotational symmetry.

    | (a) yes
    | (b) yes

14. Do the following pentominoes tesselate the plane?

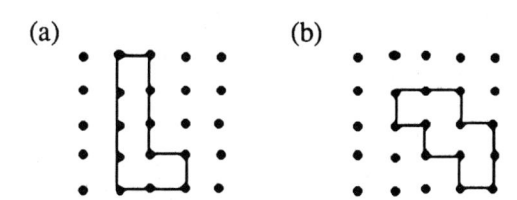

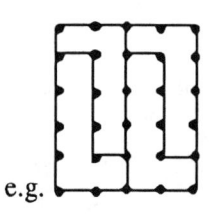

    (a)   (b)

    | (a) yes   e.g.
    | (b) no

262 CHAPTER 10 Informal Geometry

15. The dual of a tesselation is the tesselation obtained by connecting the centers of the polygons in the original tesselation that share a common side. The dual of the tesselation of squares is shown below.

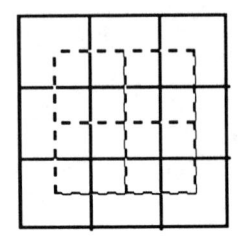

(a) The dual of the regular tesselation of squares is _____

(b) The dual of the regular tesselation of triangles is _____

(c) The dual of the regular tesselation of hexagons is _____

> (a) tesselation of squares
> (b) tesselation of hexagons
> (c) tesselation of triangles.

Section 5 Simple Closed Surfaces

Cover the right side of the page and work on the left, then check your work

1. Identify each of the following simple closed surfaces.

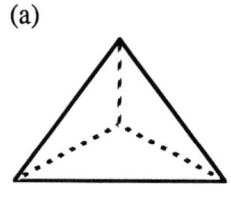

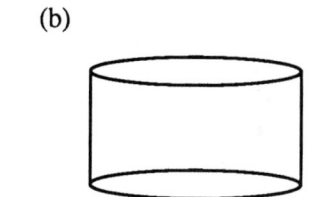

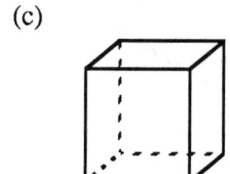

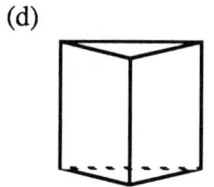

   (a)   (b)   (c)

   (d)

       (a) triangular pyramid or tetrahedron
       (b) right circular cylinder
       (c) rectangular prism
       (d) triangular prism

2. Answer each of the following true or false.

   (a) If the lateral faces of a prism are rectangles, it is a right prism.
   (b) The bases of all cones are circles.
   (c) A cylinder has only one base.

       (a) True
       (b) True
       (c) False

3. (a) What is the fewest number of faces in a pyramid?
   (b) Draw the figure from part (a).

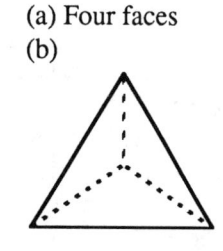

       (a) Four faces
       (b)

4. A right cylinder is cut by a plane as shown. What is the resulting cross section?

(a)   (b)

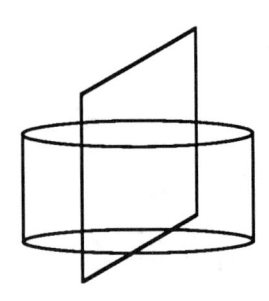

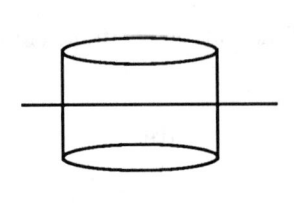

(a) rectangle
(b) circle

5. Complete the following table, indicating the number of vertices, edges, and faces.

Surface	V	E	S	V + S - E
Triangular Prism				
Quadrilateral Prism				
Pentagonal Prism				
Hexagonal Prism				
Prism with n-gon as base				

Surface	V	E	S	V + S - E
Triangular Prism	6	9	5	2
Quadrilateral Prism	8	12	6	2
Pentagonal Prism	10	15	7	2
Hexagonal Prism	12	18	8	2
Prism with n-gon as base	2n	3n	n + 2	2

6. Determine the measures of all the dihedral angles of a right prism whose bases are regular decagons.

90^0 and 144^0

7. (a) What is the cross section when a plane intersects a sphere?
   (b) If the intertersecting plane contains the center of the sphere, the cross section is called a **Great Circle**. How many great circles are there for a sphere?

(a) a circle
(b) infinite number

8. Complete the following table.

	Vertices Per	Diagonals Per	Total Number
Prism	Base	Vertex	of Diagonals
Quadrangular	4	1	4
Pentagonal	5	2	10
Hexagonal			
Heptagonal			
n-gonal			

	Vertices Per	Diagonals Per	Total Number
Prism	Base	Vertex	of Diagonals
Quadrangular	4	1	4
Pentagonal	5	2	10
Hexagonal	6	3	18
Heptagonal	7	4	28
n-gonal	n	$n-3$	$n(n-3)$

9. Which of the following patterns fold into a cube? If a pattern will fold into a cube, what letter will be opposite the * ?

(a)

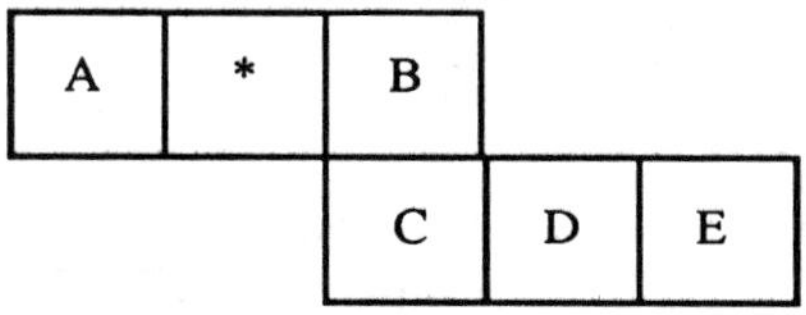

A	*	B		
		C	D	E

(b)

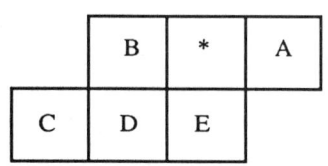

	B	*	A
C	D	E	

(c)

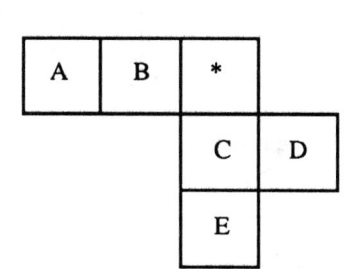

A	B	*
		C	D
		E	

(d)

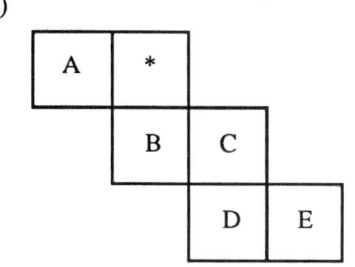

A	*	
	B	C
		D	E

(a) yes; D
(b) no
(c) no
(d) yes; D

266 CHAPTER 10 Informal Geometry

10. Consider the prism below. Its bases are regular pentagons.

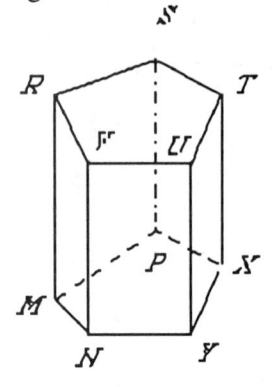

   (a) Is there a plane parallel to the plane containing points R, S, T, U, and V?
   (b) What is the measure of the dihedral angle between the plane containing points T, U, X, and Y and the plane containing points R, V, M, and N?

   (a) yes Plane containing points M, P, X, Y, and N
   (b) $108°$

11. What surface is obtained when the centers of adjacent faces are connected?

   (a)

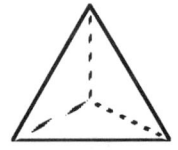

   (b)

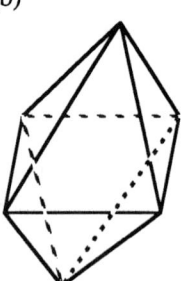

   (c)

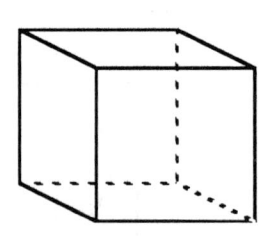

   (a) Tetrahedron
   (b) Cube
   (c) Octahedron

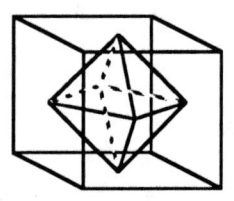

12. Complete the following.

    (a) If a prism has 47 faces, it has _____ edges?
    (b) If a pyramid has 74 edges, it has _____ faces?
    (c) Why can a prism not have exactly 68 edges?

(a) 135 edges Since it is a prism it has two bases and 45 lateral faces. Thus, it has 45 lateral edges and each base must have 45 edges.

(b) 38 faces Since it is a pyramid the number of lateral edges equals the number of edges in the base, in this case 37 + 37 = 74. Thus, there are 37 lateral faces plus the base.

(c) Since it is a prism, the number of lateral edges x equals the number of edges in each base. Thus,

$$x + x + x = 68$$
$$3x = 68$$
$$x = 22\tfrac{2}{3} \leftarrow \text{not possible}$$

13. If the figure on the left is folded, it will become which figure on the right? (You may want to make some models.)

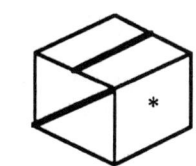

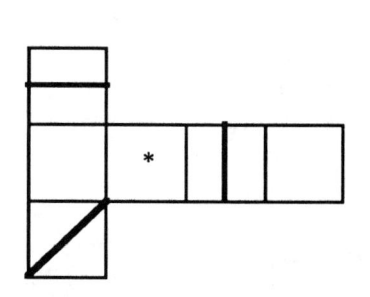

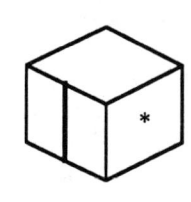

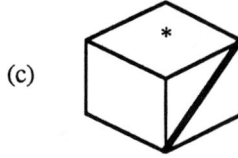

268　CHAPTER 10　　Informal Geometry

14. Which of the following cross sections are possible when a plane intersects a cube?

    (a) Square
    (b) Rectangle
    (c) Isosceles triangle
    (d) Equilateral triangle
    (e) Parallelogram
    (f) Trapezoid
    (g) Regular hexagon

    (c)

    (a) possible
    (b) possible
    (c) possible
    (d) possible

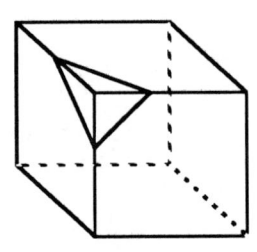

    (f) possible
    (g) not possible

15. How many axes of symmetry do the following figures have?

    (a) Right circular cylinder
    (b) Tetrahedron
    (c) Sphere

    (a) one
    (b) four
    (c) infinite

Chapter 11
Measurement and the Metric System

Section 1 Measurement and the International Metric System

Cover the right side of the page and work on the left, then check your work

1. Perform each of the following conversions.

   (a) 24.8 dm = _____ mm
   (b) 0.93 cm = _____ m
   (c) 125 dm = _____ hm
   (d) 31 km = _____ m

   > (a) 2,480 mm
   > (b) 0.0093 m
   > (c) 0.125 hm
   > (d) 31,000 m

2. Perform each of the following conversions.

   (a) 6,000 cm^2 = _____ m^2
   (b) 562 m^2 = _____ km^2
   (c) 8.7 dm^2 = _____ cm^2
   (d) 0.362 km^2 = _____ m^2

   > (a) 0.6 m^2
   > (b) 0.000562 km^2
   > (c) 870 cm^2
   > (d) 362,000 m^2

3. Convert each of the following to cubic decimeters.

   (a) 50 m^3
   (b) 4,205 mm^3
   (c) 13 km^3
   (d) 962 cm^3

   > (a) 50,000 dm^3
   > (b) 0.004205 mm^3
   > (c) 13,000,000,000,000 dm^3
   > (d) 0.962 dm^3

4. Perform each of the following conversions.

   (a) 19 ft^2 = _____ yd^2
   (b) 0.5 mi^2 = _____ ft^2
   (c) 21,000 ft^2 = _____ mi^2
   (d) 362 in^2 = _____ ft^2

270 CHAPTER 11   Measurement and the Metric System

$\quad\quad\quad\quad\quad\quad\quad\quad\quad\quad\quad$ (a) $\approx$ 2.1 yd^2
$\quad\quad\quad\quad\quad\quad\quad\quad\quad\quad\quad$ (b) 13,436,928 ft^2
$\quad\quad\quad\quad\quad\quad\quad\quad\quad\quad\quad$ $\dfrac{0.5 \text{ mi}^2}{x} = \dfrac{1 \text{ mi}^2}{(1728)^2 \text{ yd}^2}$
$\quad\quad\quad\quad\quad\quad\quad\quad\quad\quad\quad$ $x = 1{,}492{,}992 \text{ yd}^2$
$\quad\quad\quad\quad\quad\quad\quad\quad\quad\quad\quad$ $\dfrac{9 \text{ ft}^2}{1 \text{ yd}^2} = \dfrac{y}{1{,}492{,}992 \text{ yd}^2}$
$\quad\quad\quad\quad\quad\quad\quad\quad\quad\quad\quad$ $y = 13{,}436{,}928 \text{ ft}^2$
$\quad\quad\quad\quad\quad\quad\quad\quad\quad\quad\quad$ (c) 0.00078 mi^2
$\quad\quad\quad\quad\quad\quad\quad\quad\quad\quad\quad$ (d) 2.51 ft^2

5. Circle the most reasonable measure.

   (a) Length of a standard paper clip   33 mm   33 cm   33 dm
   (b) Width of a book   20 cm   20 m   20 dm
   (c) Height of a door   2 mm   2 m   2 hm
   (d) Distance between two cities   600 m   600 cm   600 km

$\quad\quad\quad\quad\quad\quad\quad\quad\quad\quad\quad$ (a) 33 mm
$\quad\quad\quad\quad\quad\quad\quad\quad\quad\quad\quad$ (b) 20 cm
$\quad\quad\quad\quad\quad\quad\quad\quad\quad\quad\quad$ (c) 2 m
$\quad\quad\quad\quad\quad\quad\quad\quad\quad\quad\quad$ (d) 600 km

6. Circle the most reasonable measure.

   (a) Area of desktop   2 m^2 2 dm^2 2 cm^2
   (b) Area of a piece of notebook paper   1,200 mm^2 1,200 m^2 1,200 cm^2
   (c) Area of a postage stamp   500 cm^2 500 mm^2 500 dm^2

$\quad\quad\quad\quad\quad\quad\quad\quad\quad\quad\quad$ (a) 2 m^2
$\quad\quad\quad\quad\quad\quad\quad\quad\quad\quad\quad$ (b) 1,200 cm^2
$\quad\quad\quad\quad\quad\quad\quad\quad\quad\quad\quad$ (c) 500 mm^2

7. For each of the following, place a decimal point in the number to make the measure reasonable.

   (a) A desk is 790 m high.
   (b) The man is 1750 cm tall.
   (c) The speed limit is 880 km/hour.
   (d) A pencil is 1900 mm long.

$\quad\quad\quad\quad\quad\quad\quad\quad\quad\quad\quad$ (a) 0.790 m
$\quad\quad\quad\quad\quad\quad\quad\quad\quad\quad\quad$ (b) 175.0 cm
$\quad\quad\quad\quad\quad\quad\quad\quad\quad\quad\quad$ (c) 88.0 km/hour
$\quad\quad\quad\quad\quad\quad\quad\quad\quad\quad\quad$ (d) 190.0 mm

8. Arrange the following in decreasing order.

   50 dm, 90 mm, 6 m, 246 cm, 5190 mm, 37 dm, 700 mm

   > 6m, 5190 mm, 50 dm, 37 dm, 246 cm, 700 mm, 90 mm

9. Choose the most appropriate metric unit (cm^2, m^2, or km^2).

   (a) Area of the face of a nickel
   (b) Area of a parking space
   (c) Area of a football field

   > (a) cm^2
   > (b) m^2
   > (c) m^2

10. At one time, the United Kingdom used the following monetary system.

    1 pound = 20 shillings   1 shilling = 12 pence (plural of penny)
    1 penny = 2 half-pennies   1 penny = 4 farthings

    (a) How many pence in a pound?
    (b) How many farthings in a pound?
    (c) How many farthings in a shilling?

    > (a) 240
    > (b) 960
    > (c) 48

11. Use dimensional analysis to make the following conversions.

    (a) 400 mi to km
    (b) 55 mi/hr to km/hr
    (c) 1,000 km to mi

    > (a) 643.74 km
    >
    > $400 \text{ mi} \left(\frac{5280 \text{ ft}}{1 \text{ mi}}\right)\left(\frac{12 \text{ in}}{1 \text{ ft}}\right)\left(\frac{2.54 \text{ cm}}{1 \text{ in}}\right)\left(\frac{1 \text{ m}}{100 \text{ cm}}\right)\left(\frac{1 \text{ km}}{1000 \text{ m}}\right)$
    >
    > ≈ 643.74 km
    >
    > (b) 88.51 km/hr
    >
    > (c) 621.37 mi $\frac{400 \text{ mi}}{643.74 \text{ km}} = \frac{x \text{ mi}}{1000 \text{ km}}$
    >
    > $x ≈ 621.37$ mi

272 CHAPTER 11   Measurement and the Metric System

12. Perform the following conversions where
    s = seconds, ms = milliseconds, cs = centiseconds

    (a) 7 s = _____ cs
    (b) 97 ms = _____ cs
    (c) 4,200 cs = _____ s

    > (a) 700 cs $\quad \dfrac{x \text{ cs}}{7 \text{ s}} = \dfrac{100 \text{ cs}}{1 \text{ s}}$
    > (b) 9.7 cs
    > (c) 42 s

13. Sound travels at a speed of 1,100 ft/sec. What is the speed of sound in mi/hr?

    > $750 \dfrac{\text{mi}}{\text{hr}}$
    >
    > $1100 \text{ ft} = 1100 \text{ ft} \left(\dfrac{1 \text{ mi}}{5280 \text{ ft}} \right) \approx 0.208 \text{ mi}$
    >
    > $0.208 \dfrac{\text{mi}}{\text{sec}} \left(\dfrac{3600 \text{ sec}}{1 \text{ hr}} \right) \approx 750 \dfrac{\text{mi}}{\text{hr}}$

14. Bill measures the length of the classroom and reports that it is 25 units long. Angela measures the length and reports that it is 40 units long. Who used the longer unit to measure the room?

    > Bill used the longer unit.

15. Light travels at a speed of 186,000 miles per second (in a vacuum.)

    (a) How far will light travel in one year? This is called a light-year.
    (b) The Sun is 93,000,000 miles from Earth. How long does it take light to travel from the Sun to the Earth?

    > (a) $5.87 \times 10^{12} \dfrac{\text{mi}}{\text{yr}}$
    >
    > $186000 \dfrac{\text{mi}}{\text{sec}} \left(\dfrac{60 \text{ sec}}{1 \text{ min}} \right) \left(\dfrac{60 \text{ min}}{\text{hr}} \right) \left(\dfrac{24 \text{ hr}}{1 \text{ day}} \right) \left(\dfrac{365 \text{ days}}{1 \text{ yr}} \right)$
    >
    > $\approx 5.87 \times 10^{12} \dfrac{\text{mi}}{\text{yr}}$
    >
    > (b) $8\dfrac{1}{3}$ minutes
    >
    > $\dfrac{x \text{ sec}}{93{,}000{,}000 \text{ mi}} = \dfrac{1 \text{ sec}}{186{,}000 \text{ mi}}$
    >
    > $x \approx 500 \text{ sec} = 8\dfrac{1}{3} \text{ minutes}$

Section 2 Congruence and Linear Measure

Cover the right side of the page and work on the left, then check your work

1. Points A, B, C, and D are located on line l below.
   If $A = {}^-2.7$, $B = {}^-1.1$, $C = 0.7$, and $D = 3.4$ find each of the following.

   ```
       A      B        C          D
   ◄──┼──●──┼──●──┼──┼──●──┼──┼──●──┼──►
      -4 -3 -2 -1  0  1  2  3  4
   ```

   (a) $m\,(\overline{AC})$
   (b) $m\,(\overline{BC})$
   (c) $m\,(\overline{BD})$
   (d) $m\,(\overline{AD})$

   | (a) 3.4
   | (b) 1.8
   | (c) 4.5
   | (d) 6.1

2. For the points A, B, C, and D from Exercise 1,

   (a) Does $m\,(\overline{AB}) + m\,(\overline{BC}) = m\,(\overline{AC})$?
   (b) Does $m\,(\overline{AC}) + m\,(\overline{BD}) = m\,(\overline{AD})$?
   (c) Does $m\,(\overline{AB}) + m\,(\overline{BC}) + m\,(\overline{CD}) = m\,(\overline{AD})$?

   | (a) yes
   | (b) no
   | (c) yes

3. Complete the following.

	Shortest Possible Length	Measurement	Largest Possible Length
(a)		15.3 m	
(b)		9.25 cm	
(c)		18 mm	
(d)		137.43 dm	

   | (a) 15.25; 15.35
   | (b) 9.245; 9.255
   | (c) 17.5; 18.5
   | (d) 137.425; 137.435

274   CHAPTER 11   Measurement and the Metric System

4. Find the perimeter of each figure where each unit is 1 cm.

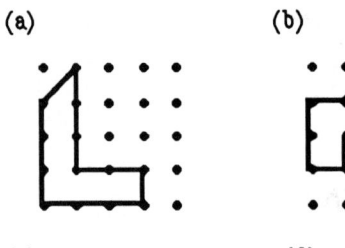

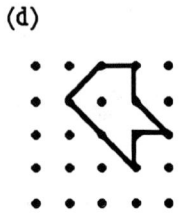

   (a)

   (b)

   (c) ▶▶ ▶ ▶ ▶ ▶

   (d)

   (a) $12 + \sqrt{2}$
   (b) $12 + 2\sqrt{2}$
   (c) $6 + 2\sqrt{2} + 2\pi$
   (d) $4 + 4\sqrt{2}$

5. Find the circumference of each circle. Use 3.14 for π.

   (a) $r = 4.5$
   (b) $d = 12$
   (c) $r = a$
   (d) $r = 2a$

   (a) 28.26
   $C = 2\pi r$
   $= 2\pi (4.5)$
   $= 2(3.14)(4.5) = 28.26$
   (b) 37.68
   (c) $2\pi a$
   (d) $4\pi a$ $C = 2\pi(2a) = 4\pi a$

6. Find the perimeter of each figure.

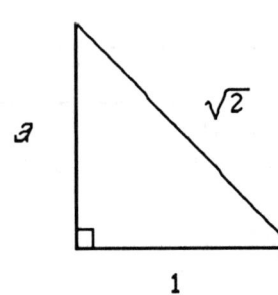

   (a) rectangle with sides 10 and 7

   (b) right triangle with legs a and 1, hypotenuse $\sqrt{2}$

(c)

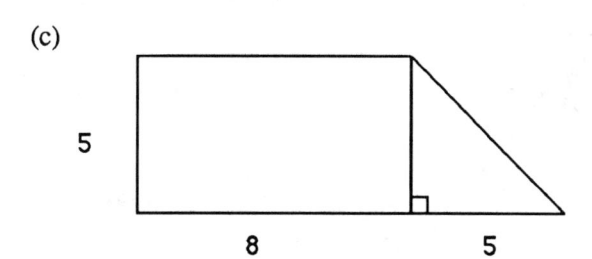

[figure: trapezoid with left side 5, bottom 8 + 5, right portion a right triangle with legs 5 and 5]

(a) 34
(b) $2 + \sqrt{2}$
$a^2 + b^2 = (\sqrt{2})^2$
$a^2 + 1^2 = 2$
$a^2 = 1$
$a = 1$
(c) $26 + \sqrt{50}$

7. Find a point that is $\frac{1}{5}$ of the distance between the points $\frac{1}{2}$ and $\frac{3}{4}$.

$\frac{11}{12}$

$\frac{3}{4} - \frac{1}{2} = \frac{1}{4}$ Distance between points

$\frac{1}{5} \cdot \frac{1}{4} = \frac{1}{20}$ $\frac{1}{5}$ of the distance

$\frac{1}{2} + \frac{1}{20} = \frac{11}{12}$

The point $\frac{1}{5}$ of the distance between

8. Find the circumference of a circle that is circumscribed by a square with side of measure 10.

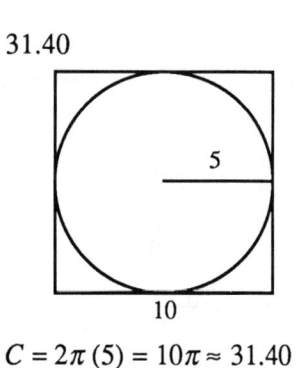

31.40

[figure: square of side 10 with inscribed circle of radius 5]

$C = 2\pi(5) = 10\pi \approx 31.40$

9. A planet that is 1,000 km in diameter has a satellite in circular orbit around it. If the satellite is 100 km above the planet, what is the length of the orbit of the satellite?

3,768 km

276 CHAPTER 11 Measurement and the Metric System

10. In Exercise 9, suppose you know the length of the satellite's orbit is 200 km longer than the circumference of the planet. How high is the satellite above the planet?

$C = 2\pi(600) = 1,200\pi \approx 3,768$ km

31.8 km
$C_{planet} = 2\pi(500) = 1,000\pi$
$\approx 3,140$ km
$C_{orbit} = 3,340$ km
$3,340 = 2\pi r$
$\dfrac{3340}{2\pi} = r$
531.8 km $\approx r$

11. Rusty Johnson's automobile has tires that are 30 inches in diameter. How many revolutions will his tires make during the Indy 500 (that is a 500 mile race)?

168,153
$C_{tire} = 2\pi(2.5) = 15.7$ ft
$\dfrac{1 \text{ rev}}{15.7 \text{ ft}} = \dfrac{x}{(500)(5280)}$
$x \approx 168.153$ revolutions

12. A rectangle whose length is 4 more than its width has a perimeter of 36. What are the dimensions of the rectangle?

width = 7
length = 11
$2w + 2(w + 4) = 36$
$2w + 2w + 8 = 36$
$4w = 28$
$w = 7$
$l = 11$

13. Find the missing parts of the following circles.
   (a)   (b)   (c)

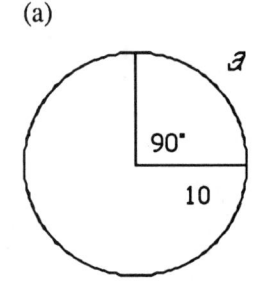

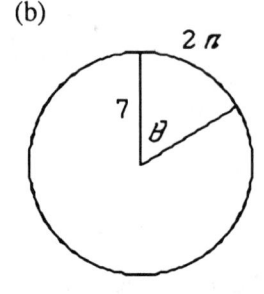

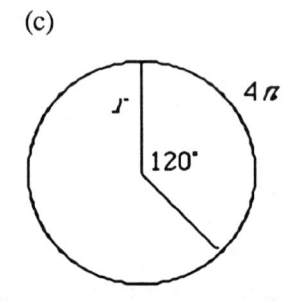

(a) 15.7
$\dfrac{90°}{360°} = \dfrac{a}{20\pi}$

Section 2   Congruence and Linear Measure   277

$a \approx 15.7$

(b) $51.4°$

$$\frac{\theta}{360°} = \frac{2\pi}{14\pi}$$

$\theta \approx 51.4°$

(c) 6

$$\frac{120°}{360°} = \frac{4\pi}{2\pi r}$$

$r = 6$

14. How much longer is the track marked by - - - - - than the track marked by _____ ? The tracks are always 2 cm apart.

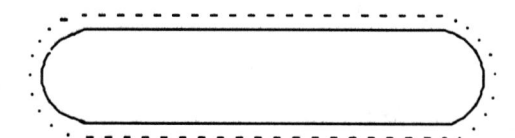

4π cm
The two ends of each track form a circle.
$C_{\text{inner track}} = 2\pi r$
$C_{\text{outer track}} = 2\pi (r + 2) = 2\pi r + 4\pi$

15. What is the length of the longest stick that could be placed in a box that has dimensions 20 cm × 20 cm × 30 cm?

41.23 cm

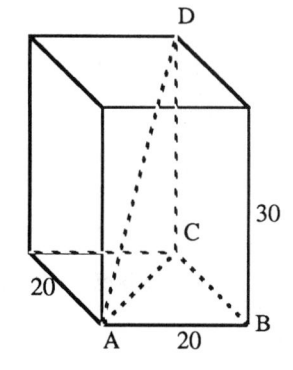

$m\,(\overline{AC})^2 = 20^2 + 20^2 = 800$
$m\,(\overline{AC}) = \sqrt{800}$ cm
$m\,(\overline{AD})^2 = (\overline{AC})^2 + 30^2$
$\phantom{m\,(\overline{AD})^2} = (\sqrt{800})^2 + 30^2$
$\phantom{m\,(\overline{AD})^2} = 800 + 900 = 1{,}700$
$m\,(\overline{AD}) = \sqrt{1700} \approx 41.23$ cm

Section 3 Two Dimensional Measure

Cover the right side of the page and work on the left, then check your work

1. Find the area of each figure where each unit is 1 cm.

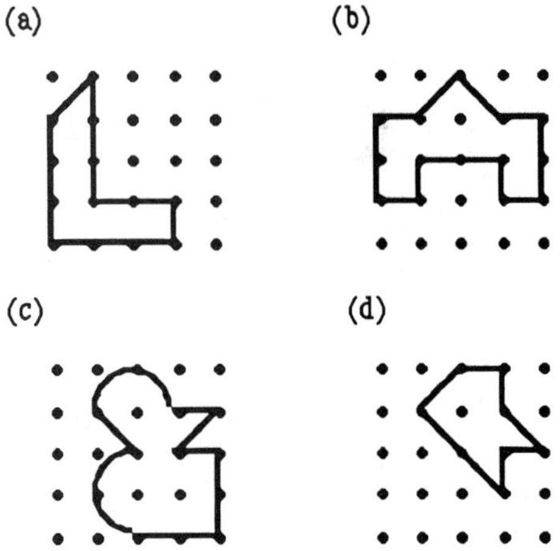

   (a) (b)

   (c) (d)

   (a) 5.5
   (b) 7
   (c) $6 + \pi$
   (d) 4

2. Find the area of each of the following rectangles.

   (a) width = 15 cm (b) width = 4 m
       length = 21 cm   length = 51 dm
   (c) width = 4.3 ft (d) width = a
       length = 6.9 ft   length = $2a + 1$

   (a) 315 cm^2
   (b) 20.4 m^2
   Area = 4 m × 51 dm
   = 4 m × 5.1 m = 20.4 m^2
   (c) 29.67 ft^2
   (d) $2a^2 + a$

3. Find the area of each of the following triangles.

   (a) triangle with legs 5 and 7
   (b) triangle with sides 8, 8, base 8
   (c) triangle with 8.4, 10, base 10

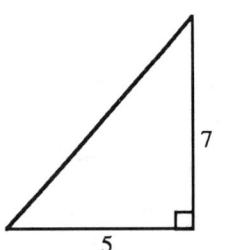

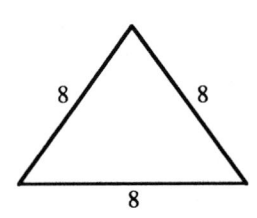

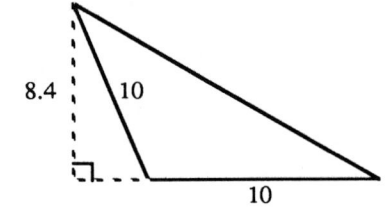

(a) 17.5
(b) $4\sqrt{48}$

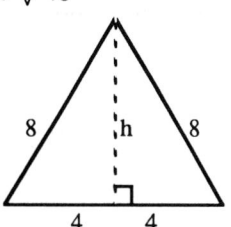

$h^2 + 4^2 = 8^2$
$h^2 + 16 = 64$
$h^2 = 48$
$h = \sqrt{48}$
(c) 42

4. Find the area of each of the following circles.

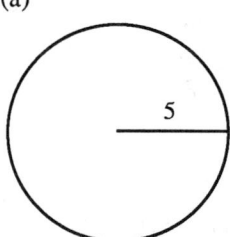

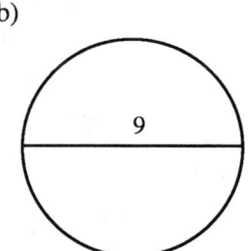

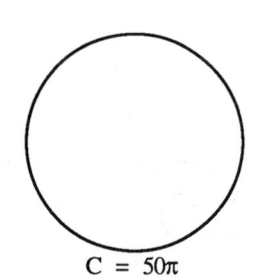

(a) [circle with radius 5]

(b) [circle with diameter 9]

(c) [circle with $C = 50\pi$]

(a) 25π
(b) 9π
(c) 625π
$2\pi r = 50\pi$
$r = 25$
$A = \pi (25)^2 = 625\pi$

5. Find the two rectangles whose sides are whole numbers and whose area and perimeter are the same numbers.

Rectangle 1 (4 by 4)
Area = Perimeter = 16
Rectangle 2 (6 by 3)
Area = Perimeter = 18

6. Find the perimeter and area for each of the following figures.

280 CHAPTER 11   Measurement and the Metric System

(a)

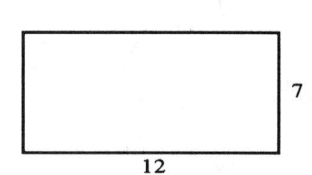

```
    ┌─────────────┐
    │             │ 7
    └─────────────┘
         12
```

(b)
```
         15
    ┌─────────┐
    │         │ 4
    └─────────┘
```

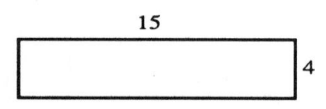

(c)
```
         /\
        /  \
    12 /    \ 12
      /_____\
         14
```

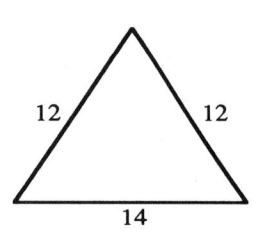

(a) Area = 84
    Perimeter = 38
(b) Area = 60
    Perimeter = 38
(c) Area = 7 $\sqrt{95}$
    Perimeter = 38

7. From Exercise 6, what can you say about the area of figures that have equal perimeters.

If two figures have equal perimeters, they do not necessarily have the same area.

8. Find the area of each shaded region.

(a)
```
         12
    ┌─────────────┐
    │╲            │
    │ ╲           │
    │  ╲          │ 10
  6 │   ╲_____│
    │             │
    └─────────────┘
         15
```

(b)

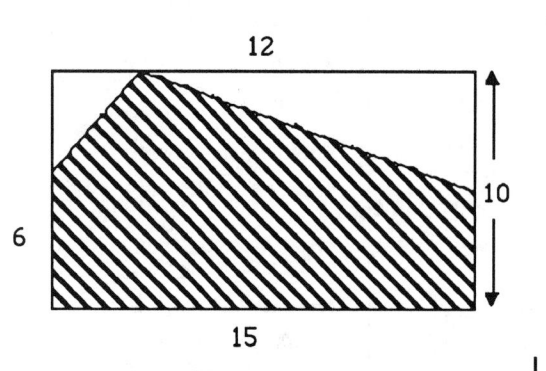

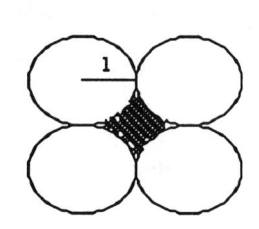

(a) 144
Area =

$15 \cdot 12 - \frac{1}{2}(5 \cdot 12) - \frac{1}{2}(3 \cdot 4)$
$= 180 - 30 - 6 = 144$

(b) $4 - \pi$

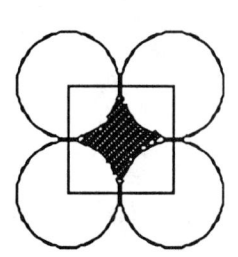

Area$_{square}$ = 2^2
Area$_{circle}$ = π
Area$_{shaded}$ = $4 - \pi$

9. Two dart boards are shown below. The area(s) inside the circle(s) are considered hits. Which board gives a player more area to hit?

(a)

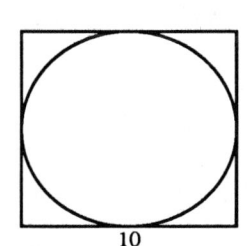

10

(b)

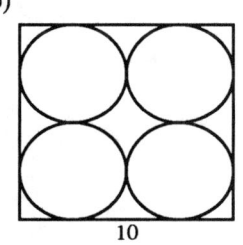

10

They are the same.
Area$_{(a)}$ = $\pi (5)^2 = 25\pi$
Area$_{(b)}$ = $4 \cdot \pi (2.5)^2 = 25\pi$

10. Find the cost of insulation for a ceiling that measures 40 dm by 70 dm if the insulation costs $1.50 per square meter.

$42
Area = 4 m × 7m = 28 m^2
Cost = $(28 \text{ m}^2)\left(\dfrac{1.50}{1 \text{ m}^2}\right)$ = $42

11. Which is the better buy, a 14-inch diameter pizza that costs $8.99 or an 18-inch diameter pizza that costs $12.99?

18-inch pizza
Area$_{14\text{-inch pizza}}$ = $\pi (7)^2 = 49\pi$ in^2

$$\text{Area}_{18\text{-inch pizza}} = \pi (9)^2 \; 81\pi \text{ in}^2$$

$$\text{Cost per ounce}_{14"} = \frac{\$8.99}{49\pi} \approx 5.8¢$$

$$\text{cost per ounce}_{18"} = \frac{\$12.99}{81\pi} \approx 5.14¢$$

12. Heron's Formula says the following.

$$\text{Area} = \sqrt{s(s-a)(s-b)(s-c)}$$

where a, b, and c are the lengths of the sides and $s = \dfrac{a+b+c}{2}$ relative to a triangle. Use this formula to find the area of the following triangles.

(a) 3 cm, 4 cm, 5 cm
(b) 5 m, 5 m, 8 m

(a) Area = 6
(b) Area = 12

13. Use Pick's formula to verify the answers for Exercise 1, parts (a), (b), and (d).

(a) $I = 0 \quad p = 13$
$A = 0 + \dfrac{13}{2} - 1 = 5.5$

(b) $I = 1 \quad p = 14$
$A = 1 + \dfrac{14}{2} - 1 = 7$

(d) $I = 1 \quad p = 8$
$A = 1 + \dfrac{8}{2} - 1 = 4$

14. Find the area of the rhombus below.

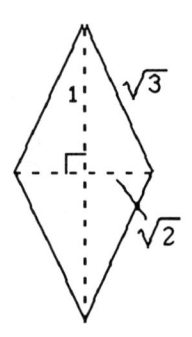

$A = 2\sqrt{2}$

15. A goat is tied to the corner of a grassy area that is in the shape of an equilateral triangle with sides of length 8 meters. How long should the rope be so that one-half of the grass is reachable by the goat?

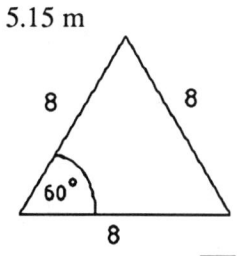

5.15 m

$\text{Area}_{\text{triangle}} = 4\sqrt{48}$

$\text{Area}_{\text{circ region}} = \dfrac{1}{6}(\pi x^2) = \dfrac{\pi x^2}{6}$

Then, $\dfrac{\pi x^2}{6} = \dfrac{4\sqrt{48}}{2}$

$x^2 = \dfrac{12\sqrt{48}}{\pi}$

$x \approx 5.15 \text{ m}$

Section 4 Surface Area and Volume

Cover the right side of the page and work on the left, then check your work

1. Find the volume and surface area for each of the following rectangular prisms with the given dimensions.

   (a) 7, 8, 9
   (b) 2, 10, 15
   (c) 7.5, 6, 14

   > (a) Volume = 504
   >   Surface area = 382
   > (b) Volume = 300
   >   Surface area = 400
   > (c) Volume = 630
   >   Surface area = 468

2. Find the volume for each of the following figures.

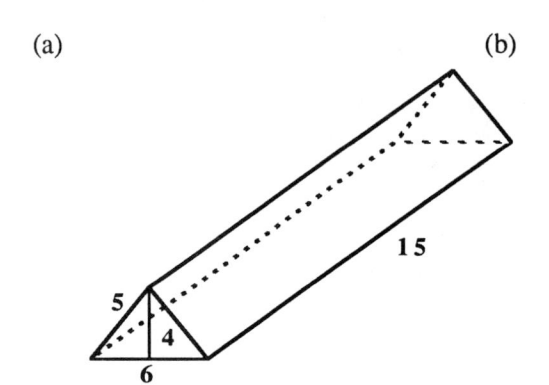

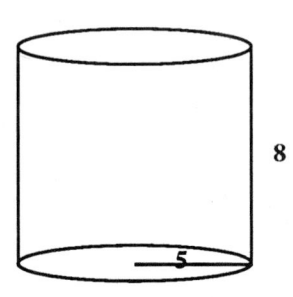

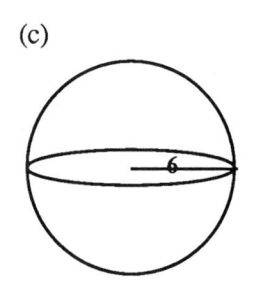

   (a)

   (b)

   (c)

   > (a) 90
   > $V = B \cdot h$
   > $= \frac{1}{2}(3 \cdot 4) \cdot 15 = 90$
   > (b) 200π
   > $V = \pi r^2 h = \pi (5)^2 \, 8$
   > $= 200\pi$

3. Find the surface area for each of the following.

   (a) Cylinder: $r = 7.6$ cm
   $h = 12.2$ cm

   (b) Triangular Prism: $B = 45$
   $h = 17$

   (c) Sphere: $r = 24$

(c) 288π
$V = \dfrac{4}{3}\pi r^3$
$= \dfrac{4}{3}\pi (6)^3 = 288\pi$

(a) 297.92π
$SA = 2\pi r^2 + 2\pi rh$
$= 2\pi (7.6)^2 + 2\pi (7.6)(12.2)$
$= 297.92\pi$
(b) 576π
$SA = 4\pi r^2$
$= 4\pi (12)^2 = 576\pi$
(c) 765
$V = 45 \cdot 17 = 765$

Find the volume and surface area for each of the following figures (4 - 6).

4.

$V = 270$
$SA = 294$
$V = B \cdot h$
$= \dfrac{1}{2}(9 + 3)5 \cdot 9 = 270$
$SA = 2 \cdot 30 + 9 \cdot 9 + 2(9 \cdot 7) + 9 \cdot 3$
$= 60 + 81 + 126 + 27 = 294$

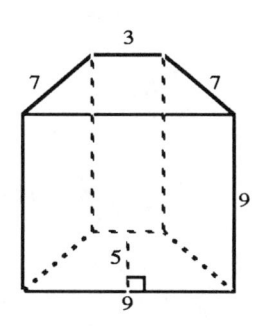

5.

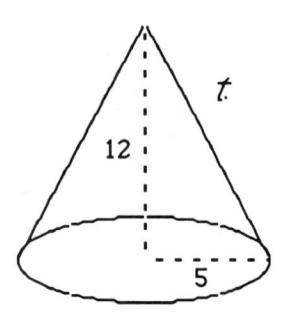

$$V = 100\pi$$
$$SA = 90\pi$$
$$V = \frac{B \cdot h}{3}$$
$$= \frac{\pi (5)^2 \cdot 12}{3} = 100\pi$$
$$SA = \pi r t + \pi r^2$$
$$= \pi (5)(13) + \pi (5)^2$$
$$= 65\pi + 25\pi = 90\pi$$

6.

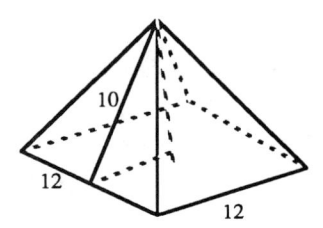

$$V = 384$$
$$SA = 384$$

7. A standard tennis ball can holds three tennis balls that each have a radius of 3.5 cm. What percent of the can is occupied by the balls?

$$66.7\%$$
$$V_{can} = \pi (3.5)^2 \cdot 21 = 257.25\pi$$
$$V_{balls} = 3\left[\frac{4}{3}\pi (3.5)^3\right] = 171.5\pi$$
$$\% = \frac{171.5\pi}{257.25\pi} = 0.667$$

8. In 1987, the Denver Mint produced 4.8 billion pennies. If the the diameter of a penny is 2 cm and it is 2 mm thick, how much copper was used to make this number of pennies?

$$301{,}000 \text{ m}^3$$
$$V_{penny} = \pi (1)^2 \cdot (0.2 \text{ cm})$$

$$= 0.2\pi \text{ cm}^3$$
$$V_{total} = (4{,}800{,}000{,}000)(0.2\pi)$$
$$= 3.01 \times 10^9 \text{ cm}^3$$
$$\approx 3.01 \times 10^5 \text{ m}^3$$
$$= 301{,}000 \text{ m}^3$$

9. How much will it cost to paint the walls and ceiling of a room that measures 12 ft square with a ceiling that is 8 ft high? The paint costs $13 per gallon and one gallon covers 250 ft² of area.

$39
$$SA = 12 \cdot 12 + 4(12 \cdot 8) = 528 \text{ ft}^2$$
$$\text{Paint needed} = \frac{528}{250} = 2.112 \approx$$
3 gallons
Cost = 3 · 13 = $39

10. A fish tank in the shape of a right rectangular prism has dimensions 40 cm, 15 cm, and 25 cm. What is the volume of the tank?

15,000 cm³
$$V = 40 \cdot 15 \cdot 25 = 15{,}000 \text{ cm}^3$$

11. Mr. Parrot wants to build a patio in the shape of a regular octagon, 12 feet on a side, with apothem of 10.

   (a) What is the area of the surface of the patio?
   (b) If the patio is 6 inches thick, what volume of concrete will be needed?
   (c) If the concrete costs $50 per cubic yard, what will be the cost of the patio?

(a) 480 ft²
$$A = \frac{pa}{2}$$
$$= \frac{(96)(10)}{2} = 480 \text{ ft}^2$$
(b) 240 ft³
$$V = B \cdot h = 480 \cdot \frac{1}{2} = 240 \text{ ft}^3$$
(c) $12,000
Cost = 240 · 5 = $12,000

12. The ratio of the volumes of two cubes is 8:27. What is the ratio of their surface areas?

$$\frac{4}{9}$$

288 CHAPTER 11   Measurement and the Metric System

$$V_1 = s_1^3$$
$$\frac{s_1^3}{s_2^3} = \frac{8}{27} \Rightarrow s_1 = \frac{2}{3} s_2$$
$$V_2 = s_2^3$$
$$SA_1 = 6\left(\frac{2}{3} s_2\right)^2 = \frac{8}{3} s_2^2$$
$$SA_2 = 6 s_2^2$$
$$\frac{SA_1}{SA_2} = \frac{\frac{8}{3} s_2^2}{6 s_2^2} = \frac{4}{9}$$

13. A spherical scoop of ice cream sits in a right circular cone. The scoop has a diameter of 10 cm. The cone has a diameter of 8 cm and a height of 12 cm. If the ice cream melts and all of it runs into the cone, will the cone overflow?

No
$$V_{scoop} = \frac{4}{3} \pi (5)^3 = 166.67\pi$$
$$V_{cone} = \pi (4)^2 \cdot 12 = 192\pi$$

14. The Great Pyramid of Egypt had a square base that was 775 feet on a side and had a height of 481 feet. What is the volume of the pyramid?

2.89×10^8 ft^3
$$V = B \cdot h = (775)^2 \cdot 481$$
$$= 2.89 \times 10^8 \text{ ft}^3$$

15. Consider a rectangular piece of paper that measures 10 by 22. The paper can be rolled into two different cylinders, one that has a height of 10 or one that has a height of 22. Which cylinder will have the larger volume?

Cylinder with Height of 10 will have larger volume.
$$V_{cyl\ 1} = \pi \left(\frac{11}{\pi}\right)^2 \cdot 10$$
$$= \frac{1210}{\pi} \approx 385.35$$
$$V_{cyl\ 2} = \pi \left(\frac{5}{\pi}\right)^2 \cdot 22 = \frac{550}{\pi} \approx 175.16$$

Section 5 Other SI Units of Measure

Cover the right side of the page and work on the left, then check your work

1. Perform the following conversions.

   (a) 5 L = _____ ml
   (b) 450 ml = _____ cl
   (c) 3.2 kl = _____ L
   (d) 53 L = _____ dl

   > (a) 5,000
   > (b) 45
   > (c) 3,200
   > (d) 530

2. Perform the following conversions.

   (a) 4678 g = _____ kg
   (b) 5.7 kg = _____ mg
   (c) 536 cg = _____ g
   (d) 39 dg = _____ hg

   > (a) 4,678
   > (b) 5,700,000
   > (c) 5.36
   > (d) 0.039

3. Convert the following temperatures from degrees Celsuis to degrees Fahrenheit.

   (a) 4°
   (b) 25°
   (c) 100°
   (d) −15°

   > (a) 39.2°
   > (b) 77°
   > (c) 212°
   > (d) 5°

4. Convert the following temperatures from degrees Fahrenheit to degrees Celsuis.

   (a) 350°
   (b) 98°
   (c) −20°
   (d) 4°

(a) 176.67°
(b) 36.67°
(c) −28.89°
(d) −15.56°

5. Choose the most reasonable measure for each of the following.

(a) Juice can   800 cl   800 ml   800 L
(b) Teaspoon   8 ml   8 L   8 cl
(c) Bathtub   10 L   20 L   200 L
(d) Bucket   4 L   4 ml   4 dl

(a) 800 ml
(b) 8 ml
(c) 200 L
(d) 4 L

6. Choose the most reasonable measure for each of the following.

(a) An adult man   75 Kg   75 g   75 dg
(b) A baseball   300 g   300 mg   300 kg
(c) A paperclip   1 mg   1 cg   1 g
(d) An automobile   1500 kg   1500 g   1500 hg

(a) 75 kg
(b) 300 g
(c) 1 g
(d) 1500 kg

7. Choose the most reasonable measure for each of the following.

(a) Comfortable room temperature   39°C   12°C   20°C
(b) Day for wearing a heavy coat   15°C   −2°C   30°C
(c) Good day for going swimming   35°C   80°C   18°C
(d) Good day for ice skating on a pond   3°C   −18°C   17°C

(a) 20°
(b) −2°C
(c) 35°C
(d) −18°C

8. An aquarium in the shape of a cylinder has a radius of 25 cm and a height of 40 cm. How much does the water in the aquarium weigh?

> 78.5 kg
> $V = \pi (25)^2 (40) \approx 78{,}500 \text{ cm}^3$
> Weight $= 78{,}500 \text{ g} = 78.5 \text{ kg}$

9. Complete the following chart for right rectangular prisms.

	Length	Width	Height	Volume in cm³	Volume in dm³	Volume in ml	Volume in L
(a)	30 cm	20 cm	10 cm				
(b)	10 cm	5 dm	2 dm				
(c)	3 dm	1 dm				6000 l	

> | | Length | Width | Height | Volume in cm³ | in dm³ | in ml | in L |
> |---|---|---|---|---|---|---|---|
> | (a) | 30 cm | 20 cm | 10 cm | 6,000 | 6 | 6,000 | 6 |
> | (b) | 10 cm | 5 dm | 2 dm | 10,000 | 10 | 10,000 | 10 |
> | (c) | 3 dm | 1 dm | 1.5 dm | 4,500 | 4.5 | 4,500 | 4.5 |

10. For each of the following, place a decimal point in the number to make the measure reasonable.

   (a) A coffee cup holds 3000 ml.
   (b) A quart milk container holds 10 L
   (c) A brick weighs about 20 kg
   (d) A nickel weighs about 5 dag

> (a) 300.0
> (b) 1.0
> (c) 2.0
> (d) 0.5

11. A right cylindrical container is to be made to hold 1 L of juice. What should the radius of the can be if its height is 10 cm?

> 5.64 cm
> $1 \text{ L} = 1000 \text{ ml} = \pi r^2 \cdot 10$
> $r^2 = \dfrac{1000}{10 \, \Pi}$
> $r^2 \approx 31.847$
> $r \approx 5.64 \text{ cm}$

292 CHAPTER 11   Measurement and the Metric System

12. Order the following from smallest to largest.

    4 kg, 3672 g, 42 cg, 897 mg, 26 dg

    | 42 cg, 897 mg, 26 dg, 3,672 g, 4 kg

13. How much liquid (in ml) can be held in a straw that is 30 cm long and 4 mm in diameter?

    | 15 cm^3 or 15 ml
    | $V = \pi (0.4 \text{ cm})^2 \, 30 \text{ cm} \approx 15 \text{ cm}^3$

14. Choose the most appropriate metric unit (kg, g, mg)

    (a) weight of a pencil
    (b) weight of an aspirin tablet
    (c) weight of a chair
    (d) weight of a can of soup.

    | (a) g
    | (b) mg
    | (c) kg
    | (d) g

15. The human heart pumps 65 milliliters of blood per hearbeat. If the heart beats 68 beats per minute,

    (a) how much blood is pumped per minute?
    (b) how much blood is pumped per hour?
    (c) how long will it take to pump on liter of blood?

    | (a) 4.420 ml
    | $\dfrac{65 \text{ ml}}{1 \text{ beat}} = \dfrac{x \text{ ml}}{68 \text{ beats}}$
    | $x = 4{,}420$ ml
    | (b) 265.2 L
    | $\dfrac{4{,}420 \text{ ml}}{1 \text{ min}} = \dfrac{x \text{ ml}}{60 \text{ min}}$
    | $x = 265{,}200$ ml $= 265.2$ L
    | (c) 13.5 seconds
    | $\dfrac{4420 \text{ ml}}{1 \text{ min}} = \dfrac{1000 \text{ ml}}{x}$
    | $x = 0.226$ min ≈ 13.5 seconds

Chapter 12
Using Transformations in Geometry

Section 1 Congruence and Triangles

Cover the right side of the page and work on the left, then check your work

1. Determine whether each pair of triangles is congruent.

   (a)

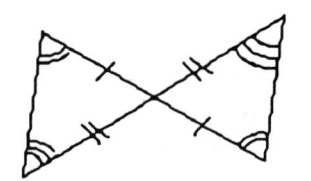

   (b)

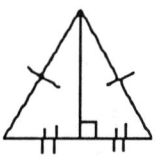

   (c)

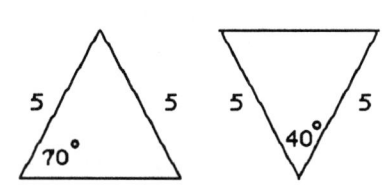

   (d)

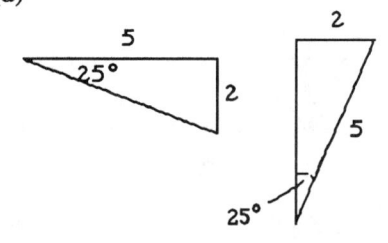

   (a) congruent - AAS or SAS
   (b) congruent - SSS or SAS
   (c) congruent - ASA or SAS
   (d) not congruent

2. Given that $\triangle ABC \cong \triangle DEF$, complete the following.

   (a) $\overline{AB} \cong$ _____
   (b) $\overline{CB} \cong$ _____
   (c) $\angle CAB \cong$ _____
   (d) $\angle DEF \cong$ _____

   (a) $\overline{DE}$
   (b) $\overline{FE}$
   (c) $\angle FDE$
   (d) $\angle ABC$

3. For the triangles shown, answer the following questions.

   (a) Which side of $\triangle DEB$ corresponds to $\overline{BC}$?

294 **CHAPTER 12** Using Transformations in Geometry

(b) Which angle of △ ABC corresponds to ∠ DEF ?
(c) If $m(\angle BAC) = 30°$, then what is the measure of ∠ BDF ?

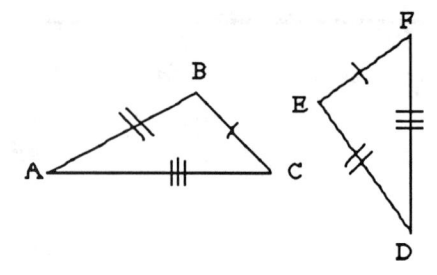

(a) $\overline{EF}$
(b) ∠ ABC
(c) 30°

4. Verify that the following pair of triangles are congruent.

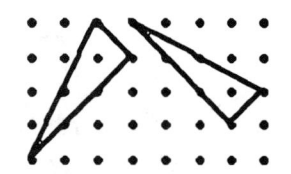

Congruent by SSS

5. For which of the pairs of triangles given is the information inadequate to determine whether they are congruent?

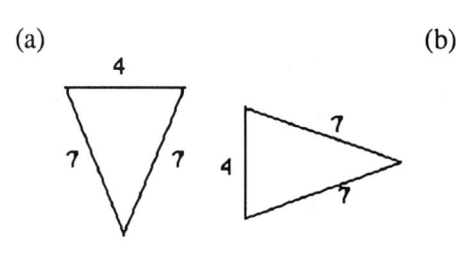

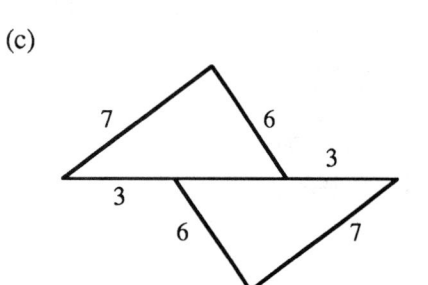

(a)

(b)

(c)

(a) Adequate information
(b) Inadequate information

	(c) Adequate information
6. Polygon *ABCD* is a square. (a) What relationship exists between point *E* and the diagonals $\overline{AC}$ and $\overline{BD}$? (b) Prove your answer to (a)	

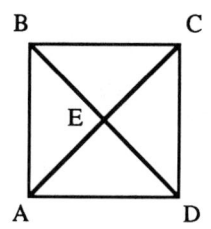

B   C
[square with diagonals crossing at E]
A   D

	(a) *E* is the midpoint of each diagonal (b) $\overline{AB} \cong \overline{CD}$ sides of the square $\angle BEA \cong \angle DEC$ vertical angles $\angle ABE \cong \angle CDE$ alternate interior angles $\triangle BEA \cong \triangle DEC$ AAS $\overline{BE} \cong \overline{DE}$ corresponding parts of congruent triangles. In a similar way, we can show $\overline{CE} \cong \overline{AE}$
7. For Exercise 6, (a) What are the measures of $\angle AEB$ and $\angle DEA$? (b) Prove your answer to (a).	(a) 90° (b) From 6, we can show $\triangle AEB \cong \triangle AED$ since we showed $\triangle AEB \cong \triangle DEC$ and $\triangle DEC \cong \triangle AED$ (Transitive) Then we have $\angle AEB \cong \angle DEA$ by corresponding parts. We also see that $\angle AEB$ and $\angle DEA$ are supplementary. Thus, $\angle AEB \cong \angle DEA = 90°$
8. Could the ASA congruence property be used to show the following pairs of triangle are congruent?	

296 CHAPTER 12   Using Transformations in Geometry

(a)

(b)

(a) yes
(b) no

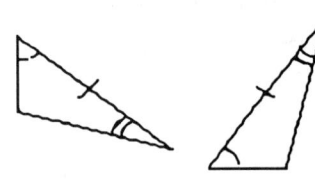

9. Prove that $\triangle ABC \cong \triangle DEF$.

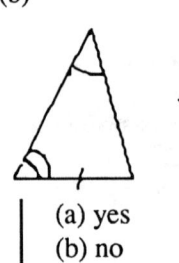

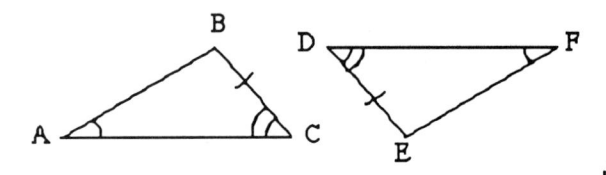

$\angle A \cong \angle F$
$\angle C \cong \angle D$
$\overline{BC} \cong \overline{DE}$
$\triangle ABC \cong \triangle DEF$ AAS

10. How could the following diagram be useful in finding the distance m $\overline{(AB)}$, across the lake? We know that $\angle DBA \cong \angle BCA$ and $\overline{DC} \cong \overline{BC}$.

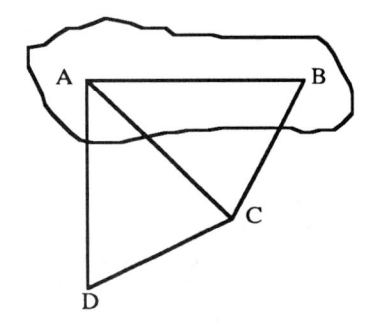

Given: $\angle DCA \cong \angle BCA$
$\overline{DC} \cong \overline{BC}$

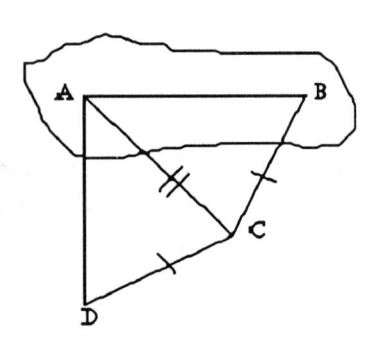

We also know, $\overline{AC} \cong \overline{AC}$
Then, $\triangle DAC \cong \triangle BAC$ (SAS)
And, $\overline{AB} \cong \overline{AD}$ (corresponding parts)
We can measure $\overline{AD}$ much more easily than $\overline{AB}$.

11. Given: $\overline{DC} \perp$ plane α at C
    $\overline{CA}$ and $\overline{CB}$ are in α
    $\overline{AD} \cong \overline{BD}$
    (a) Prove: $\angle DAB \cong \angle DBA$
    (b) Prove: $\triangle DCA \cong DCB$

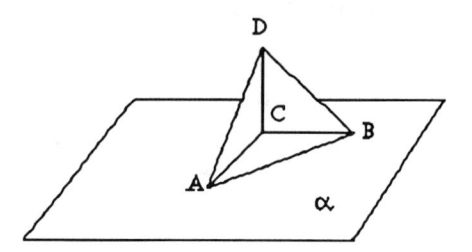

(a) $\overline{AD} \cong \overline{BD}$ - is given
Thus, $\triangle ADB$ is isosceles
And, $\angle DAB \cong \angle DBA$ - Angles opposite congruent sides are congruent.

(b) $\overline{AD} \cong \overline{BD}$ - given
$\angle DCA \cong \angle DCB$ - both are right angles
$\angle CDA \cong \angle CDB$ - corresponding parts from (a)
Then, $\triangle DCA \cong \triangle DCB$ (ASA)

12. Prove that $\triangle RST$ is isosceles.

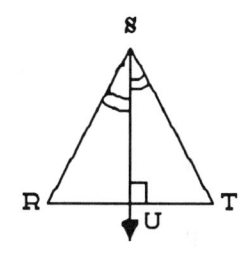

$\angle RSU \cong \angle TSU$ - is given
$\angle SUR \cong \angle SUT$ - both are right angles
$\overline{SU} \cong \overline{SU}$ - reflexive
$\triangle RSU \cong \triangle TSU$ - ASA

298   CHAPTER 12   Using Transformations in Geometry

$\overline{RS} \cong \overline{TS}$ - corresponding parts
$\triangle RST$ is isosceles - definition of isosceles triangle

13. In $\triangle GUI$ and $\triangle TAR$, $\overline{GU} \cong \overline{AR}$ and $\angle U \cong \angle A$. Name the other pair(s) of parts you would need to show congruent in order to show that the triangles are congruent by the

   (a) *SAS* postulate
   (b) *ASA* theorem.

(a) $\overline{UI} \cong \overline{AT}$
(b) $\angle G \cong \angle R$

14. Given: $\angle 3 \cong \angle 4$
    $\angle 4 \cong \angle 5$
    $\angle LU \cong \angle I$
    $\angle 1 \cong \angle 2$

    Prove: $\triangle TUL \cong \triangle PIL$

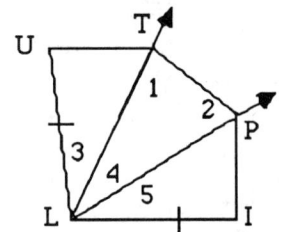

$\overline{LU} \cong \overline{LI}$ - given
$\angle U \cong \angle I$ - given
$\angle 3 \cong \angle 4$ and $\angle 4 \cong \angle 5$ - given
$\angle 3 \cong \angle 5$ - transitive
$\triangle TUL \cong \triangle PIL$ - ASA

15. Show that the following figure is a parallelogram.

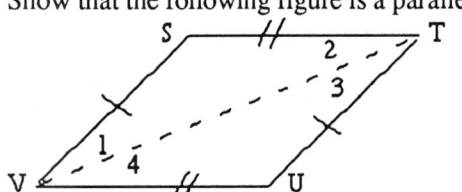

$\overline{RS} \cong \overline{TU}$ - given
$\overline{ST} \cong \overline{UR}$ - given
$\overline{RT} \cong \overline{RT}$ - reflexive
$\triangle TSR \cong \triangle RUT$ - SSS
$\angle 1 \cong \angle 3$ and $\angle 2 \cong \angle 4$ - corresponding parts
$\overline{ST} \parallel \overline{UR}$ and $\overline{SR} \parallel \overline{UT}$ - lines cut by transversal ($\overline{RT}$) for alternate interior angles that are congruent.

Section 2 Congruence and Right Triangles

Cover the right side of the page and work on the left, then check your work

1. Which of the following pairs of right triangles are congruent?

   (a)

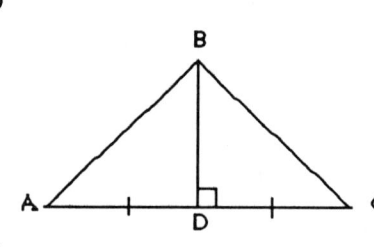

   (b)

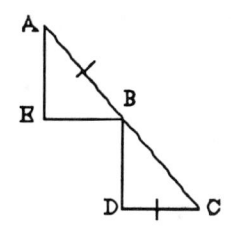

   (c)

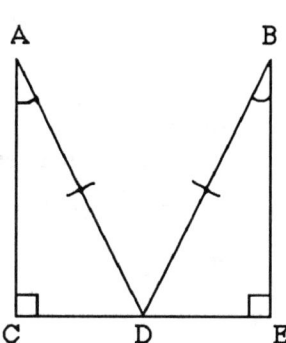

   (a) Congruent by SAS
   (b) Not congruent
   (c) Congruent by AAS

2. Find the length of the missing side on the following triangles.

   (a)

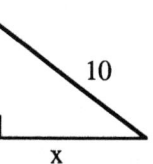

   (b)

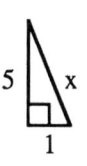

   (c)

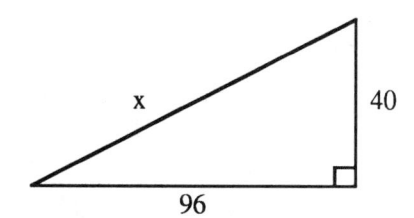

   (a) $x = 8$
   $x^2 + 6^2 = 10^2$
   $x^2 + 36 = 100$
   $x^2 = 64$
   $x = 8$

300   CHAPTER 12   Using Transformations in Geometry

(b) $x = \sqrt{26}$
$1^2 + 5^2 = x^2$
$1 + 25 = x^2$
$\sqrt{26} = x$
(c) $x = 104$

3. Why are the following triangles congruent?

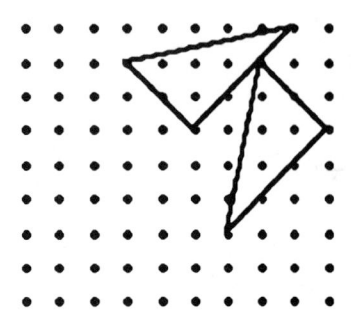

The measure of the three sides of one triangle are congruent to three sides of the other triangle (SSS).

4. Which of the following are sides of a right triangle.

(a) 7, 24, 25
(b) 8, 9, 15
(c) 10, 12, 16
(d) 28, 21, 25

(a) yes
(b) no
(c) no
(d) yes

5. Show that $\angle BXA \cong \angle CXA$

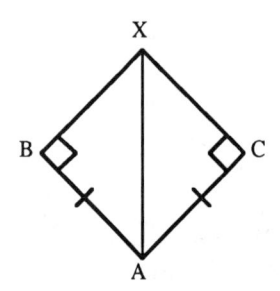

$\overline{AB} \cong \overline{AC}$	Given
$\overline{XA} \cong \overline{XA}$	Reflexive
$\triangle AXB \cong \triangle AXC$	SSS
$\angle BXA \cong \angle CXA$	Corresponding

Section 2   Congruence and Right Triangles

parts of congruent triangles

6. Explain why quadrilateral *ABCD* is a trapezoid.

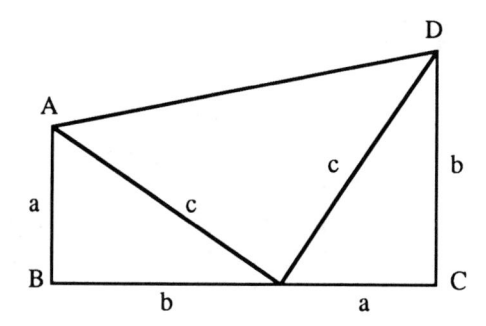

$\overline{BC}$ is a transversal for $\overline{AB}$ and $\overline{PC}$ (see drawing)

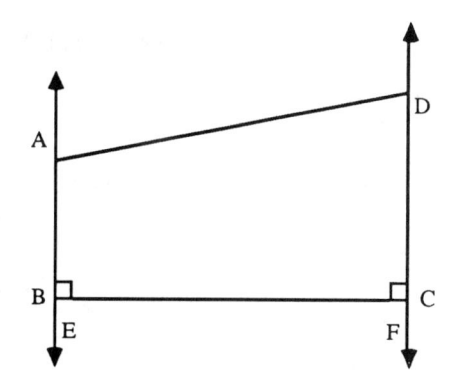

∠*EBC* is adjacent and supplementary to ∠*ABC*
Thus, ∠*EBC* is a right angle
∠*EBC* ≅ ∠*DCB* - both are right angles
AB ∥ *DC* - alternate interior angles are congruent

7. Show how you would construct a segment of length $\sqrt{17}$.

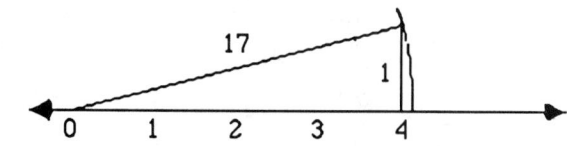

8. Show that $\overline{AB} \cong \overline{AC}$.

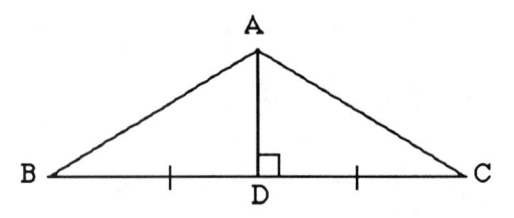

302 **CHAPTER 12**   Using Transformations in Geometry

$\overline{BD} \cong \overline{CD}$ — Given
$\angle ADB \cong \angle ADC$ — Both are right angles
$\overline{AD} \cong \overline{AD}$ — Reflexive
$\triangle ADB \cong \triangle ADC$ — SAS
$\overline{AB} \cong \overline{AC}$ — Corresponding parts of congruent triangles

9. How much distance will Joey save in going from his house to the store if he cuts across the vacant lot rather than stay on the sidewalk?

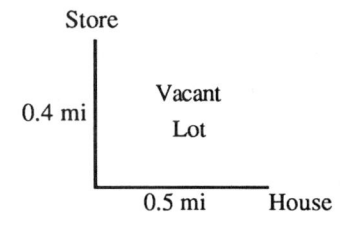

0.26 miles
Distance via sidewalk = 0.4 + 0.5 = 0.9 miles
Distance across lot = x
Then, $x^2 = (0.4)^2 + (0.5)^2$
$= 0.16 + 0.25 = 0.41$
$x \approx 0.64$ miles
Difference = $0.9 - 0.64 = 0.26$ miles

10. True or false? "The area of the semicircle on the hypotenuse of a right triangle equals the sum of the areas of the semicircles on the legs of the triangle." Use the example below.

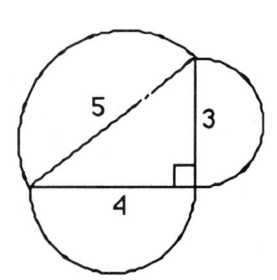

True

Area of Semicircle$_3$ = $\frac{1}{2}\pi (1.5)^2$

= 1.125π

Area of Semicircle$_4$ = $\frac{1}{2}\pi (2)^2 = 2\pi$

Area of Semicircle$_5$ = $\frac{1}{2}\pi (2.5)^2 =$ 3.125π

$1.125\pi + 2\pi = 3.125\pi$

11. A railroad track is made from two pieces of rail, each 1,000 meters long. On a hot day, the rails expand so that each piece is 1,001 meters long, causing the track to buckle as shown. Find h.

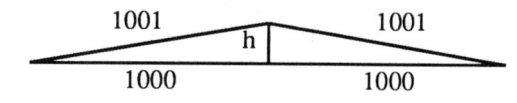

$h = 44.7$ m
$h^2 + 1000^2 = 1001^2$
$h^2 + 1,000,000 = 1,002,001$
$h^2 = 2001$
$h \approx 44.7$ m

12. A toy maker wants to cut the plastic rectangle □EFGH into four right triangles and a rectangle as shown. How long is $\overline{FA}$? How long is $\overline{CE}$?

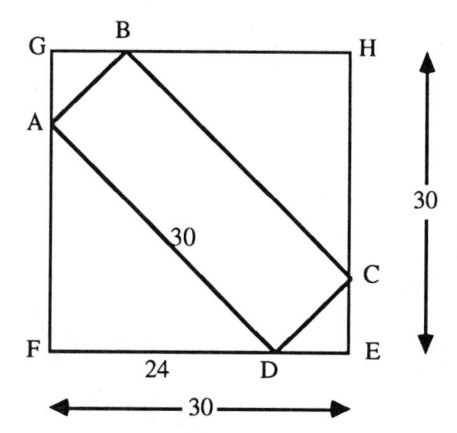

$m\ (\overline{FA}) = 18$
$[m\ (\overline{FA})]^2 + 24^2 = 30^2$
$[m\ (\overline{FA})]^2 = 324$
$m\ (\overline{FA}) = 18$

304 CHAPTER 12 Using Transformations in Geometry

13. For the following square □ABCD, find the length of $\overline{AB}$.

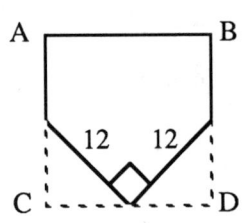

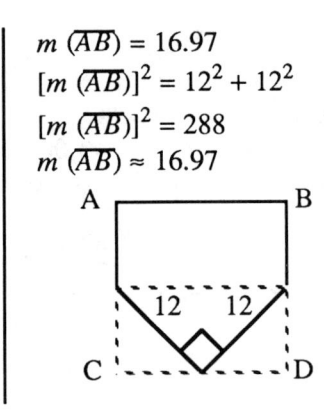

$m\ (\overline{AB}) = 16.97$
$[m\ (\overline{AB})]^2 = 12^2 + 12^2$
$[m\ (\overline{AB})]^2 = 288$
$m\ (\overline{AB}) \approx 16.97$

14. The figure below shows a tetrahedron RSTU formed by four vertices of a right rectangular prism.

   (a) Determine the lengths of the six edges of the tetrahedron.
   (b) Are the four faces of the tetrahedron congruent to each other?

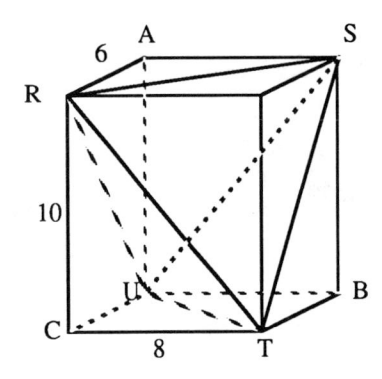

(a) $m\ (\overline{RS}) = m\ (\overline{TU}) = 10$
$m\ (\overline{ST}) = m\ (\overline{RU}) = \sqrt{136}$
$m\ (\overline{RT}) = m\ (\overline{SU}) = \sqrt{164}$
(b) yes by SSS

15. It can be shown that in a right triangle with angle measures of $30°$, $60°$, and $90°$ that the measure of the side opposite the $30°$ angle is one-half the measure of the hypotenuse. Find the measures of the sides of the following triangles.

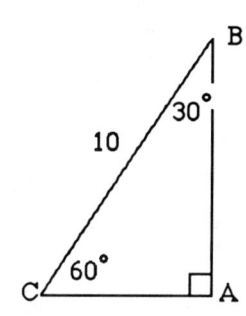

(a)

(b)

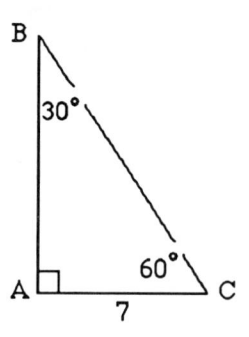

(a) $m\,(\overline{BC}) = 10$
$m\,(\overline{AC}) = 5$
$m\,(\overline{AB}) = \sqrt{75}$

(b) $m\,(\overline{AC}) = 7$
$m\,(\overline{BC}) = 14$
$m\,(\overline{AB}) = \sqrt{147}$

Section 3 Justification of Constructions

Cover the right side of the page and work on the left, then check your work

1. Construct a triangle with the three sides given.

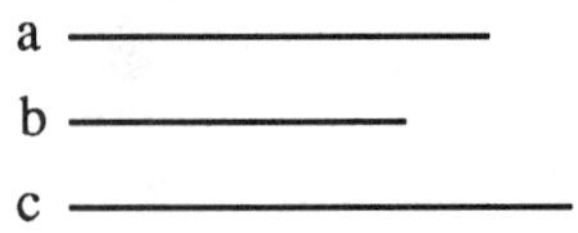

    a ─────────
    b ──────
    c ────────────

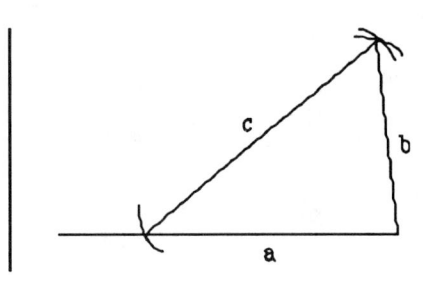

2. Construct a triangle congruent to △ABC using SAS.

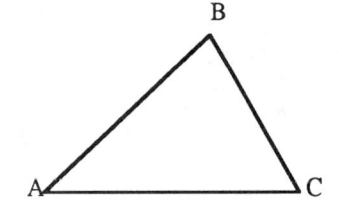

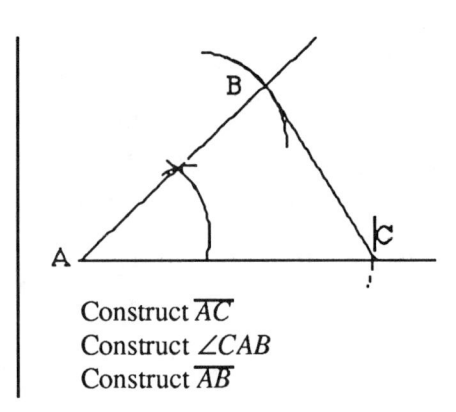

    Construct $\overline{AC}$
    Construct $\angle CAB$
    Construct $\overline{AB}$

3. Construct the three angle bisectors for the triangle in Exercise 2.

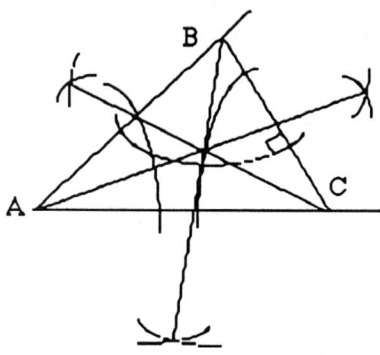

4. Given the two angles below, construct each of the following.

   (a) An angle with measure $a + b$
   (b) An angle with measure $2b$
   (c) An angle with measure $a - b$

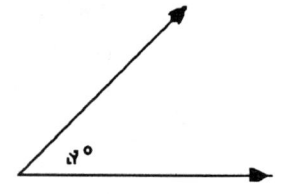

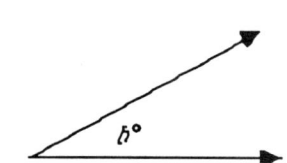

Answers follow:

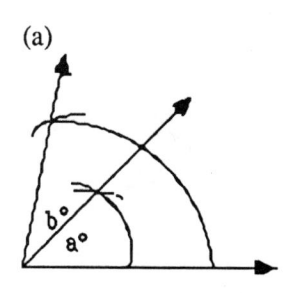

(a)   (b)   (c)

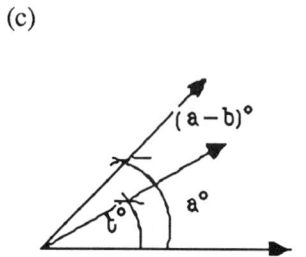

5. Draw a convex quadrilateral. Bisect all four sides and connect these midpoints to form another quadrilateral. Repeat for several quadrilaterals. Do you find a pattern?

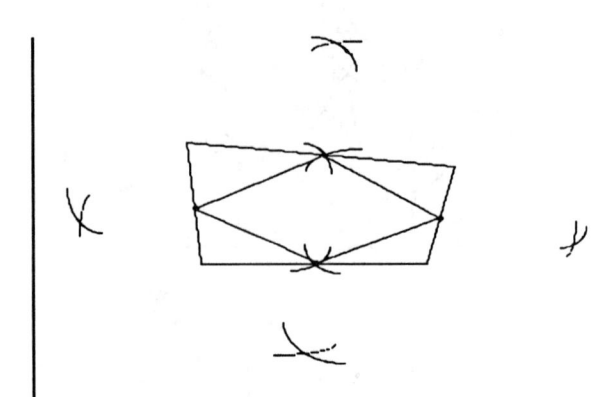

308 CHAPTER 12 Using Transformations in Geometry

Quadrilateral formed by the midpoints is a parallelogram.

6. Construct the perpendicular bisector of the segment below.

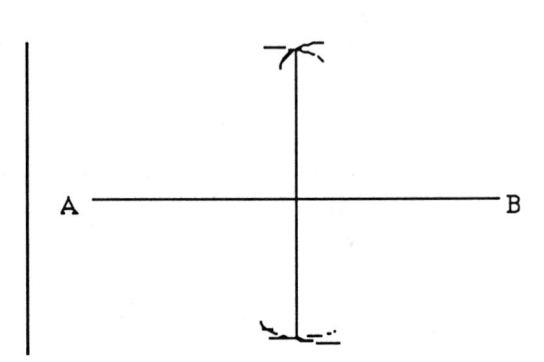

A ——————————— B

7. Construct a circle circumscribed about the triangle below.

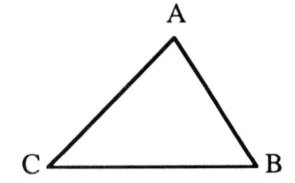

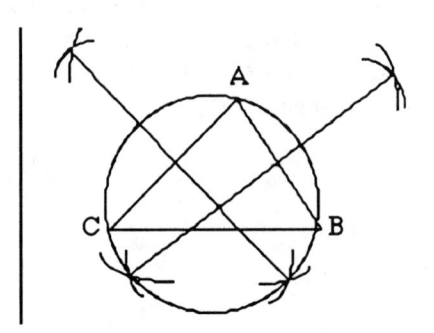

8. Construct an inscribed circle to the triangle in Exercise 7.

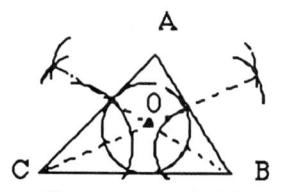

Construct angle bisector for $\angle B$
Construct angle bisector for $\angle C$
O is center of circle
Construct line through $O \perp BC$

9. Consider the following figure.

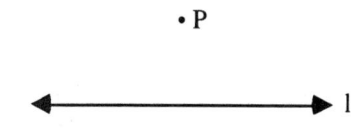

   • P

   ←——————→ l

   (a) Construct the line that contains P and is perpendicular to l. Call this line m.
   (b) Construct a line that contains P and is perpendicular to m. Call this line n.
   (c) What is the relationship between lines n and l?

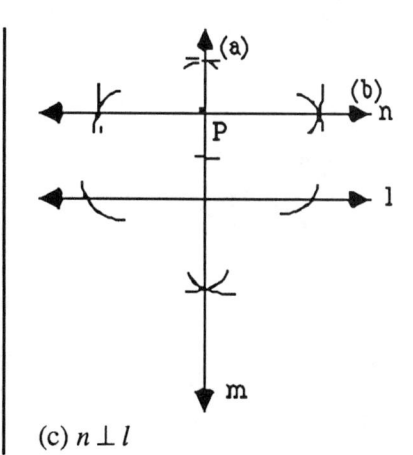

   (c) $n \perp l$

10. Construct two angles with their respective sides perpendicular. What can you hypothesize about the measures of the two angles?

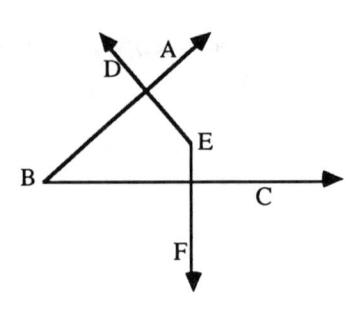

$\overline{AB} \perp \overline{DE}$ and $\overline{BC} \perp \overline{EF}$
$m(\angle ABC) + m(\angle DEF) = 180°$

11. Construct each of the following.

    (a) A square, given one of its sides
    (b) A square, given one of its diagonals

    (a) Copy a side and construct perpendicular lines

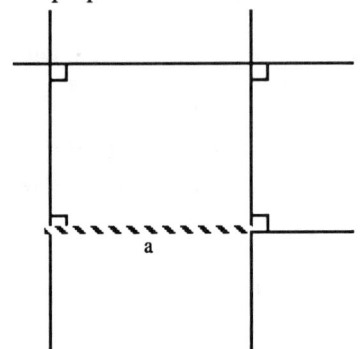

    (b) Diagonals are perpendicular and the same length

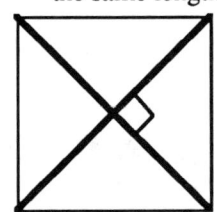

12. Construct a circle, several chords, and a perpendicular bisector for each chord. Make a conjecture about the perpendicular bisector of a chord and the center of the circle.

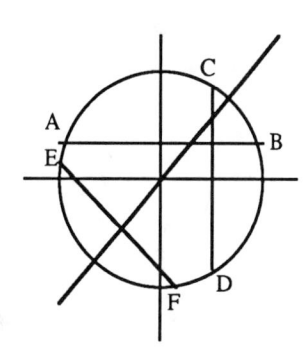

Section 3   Justifications of Constructions   311

Perpendicualr bisector will contain the center of the circle.

13. Find the center of the circle below.

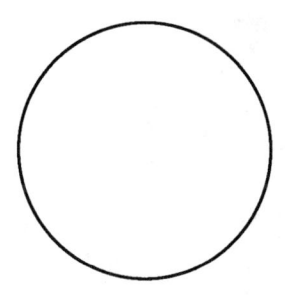

Use the results from Exercise 12.

14. The ACME Roller Skate Company wants to build its factory so that it is equidistant from three neighboring communities (point A, B, C in the diagram). How can the location of the factory be found?

City A

• City B

City C

(1) Connect points A, B, and C to form a triangle.
(2) Find perpendicular bisectors of $\overline{AB}$ and $\overline{AC}$.
(3) Consider point of intersection from (2).

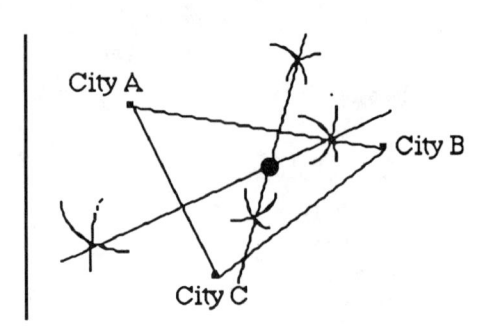

15. It is not possible to construct a regular n-gon if n

   (i) has an odd prime <u>not</u> of the form $2^k + 1$ in its prime factorization, or
   (ii) has the square of <u>any</u> odd prime in its prime factorization.

   (a) Is it possible to construct a 17-gon?
   (b) Is it possible to construct a 18-gon?
   (c) Which n-gons (for $3 \leq n \leq 30$) are constructible?

(a) yes
$17 = 1 \cdot 17$ $(17 = 2^4 + 1)$
(b) no
$18 = 2 \cdot 3^2$ ← square of an odd prime
(c) $n = 3, 4, 5, 6, 8, 10, 12, 15, 16, 17, 20, 24$

Section 4 Similar Geometric Figures

Cover the right side of the page and work on the left, then check your work

1. Assume that the following pairs of figures are similar, and find x.

   (a)

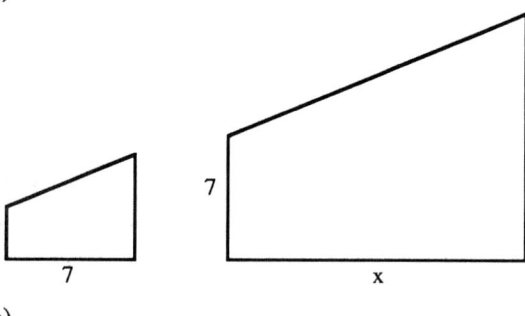

   (b)

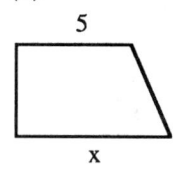

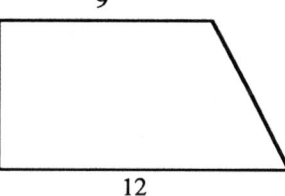

   (a) $x = 16\frac{1}{3} = 16.\overline{3}$
   $\dfrac{3}{7} = \dfrac{7}{x}$
   $3x = 49$
   $x = 16\frac{1}{3}$

   (b) $x = 6\frac{2}{3} = 6.\overline{6}$

2. The following pair of figures are similar. Find the area of the smaller figure.

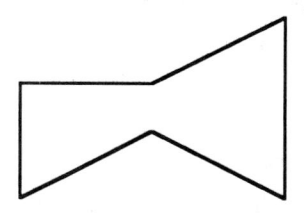

   Perimeter = 10 cm
   Area = 5 cm^2

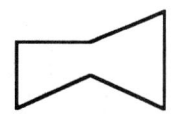

   Perimeter = 6 cm
   Area = ?

   Area = 1.8 cm^2

314 CHAPTER 12   Using Transformations in Geometry

> Scale factor for perimeter = $\frac{6}{10}$
>
> Scale factor for area = $(\frac{6}{10})^2 = \frac{36}{100}$

3. Make a quadrilateral similar to the one given, with a scale factor of 2.

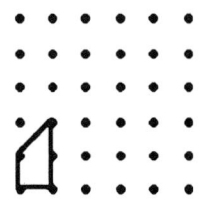

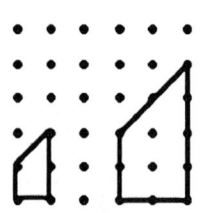

4. Two right triangles are similar, and the lengths of the sides of one triangle are one-half the lengths of the corresponding sides from the other triangle. How do the areas of the triangles compare?

> Areas are in ratio of 1 to 4.

5. A large egg is 7 cm long and a small egg is 5 cm long. If we assume that all the eggs are similar in shape, then which is the better buy, a dozen small eggs for 70¢ or a dozen large eggs for $1.05?

> Dozen large eggs
>
> Scale factor for length = $\frac{7}{5}$
>
> Scale factor for volume = $\left(\frac{7}{5}\right)^3 = 2.7$
>
> Scale factor for price = $\frac{105}{70} = 1.5$

6. Find x and y for each of the following.
   (a)

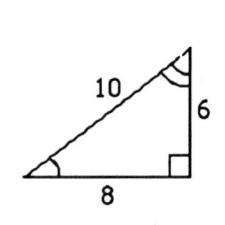

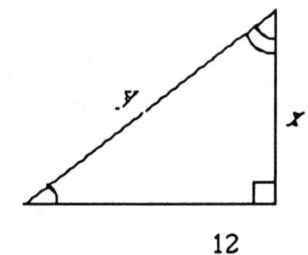

(b)

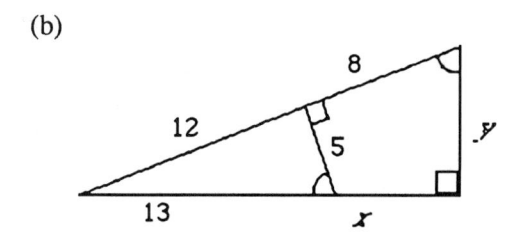

(a) $x = 9$
$y = 15$
(b) $x = 5.46$
$\dfrac{x + 13}{12} = \dfrac{70}{13}$
$x \approx 5.46$
$y = 7.69$
$\dfrac{y}{20} = \dfrac{5}{13}$
$y \approx 7.69$

7. Fill in the following chart.

Object	Shadow Length (L)	Height (h)	Ratio $\left(\dfrac{L}{h}\right)$
Flagpole	10 m	8 m	
Boy		152 cm	5:4
Shed	4 m		5:4
Bush	110 cm	88 cm	

Object	Shadow Length (L)	Height (h)	Ratio $\left(\dfrac{L}{h}\right)$
Flagpole	10 m	8 m	5:4
Boy	190 cm	152 cm	5:4
Shed	4 m	3.2 m	5:4
Bush	110 cm	88 cm	5:4

8. To determine the width of a river, Mr. Stream and his seventh grade class sighted right triangles $\triangle ABC$ and $\triangle ADE$ and made the following measurements: $\overline{BC} = 30$ m, $\overline{DE} = 70$ m, $\overline{AC} = 62$ m. Find the length x.

$x = 82.67$ m

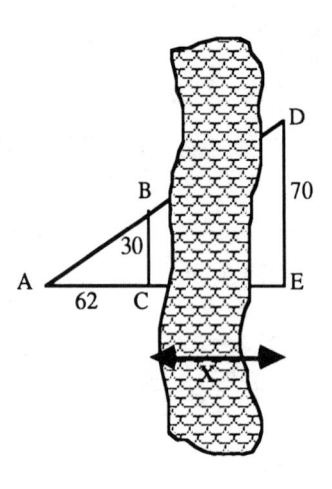

$$\frac{62 + x}{62} = \frac{70}{30}$$

$x \approx 82.67$

9. Explain why each of the following pairs of triangles are similar.

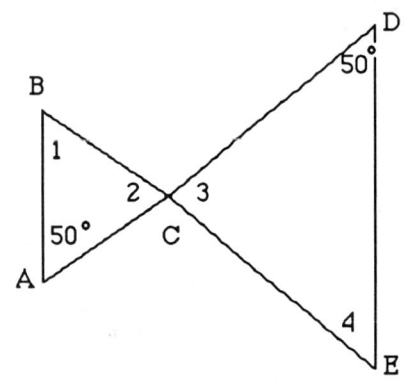

(a)

(b)

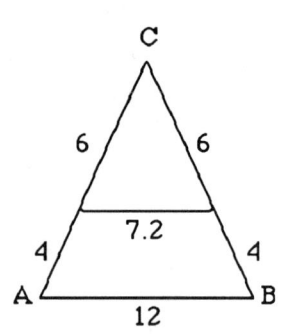

(a) Corresponding pairs of angles are congruent.

$\angle 2 \cong \angle 3$ Vertical angles
$\angle 1 + \angle A + \angle 2 = 180°$
$\angle 4 + \angle D + \angle 3 = 180°$
$\angle 1 + \angle A + \angle 2 = \angle 4 + \angle D + \angle 3$
$\angle 1 + \angle A + \angle 2 = \angle 4 + \angle A + \angle 2$
$\angle 1 = \angle 4$

(b) Corresponding sides are proportional. Ratio of 6 : 10.

10. Allen is 6 feet tall. He wants to locate a tree that is 150 feet tall. If his shadow is 10 feet long, what length shadow for a tree should he look for?

$$\text{Shadow} = 250 \text{ ft}$$
$$\frac{6}{10} = \frac{150}{s}$$
$$s = 250 \text{ ft}$$

11. Prove: Two isosceles triangles are similar if their nonbase angles are congruent.

   Answer follows:

$m(\angle B) = m(\angle E)$	Given
$m(\angle B) + m(\angle A) + m(\angle C) = m(\angle E) + m(\angle D) + m(\angle F)$	Sum of angle measures for a triangle = 180^0
$m(\angle A) = m(\angle C)$ and $m(\angle D) = m(\angle F)$	Base angles of isosceles triangle
$m(\angle B) + 2m(\angle A) = m(\angle E) + 2m(\angle D) = 180^0$	Substitution
$m(\angle B) + 2m(\angle A) = m(\angle B) + 2m(\angle D)$	Substitution
$2m(\angle A) = 2m(\angle D)$	Subtract $m(\angle B)$
$m(\angle A) = m(\angle D)$	Divide by 2
$\angle A \cong \angle D$	Definition of $\cong$ angles

   In a similar way we can show that $\angle C \cong \angle F$
   Thus, $\triangle ABC \cong \triangle DEF$

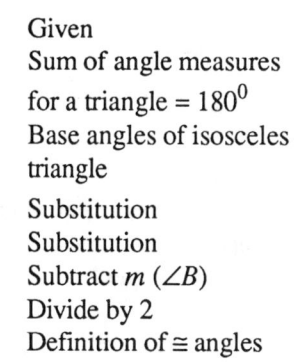

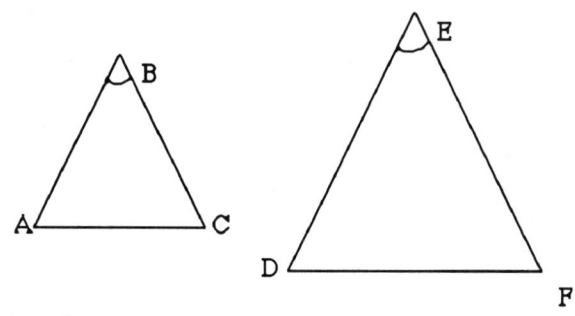

12. The sides of a pentagon are 4, 6, 8, 9, and 10 cm. If the longest side of a similar pentagon is 8 cm, find the lengths of the other four sides.

    3.2, 4.8, 6.4, 7.2, and 8

13. A pyramid in Egypt has a square base that measures 220 meters on a side and a height of 70 meters. If a scale model of the pyramid is to be built using a scale of 5 cm = 1 m, what are the dimensions of the model?

side = 1,100 cm
height = 350 cm

$$\frac{5 \text{ cm}}{1 \text{ m}} = \frac{s}{220 \text{ m}} \qquad \frac{5 \text{ cm}}{1 \text{m}} = \frac{h}{70 \text{ m}}$$

$s = 1{,}100$ cm $\qquad h = 350$ cm

14. A cylinder has a radius of 10 cm and a height of 20 cm. A similar cylinder has a surface area of 150π cm^2. What are the dimensions of the second cylinder?

$r = 5$ cm
$h = 10$ cm $= 600\pi$
SA $= 2\pi(10)^2 + 2\pi(10)(20)$
ratio of surface areas $= \dfrac{150\pi}{600\pi} = \dfrac{1}{4}$
ratio of dimensions $= \dfrac{1}{2}$

15. Give five examples from the physical world that imply similarity but not congruence.

Possible answers:

Photograph
Model airplane
Blueprints
Road map
Model electric train

Section 5 Topological Equivalence

Cover the right side of the page and work on the left, then check your work

1. Which of the following pairs of figures are topologically equivalent?

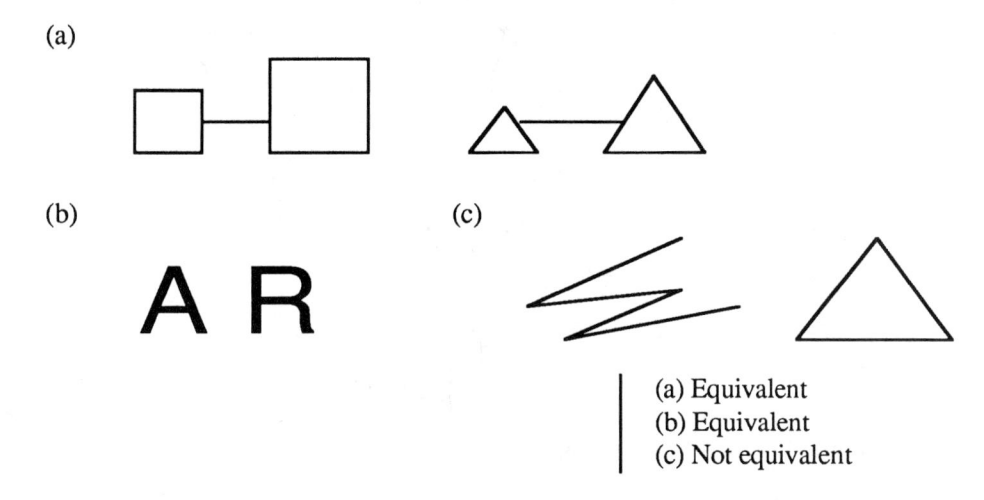

   (a)

   (b)

   A R

   (c)

       (a) Equivalent
       (b) Equivalent
       (c) Not equivalent

2. Complete the figure on the right so that it is topologically equivalent to the figure on the left.

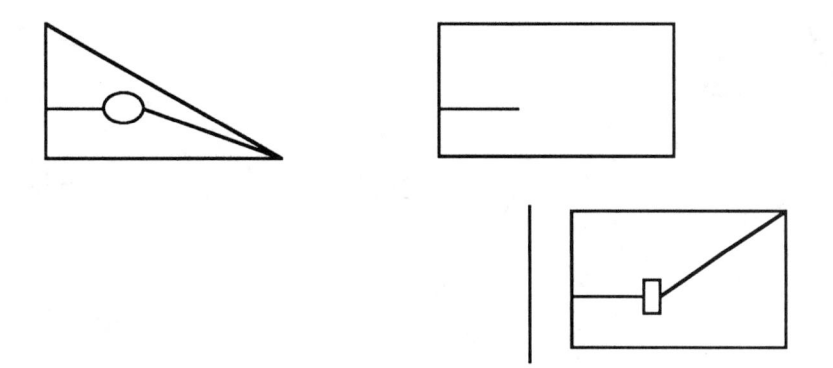

3. Are the two figures below topologically equivalent? Why or why not?

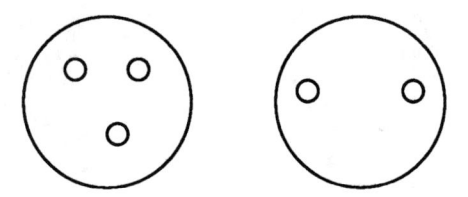

   Not equivalent. The figure on the left has three holes while the figure on the right has two holes.

320 CHAPTER 12 Using Transformations in Geometry

4. Which two of these figures are topologically equivalent?

(a)

(b)

(c)

(d)

| (a) and (d)
| (b) and (c)

5. Which of the figures below are topologically equivalent to the given figure?

(a)

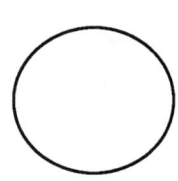

(b)

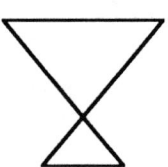

(c)

(d)

(e)

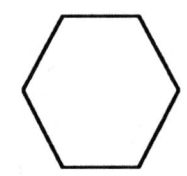

| (a), (d), (e)

Section 5   Topological Equivalence   321

6. How many cuts could be made on each figure without cutting the figure into two pieces?

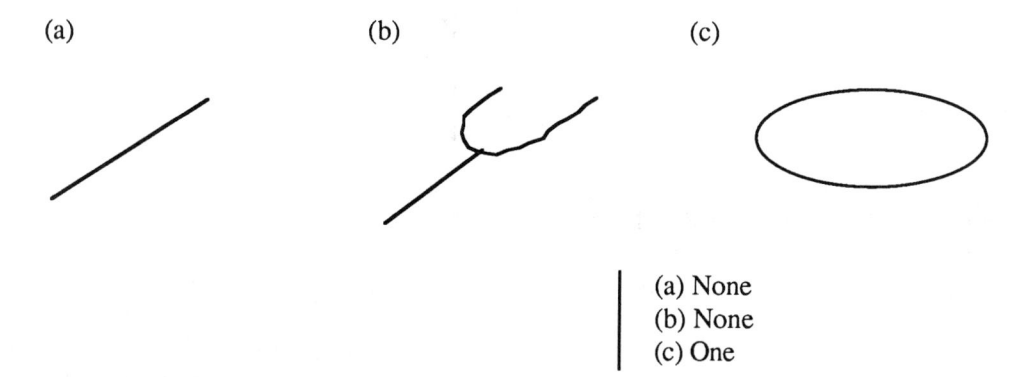

(a)   (b)   (c)

(a) None
(b) None
(c) One

7. Figures can be classified according to their <u>genus</u>, the number of cuts that can be made without cutting the figure into two pieces. What is the genus of the figures in Exercise 6?

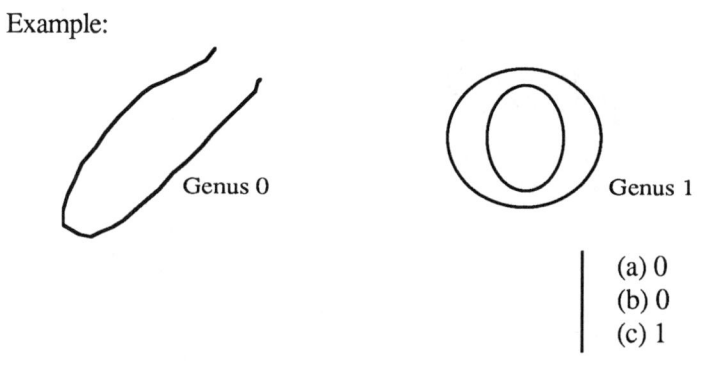

Example:

Genus 0   Genus 1

(a) 0
(b) 0
(c) 1

8. Draw three figures that have genus of 1.

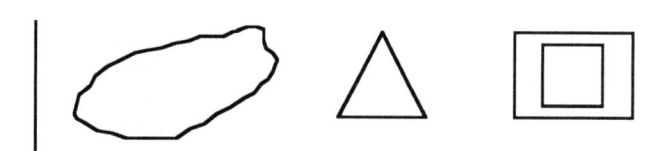

9. Draw three figures that have genus of 2.

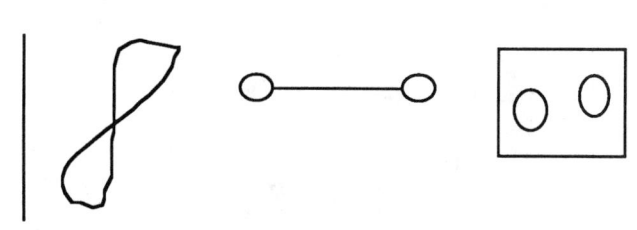

10. Draw a figure that has genus of 3.

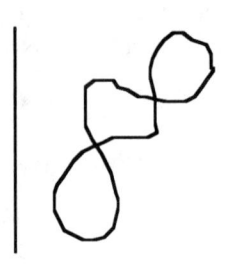

11. From the results of Exercises 8 - 11, what generalization can you make concerning the genus of a figure and the number of "holes" in the figure?

The genus of a figure appears to be directly related to the number of "holes" the figure has.

Section 6 Traversing a Network

Cover the right side of the page and work on the left, then check your work

1. Complete the table for networks (a) through (e)

Network	Number of Even Vertices	Number of Odd Vertices	Traversable (Yes or No)
(a)			
(b)			
(c)			
(d)			
(e)			

   (a)   (b)   (c)

   (d)   (e)

Network	Number of Even Vertices	Number of Odd Vertices	Traversable (Yes or No)
(a)	2	4	no
(b)	4	0	yes
(c)	1	2	yes
(d)	0	4	no
(e)	0	5	no

2. Classify each of the traversable paths from Exercise 1 as an Euler circuit or an Euler path.

   (b) Euler circuit
   (c) Euler path

3. For each traversable path from Exercise 1, find a path.

324 CHAPTER 12   Using Transformations in Geometry

(b)

(c)

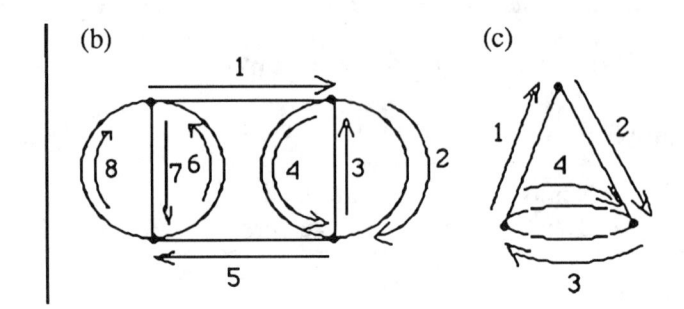

4. Is it possible to take an entire trip through the houses whose floor plans are shown below and pass through each door once and only once?

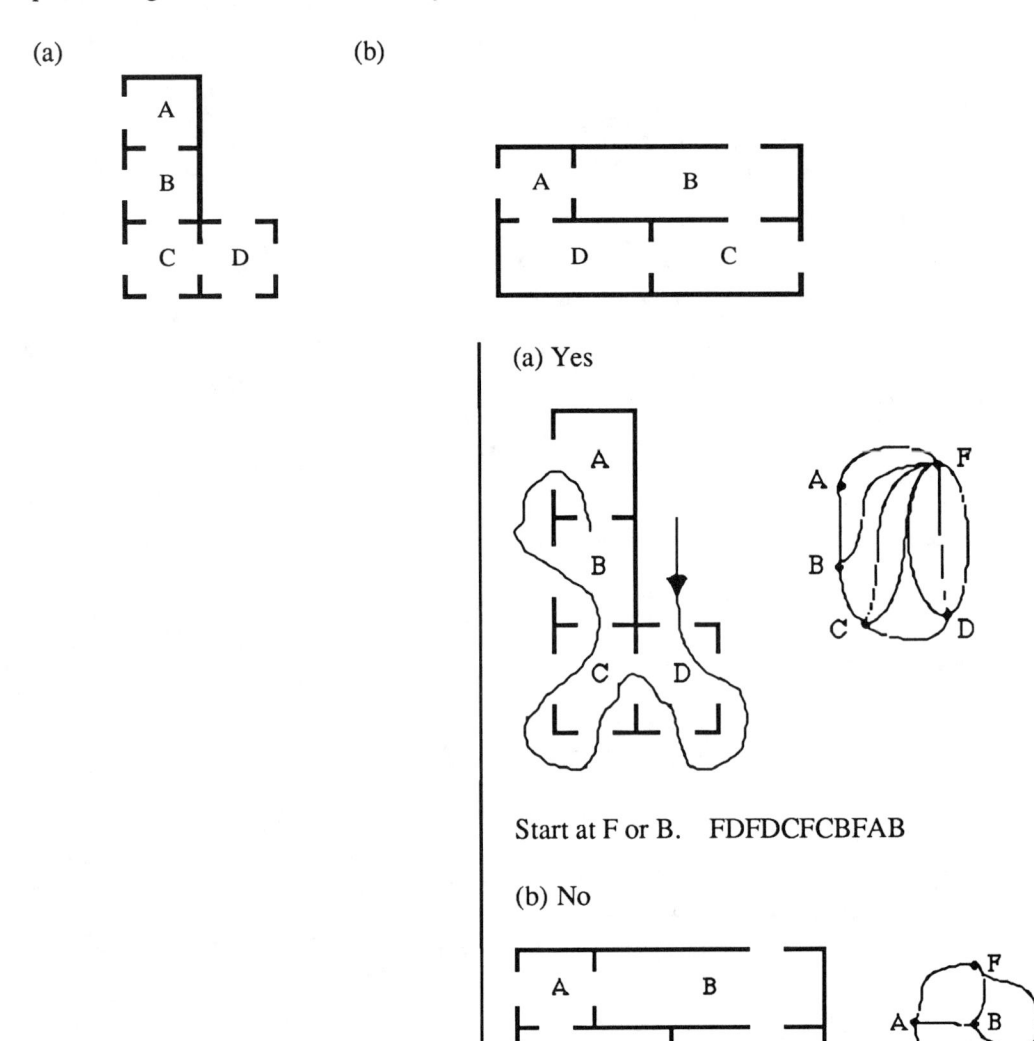

(a)

(b)

(a) Yes

Start at F or B. FDFDCFCBFAB

(b) No

Not traversable.

5. Below is a drawing representing the floor plan of a historic house. If you must start at the entrance, could you see every room but not go through any door twice?

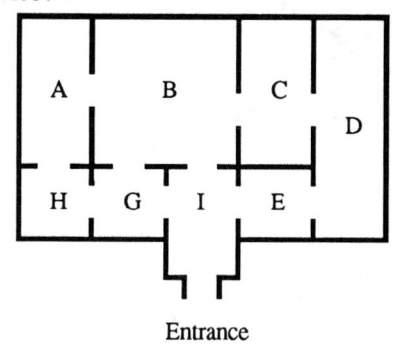

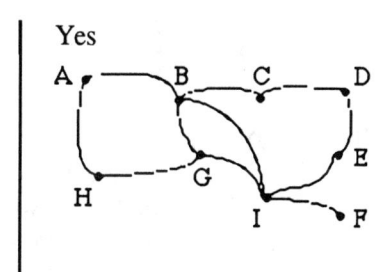

Yes

Start at F.   FIEDCBIGBAHG

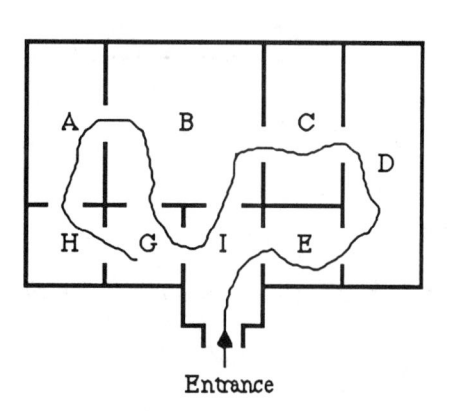

6. For the networks in Exercise 1, complete the following table where R is the number of interior and exterior regions, V is the number of vertices, and A is the number of arcs.

   Network   R V   A $V + R - A$
   (a)
   (b)
   (c)
   (d)
   (e)

Network	R	V	A	$V + R - A$
(a)	6	6	10	2
(b)	6	4	8	2
(c)	3	3	4	2
(d)	4	4	6	2
(e)	5	4	7	2

326　CHAPTER 12　　Using Transformations in Geometry

7. Are the following rooms traversable?

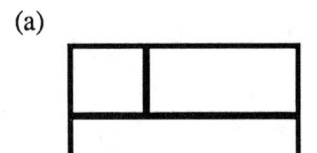

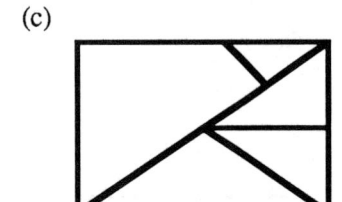

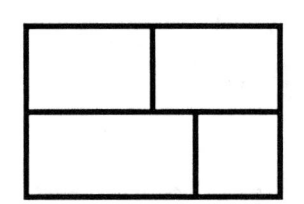

(a)　　　　　　　　　(b)

(c)　　　　　　　　　(d)

(a) Yes.　All even vertices
(b) No.　More than two odd vertices
(c) No.　More than two odd vertices
(d) Yes.　Two odd vertices

8. Find a path for each traversable drawing in Exercise 7.

Answers follow:
(a)　　　　　　　　　(d)

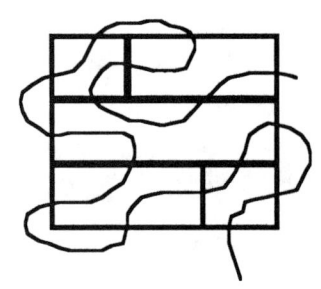

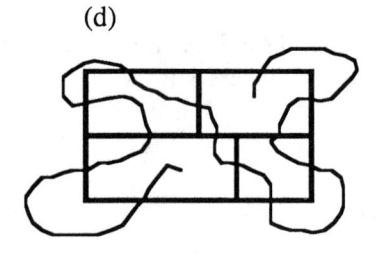

9. Use each drawing in Exercise 7 as a map. Use the Largest First Algorithm to color the map.

(a)

(b)

(c)

(d)

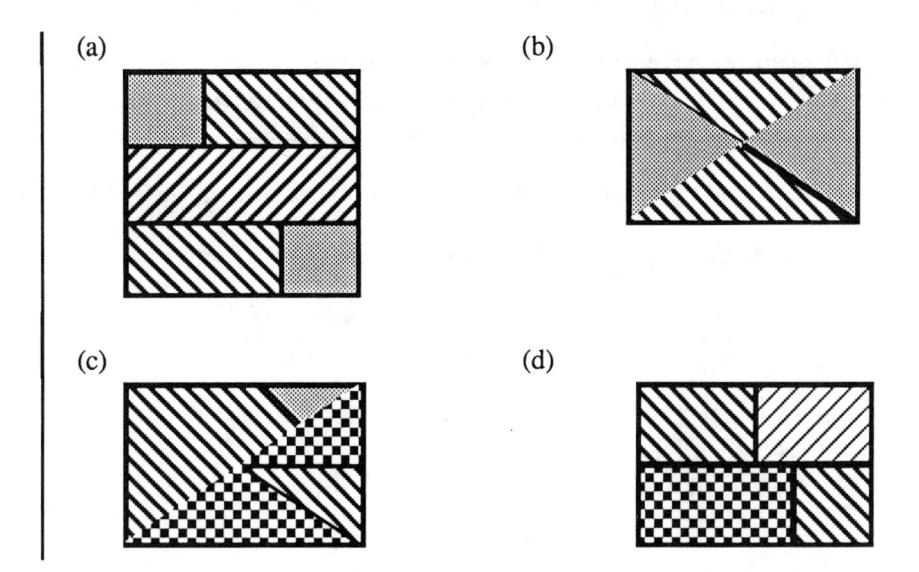

10. Below is a simplified map of a city. Subway connections are shown for each borough. Is it possible to travel the system and use each subway exactly once? Use a network diagram to decide.

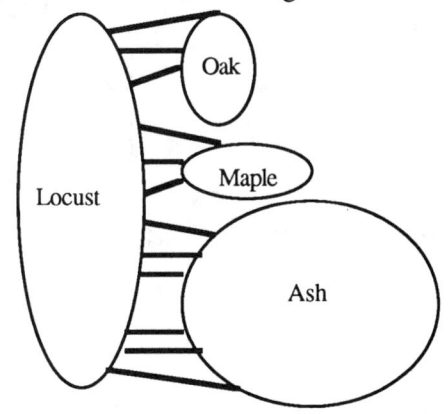

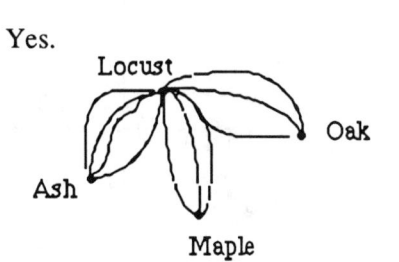

Yes.

There are two odd vertices. You can begin at Oak or Maple. You could go in the following order.
Oak, Locust, Oak, Locust, Ash, Locust, Ash, Locust, Maple, Locust, Maple

Chapter 13
Coordinate Geometry and Transformations

Section 1 Coordinate Geometry

Cover the right side of the page and work on the left, then check your work

1. Plot the following points on a coordinate system.

   (a) (3, 4)
   (b) (5, ⁻2)
   (c) (⁻3, ⁻4)
   (d) (⁻1, 5)

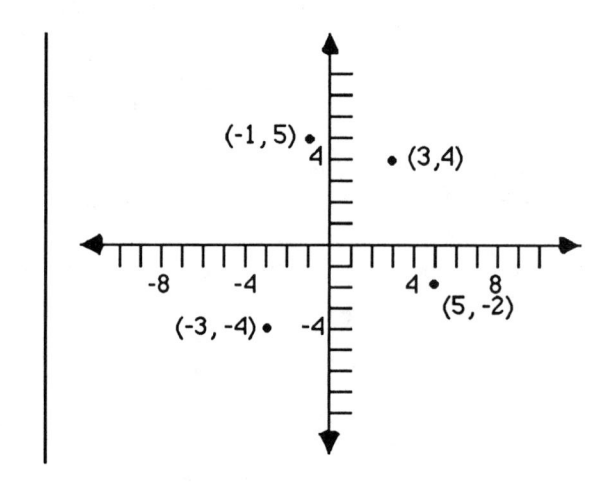

2. Give the coordinates for each of the points A, B, C, D, and E.

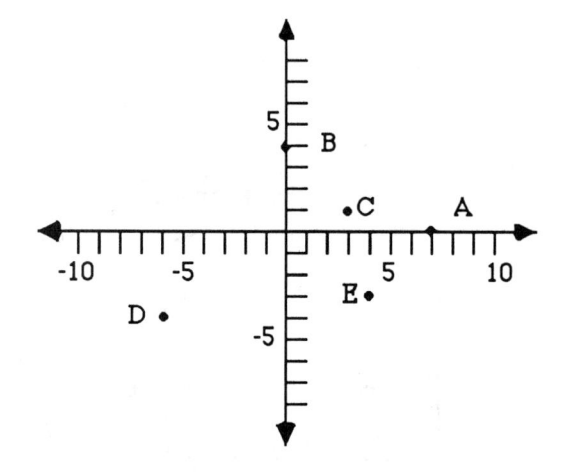

   A (7, 0)
   B (0, 4)
   C (3, 1)

Section 1 Coordinate Geometry 329

$D\ (^-6,\ ^-4)$
$E\ (4,\ ^-3)$

3. Describe all the points contained in the shaded region.

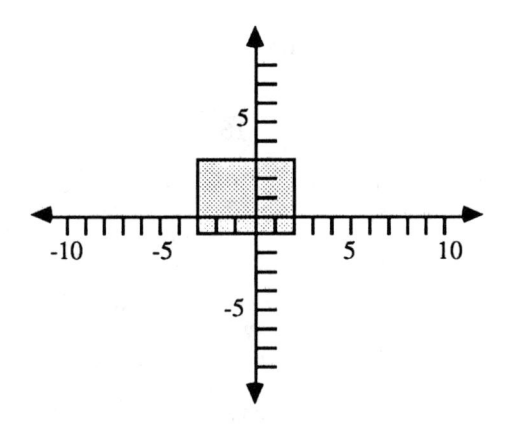

Points whose x-coordinates are between $^-3$ and 2 and whose y-coordinates are between $^-1$ and 3.

4. Find the distance between the given points.

   (a) $A\ (4,\ 5)$ and $B\ (4,\ 9)$
   (b) $A\ (5,\ 2)$ and $B\ (^-3,\ 4)$
   (c) $A\ (^-1,\ 2)$ and $B\ (2,\ ^-4)$
   (d) $A\ (2.2,\ ^-3.5)$ and $B\ (^-1.4,\ 2.6)$

(a) 4
(b) $\sqrt{68} \approx 8.25$
$$d = \sqrt{(5-(^-3))^2 + (2-4)^2}$$
$$= \sqrt{8^2 + (^-2)^2} = \sqrt{68}$$
(c) $\sqrt{45} \approx 6.71$
(d) $\sqrt{49.46} \approx 7.01$

5. Find the midpoint of the segment joining each pair of points in Exercise 4.

(a) $M\ (4,\ 7)$
(b) $M\ (1,\ 3)$
$$M = (\frac{5 + (^-3)}{2},\ \frac{2 + 4}{2})$$
$$= (\frac{2}{2},\ \frac{6}{2}) = (1,\ 3)$$

330　CHAPTER 13　Coordinate Geometry and Transformations

(c) $M\left(\frac{1}{2}, ^-1\right)$

(d) $M\ (0.35,\ ^-0.45)$

6. Show that $\triangle ABC$ is an isosceles triangle:
   A $(^-2, ^-5)$, B $(1, ^-1)$, C $(5, 2)$.

   Is isosceles.
   $d_{AB} = \sqrt{(^-2 - 1)^2 + (^-5 - (^-1))^2}$
   $= \sqrt{9 + 16} = 5$
   $d_{BC} = \sqrt{(1 - 5)^2 + (^-1 - 2)^2}$
   $= \sqrt{16 + 9} = 5$
   $d_{AC} = \sqrt{(^-2 - 5)^2 + (^-5 - 2)^2}$
   $= \sqrt{49 + 49} = \sqrt{98}$

7. Show that $\triangle RST$ is a right triangle: $R\ (0, 6)$, $S\ (^-3, 0)$, $T\ (9, ^-6)$.

   Is a right triangle
   $d_{RS} = \sqrt{(0 - (^-3))^2 + (6 - 0)^2}$
   $= \sqrt{9 + 36} = \sqrt{45}$
   $d_{ST} = \sqrt{(^-3 - 9)^2 + (0 - (^-6))^2}$
   $= \sqrt{144 + 36} = \sqrt{180}$
   $d_{RT} = \sqrt{(0 - 9)^2 + (6 - (^-6))^2}$
   $= \sqrt{81 + 144} = \sqrt{225}$
   $(d_{RT})^2 = 225$
   $(d_{RS})^2 = 45$
   $(d_{ST})^2 = 180$
   $(d_{RT})^2 = (d_{RS})^2 + (d_{ST})^2$

8. Find the perimeter and area of the triangle with vertices $(2, 2)$, $(2, 10)$, and $(7, 6)$.

   $P = 8 + 2\sqrt{41}$
   $A = 20$
   $d_1 = \sqrt{(2 - 2)^2 + (2 - 10)^2} = 8$
   $d_2 = \sqrt{(2 - 7)^2 + (10 - 6)^2}$
   $= \sqrt{25 + 16} = \sqrt{41}$
   $d_3 = \sqrt{(2 - 7)^2 + (2 - 6)^2} = \sqrt{41}$
   $P = d_1 + d_2 + d_3 = 8 + 2\sqrt{41}$
   Area $= \frac{1}{2} \cdot 8 \cdot h$
   Area $= \frac{1}{2}(8)(5) = 20$

Section 1   Coordinate Geometry   331

9. One endpoint of a segment is (3, ⁻1). The midpoint of the segment is (⁻2, 5). Find the coordinates of the other endpoint.

$M\ (^-7,\ 11)$
$M = \left(\dfrac{3+x}{2}, \dfrac{-1+y}{2}\right) = (^-2,\ 5)$
$\dfrac{3+x}{2} = ^-2 \qquad \dfrac{-1+y}{2} = 5$
$3 + x = ^-4 \qquad ^-1 + y = 10$
$x = ^-7 \qquad\qquad y = 11$

10. Use the distance formula to show that the following points lie on the same line.
    A (1,2), B (⁻1, ⁻4), C (3, 8)

Points are collinear
$d_{AB} = \sqrt{40} \quad d_{BC} = \sqrt{160}$
$d_{AC} = \sqrt{40}$
$d_{AB} + d_{AC} = 2\sqrt{40} = \sqrt{4 \cdot 40} = \sqrt{160} = d_{BC}$

11. Using the following figure, show that the midpoint M of the hypotenuse of the right triangle is equidistant from the vertices of the triangle.

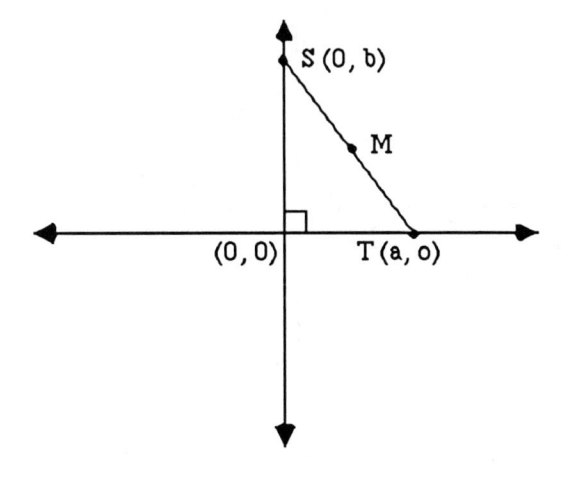

S (0, b)

M

(0, 0)   T (a, 0)

$M = \left(\dfrac{a}{2},\ \dfrac{b}{2}\right)$

$$d_{MS} = \sqrt{\left(\frac{a}{2}-0\right)^2 + \left(b-\frac{b}{2}\right)^2}$$

$$= \sqrt{\frac{a^2}{4} + \frac{b^2}{4}} = \frac{\sqrt{a^2+b^2}}{2}$$

$$d_{MT} = \sqrt{\left(\frac{a}{2}-a\right)^2 + \left(\frac{b}{2}-0\right)^2}$$

$$= \sqrt{\frac{a^2}{4} + \frac{b^2}{4}} = \frac{\sqrt{a^2+b^2}}{2}$$

$$d_{MO} = \sqrt{\left(\frac{a}{2}-0\right)^2 + \left(\frac{b}{2}-0\right)^2}$$

$$= \sqrt{\frac{a^2}{4} + \frac{b^2}{4}} = \frac{\sqrt{a^2+b^2}}{2}$$

12. Given the points $A\ (3, 4)$ and $B\ (6, 2)$, find another point C so that $\triangle ABC$ is a right triangle.

Plot the points. Choose point $(3, 2)$ as the other vertex.

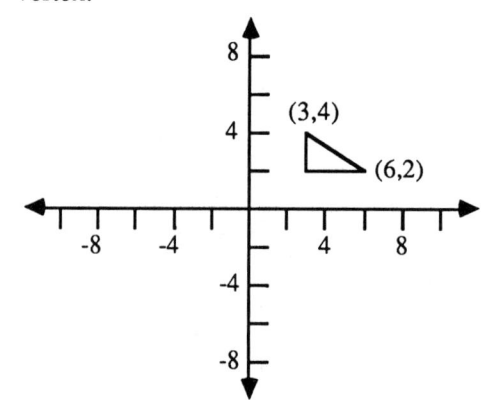

13. Which of the following properties are true for the trapezoid shown below.

   (a) The diagonals have the same length.
   (b) The diagonals are perpendicular to each other.
   (c) The diagonals bisect each other.

(a) True
(b) False
(c) True

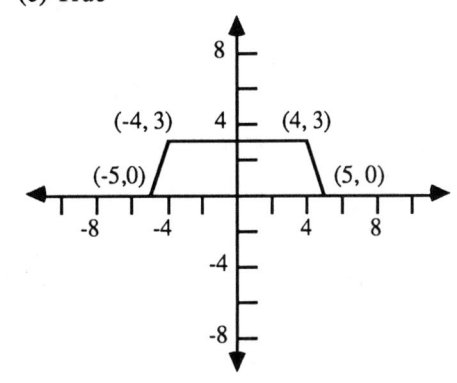

14. Consider the following diagram showing distances between cities. Use coordinates to show which path from city A to city D, through city B or city C, is shorter.

B is 40 miles west of A
B is 180 miles north of A
C is 230 miles east of A
C is 40 miles south of A
D is 360 miles east of A
D is 200 miles north of A

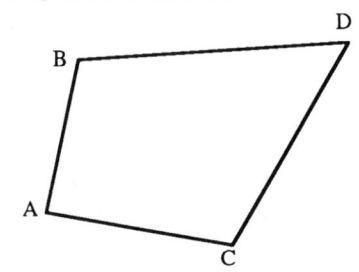

Place the diagram on a coordinate system with city A as the origin.

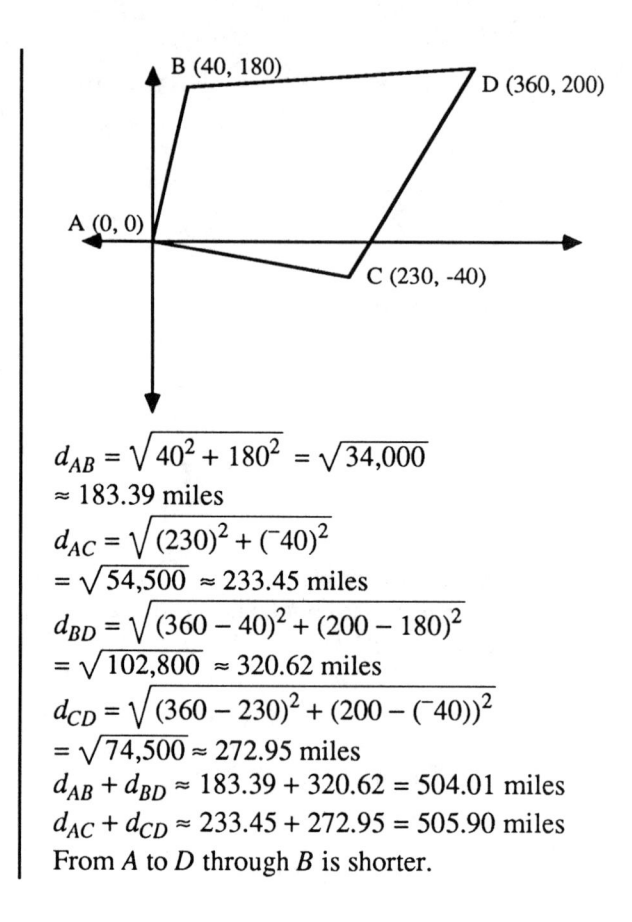

334   **CHAPTER 13**   Coordinate Geometry and Transformations

B (40, 180)
D (360, 200)
A (0, 0)
C (230, -40)

$d_{AB} = \sqrt{40^2 + 180^2} = \sqrt{34{,}000}$
≈ 183.39 miles

$d_{AC} = \sqrt{(230)^2 + (^-40)^2}$
$= \sqrt{54{,}500} \approx 233.45$ miles

$d_{BD} = \sqrt{(360 - 40)^2 + (200 - 180)^2}$
$= \sqrt{102{,}800} \approx 320.62$ miles

$d_{CD} = \sqrt{(360 - 230)^2 + (200 - (^-40))^2}$
$= \sqrt{74{,}500} \approx 272.95$ miles

$d_{AB} + d_{BD} \approx 183.39 + 320.62 = 504.01$ miles
$d_{AC} + d_{CD} \approx 233.45 + 272.95 = 505.90$ miles

From A to D through B is shorter.

15. Find the distance from the midpoint of the segment joining $A\ (1, 3)$ and $B\ (3, 5)$ to the midpoint of the segment joining $C\ (4, 6)$ and $D\ (^-2,\ 5)$.

Distance from $M_1\ (2, 4)$ to $M_2\ (1, 5.5) = \sqrt{3.25}$

Section 2 An Introduction to Reflections or "Flips"

Cover the right side of the page and work on the left, then check your work

1. Find the images of the given figures or points after a reflection about line *l*.

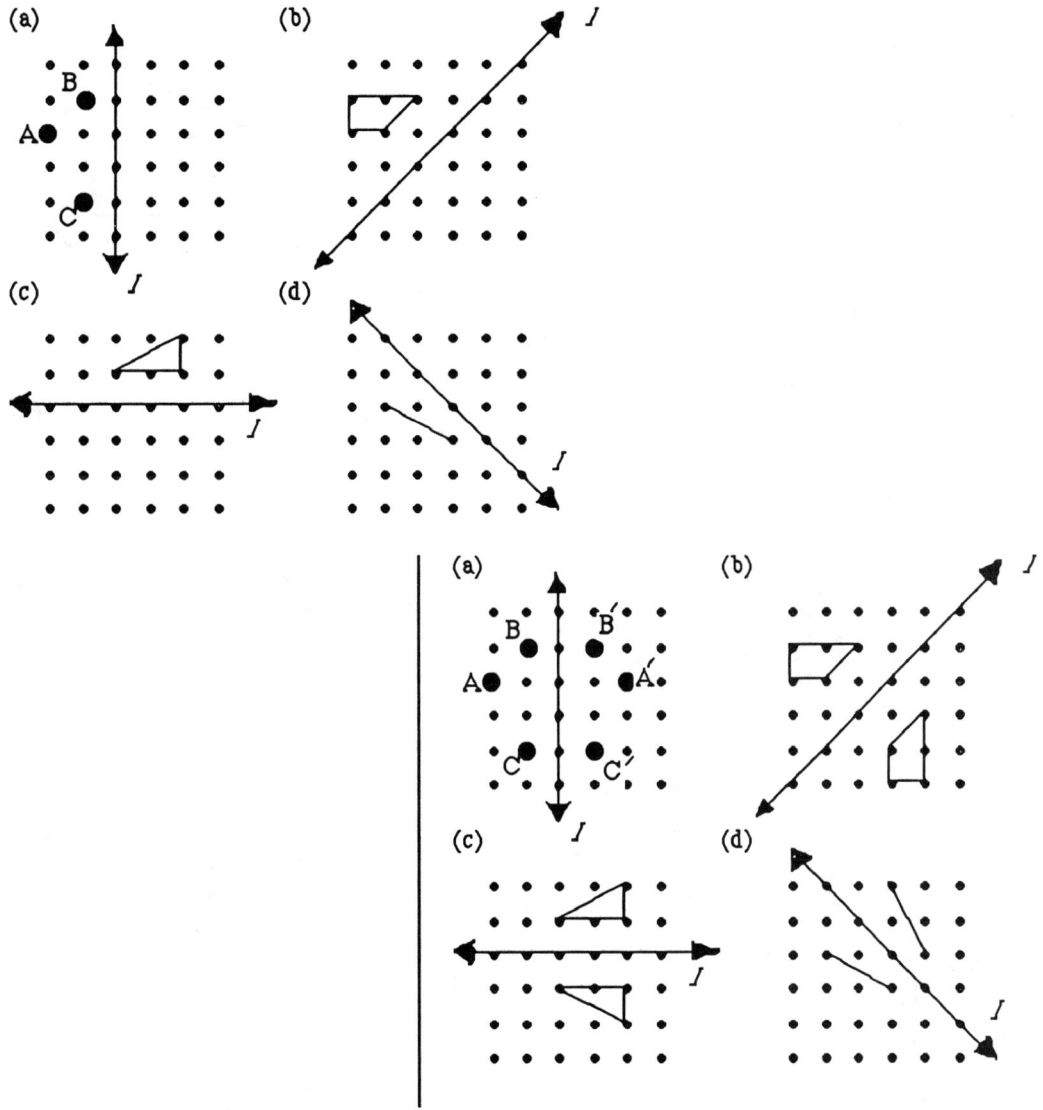

    (a)   (b)

    (c)   (d)

    (a)   (b)

    (c)   (d)

2. Find the reflection image of △ ABC with respect to *l* by

    (a) using tracing paper and folding.
    (b) using geometric construction

336 CHAPTER 13 Coordinate Geometry and Transformations

(b)

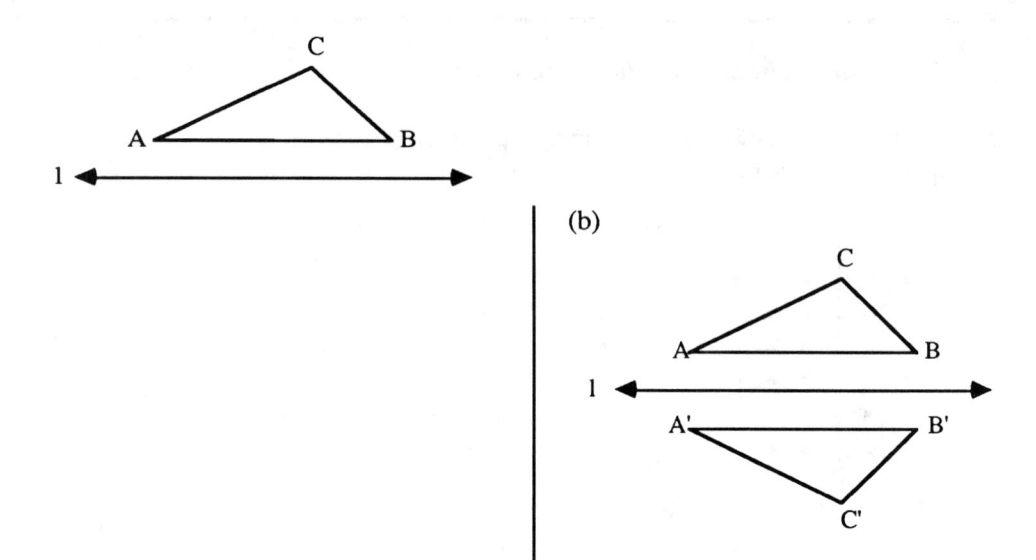

3. Find the image of the figure after a reflection about line *l*.

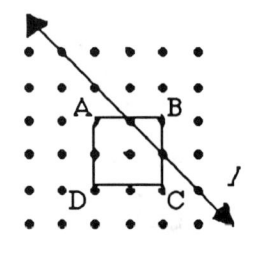

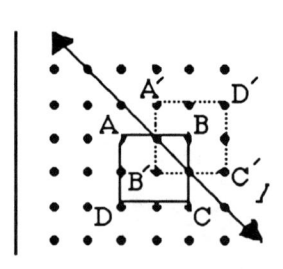

Copy the drawings that follow. Then reflect △ABC over line *l*.

4.

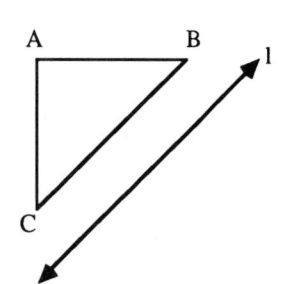

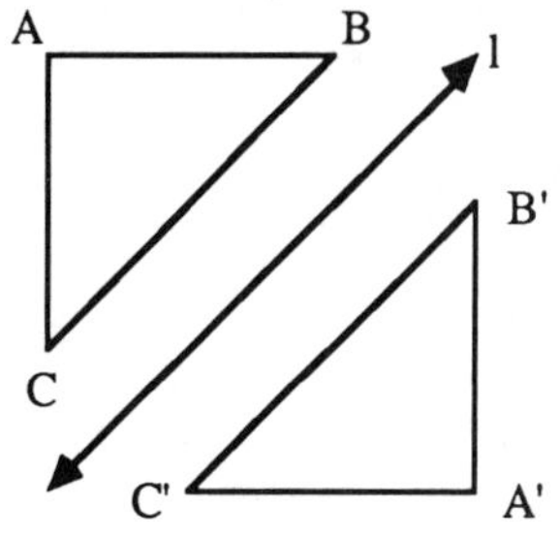

Section 2 An Introduction to Reflections or "Flips" **337**

5.

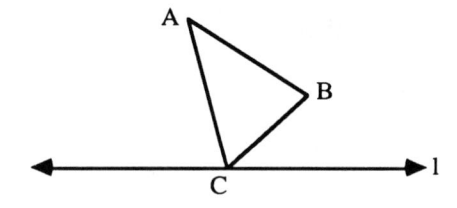

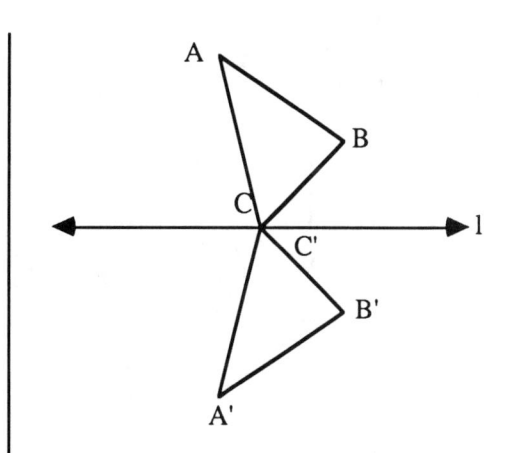

6.

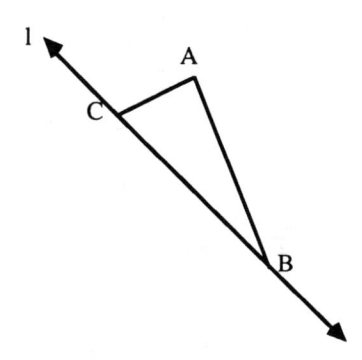

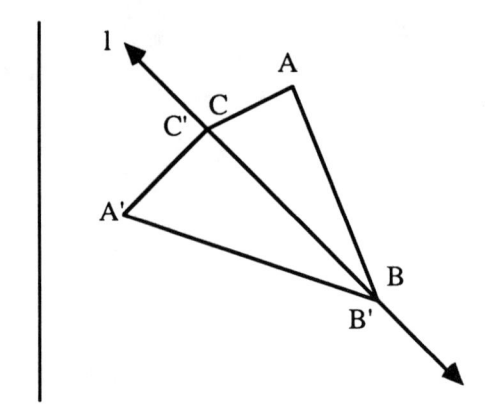

7.

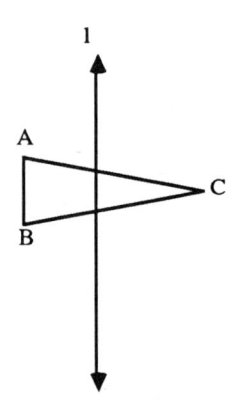

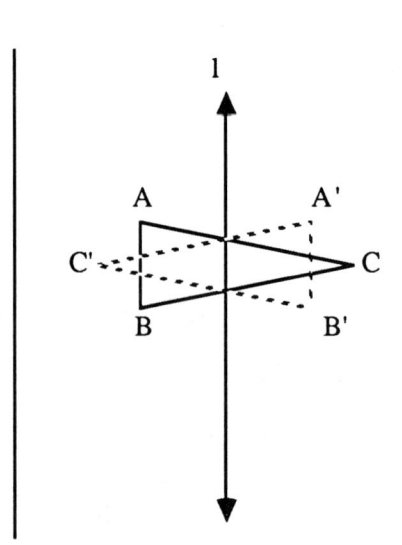

8. Which of the following figures have a reflecting line so that the image over the line is the figure itself?

   (a) circle
   (b) rectangle
   (c) isosceles triangle
   (d) arc

(e) scalene triangle

(a) Yes Any line passing through the center of the circle
(b) Yes Line must be the perpendicular bisector of opposite sides
(c) Yes Line must be the perpendicular bisector of the side that is not congruent to the other two sides
(d) Yes
(e) No

9. Show the reflection of each figure over the line *l*.

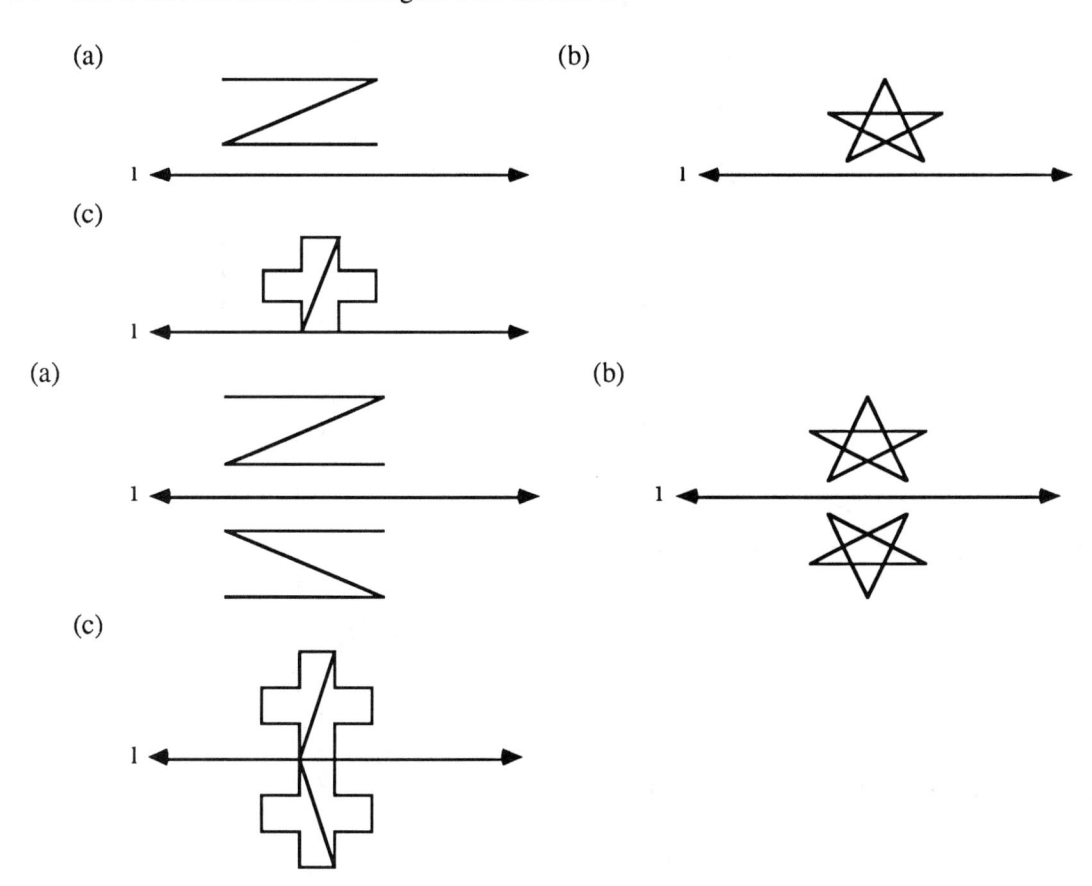

   (a)

   (b)

   (c)

(a)

(b)

(c)

10. Reflect ∠ ABC over line *l* to obtain ∠A'B'C'. Then reflect ∠A'B'C' over line *m* to obtain ∠A"B"C". Lines *l* and *m* are parallel.

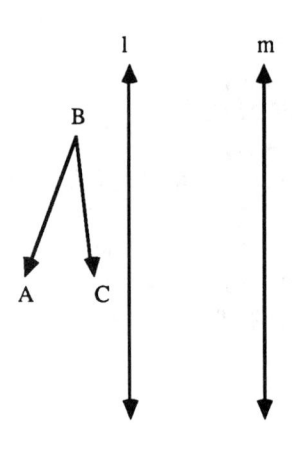

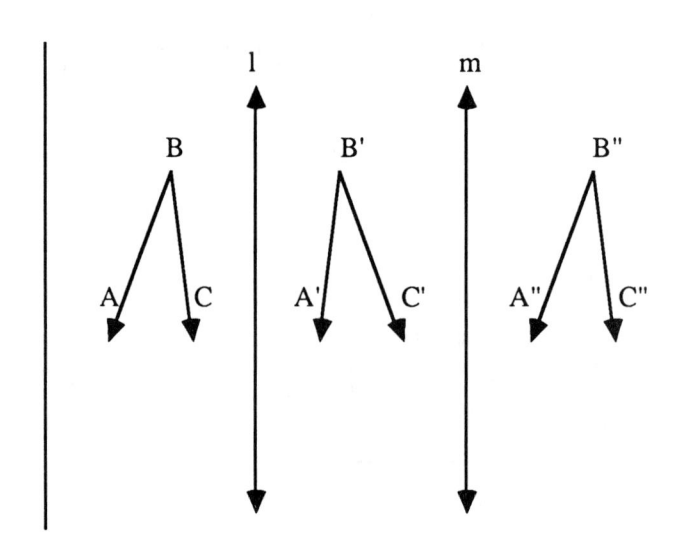

340 CHAPTER 13 Coordinate Geometry and Transformations

11. How do ∠ABC and ∠A″B″C″ from Exercise 10 compare?

    They are congruent.

12. Find the image coordinates of the image reflected about the x-axis. Draw the figures.

    A (3, 2), B (2, 5), C (8, 1), D (6, 7)

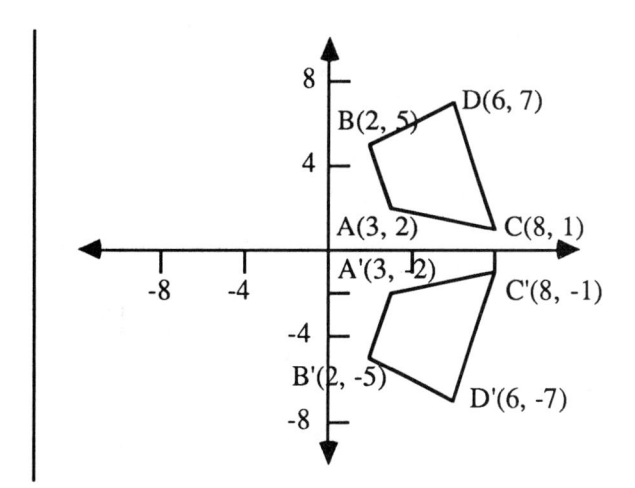

13. Find the coordinates of the image reflected about the y-axis. Draw the figures.

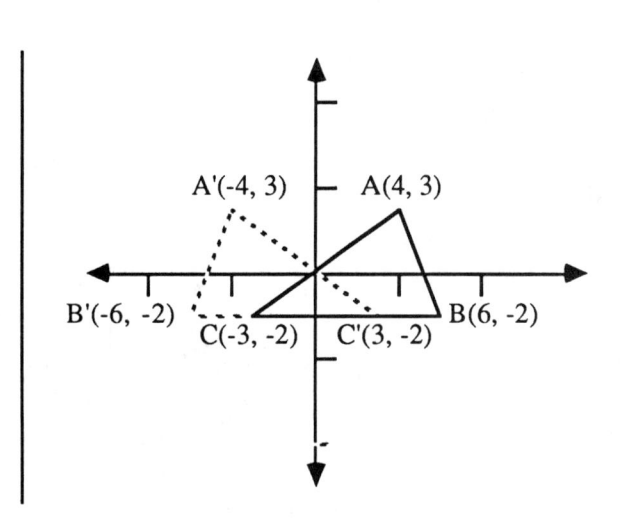

    A (4, 3), B (6, ⁻2), C (⁻3, ⁻2)

14. Reflect the figure from Exercise 12 about the line $y = x$.

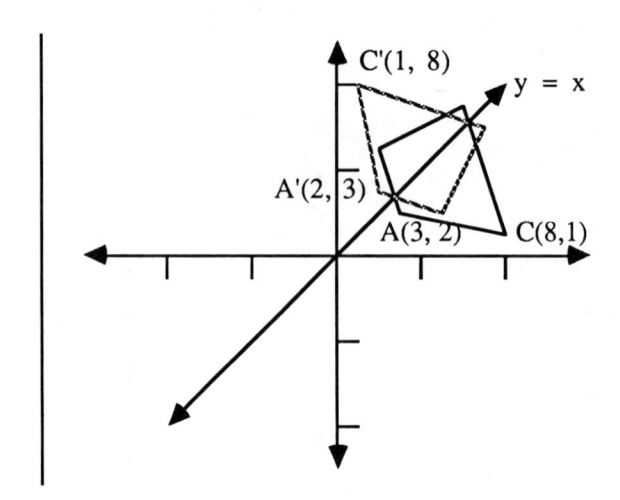

342 CHAPTER 13 Coordinate Geometry and Transformations

15. Reflect the figure from Exercise 13 about the line $y = -x$.

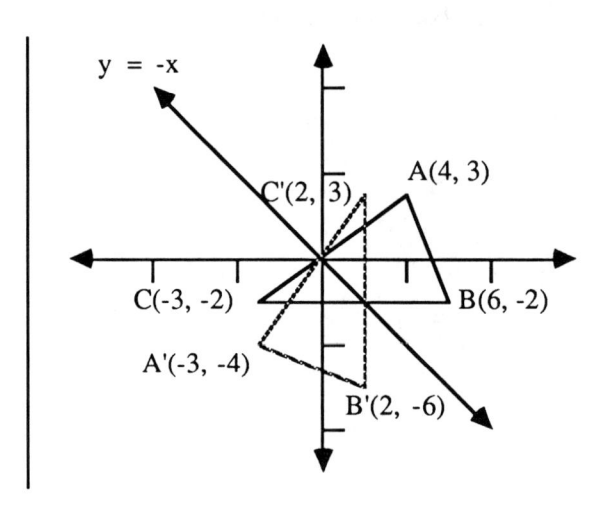

Section 3 Slides or Translations

Cover the right side of the page and work on the left, then check your work

1. Perform the translations $T_{\overrightarrow{AB}}$ on each of the following figures.

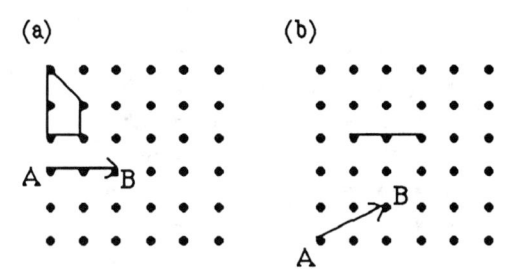

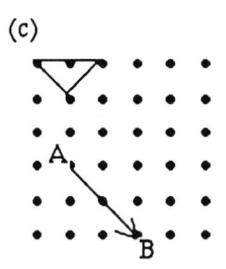

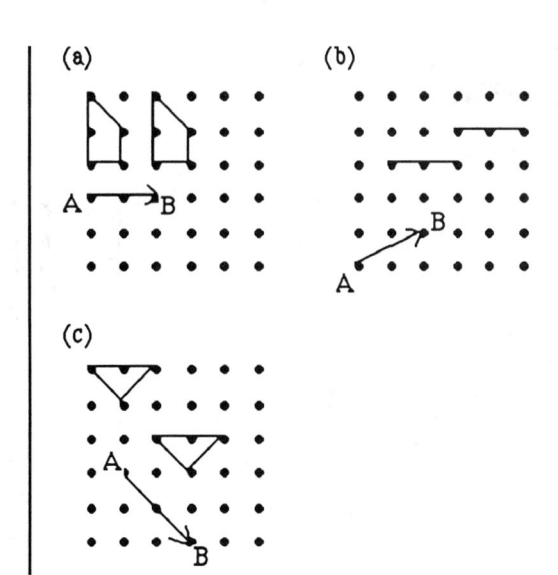

   (a)

   (b)

   (c)

2. Perform the translation $T_{\overrightarrow{RS}}$ on the geometric figures from Exercise 1.

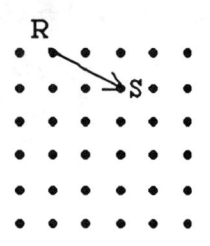

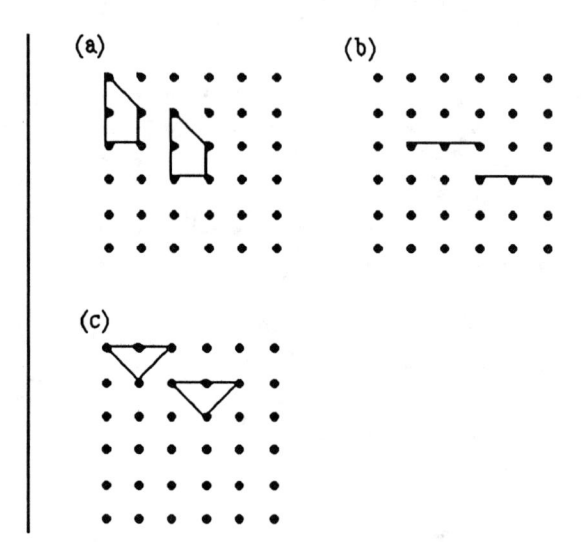

3. Perform $T_{1,-2}$ on the figures from Exercise 1.

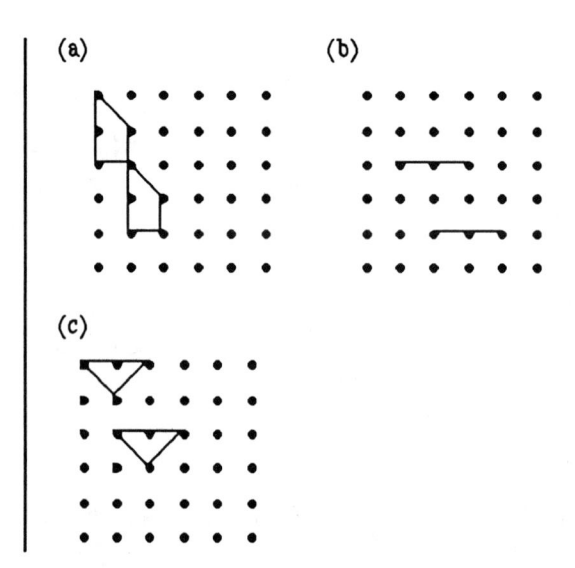

4. Use construction to find the slide image of $\overline{RS}$ under the translation $T_{\overrightarrow{AB}}$.

Section 3 Slides or Translations

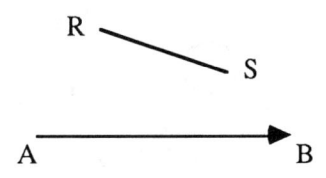

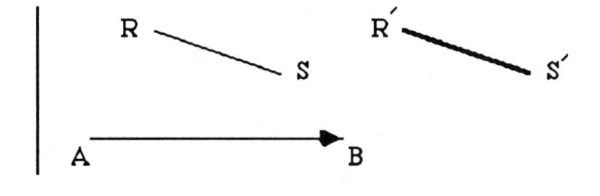

5. Draw the slide image of each of the following.
   (a)

   (b)

   (c)

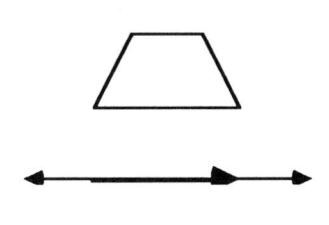

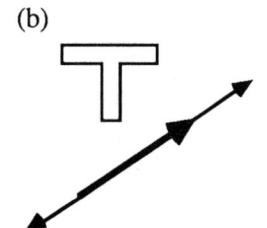

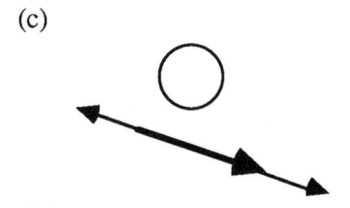

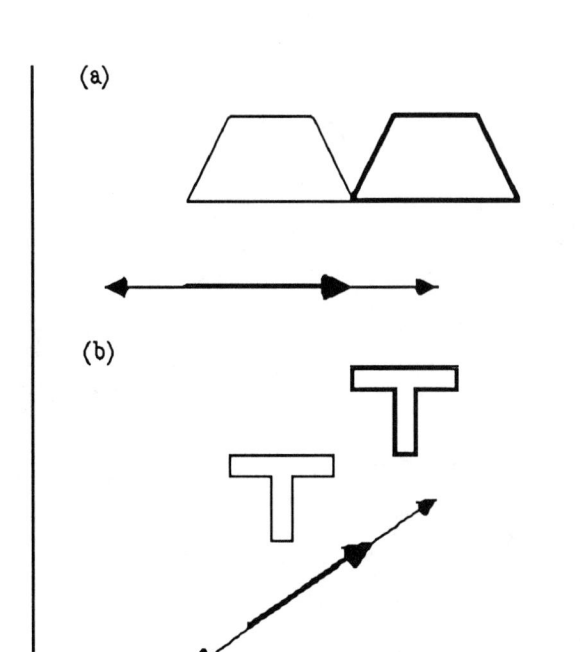

(a)

(b)

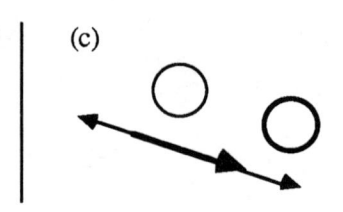

6. Slide each figure 3 centimeters in the direction of $\overline{AB}$.

(a)

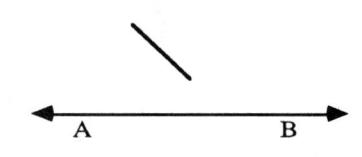

(b)

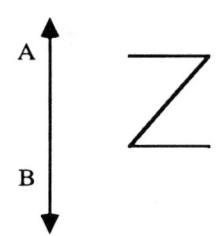

(c)

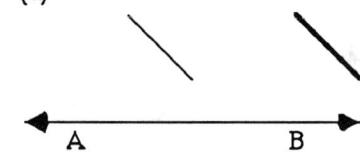

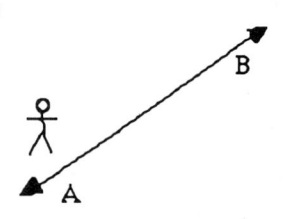

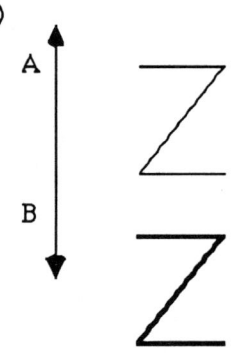

Section 3   Slides or Translations   347

(c)

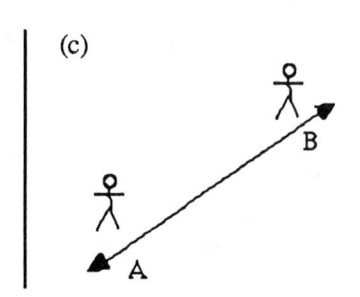

7. True or false.

   (a) If $T_{\overrightarrow{AB}}$ is a translation and if line l is parallel to AB, then the image of l under $T_{\overrightarrow{AB}}$ is l.
   (b) The slide image of a triangle can have at most one fixed point.
   (c) Translations preserve orientation.

   (a) True
   (b) False. Will have no fixed points.
   (c) True

8. Suppose $T_{x,y}$ is a translation and (3, 6) is translated to (2, 3). Find the coordinates of each of the following under this translation.

   (a) (⁻3, ⁻2)
   (b) (5, 1)
   (c) (0, 0)

   (a) (⁻4, 1)
   (b) (4, 4)
   (c) (⁻1, 3)

9. Describe in words each of the indicated translations.

   (a) $T_{\overrightarrow{AB}}$
   (b) $T_{\overrightarrow{RS}}$
   (c) $T_{\overrightarrow{XY}}$

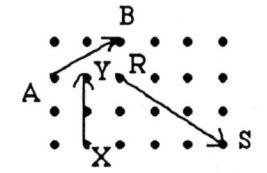

   (a) Move two units to the right and one unit up.

348 CHAPTER 13 Coordinate Geometry and Transformations

(b) Move three units to the right and three units down.
(c) move two units up.

10. Below are coordinates of image points under the translation in Exercise 8. Find the coordinates of the points that are translated to the following image points.

(a) (7, 6)
(b) (⁻3, ⁻5)
(c) (6, ⁻2)

(a) (8, 3)
(b) (⁻2, ⁻8)
(c) (7, ⁻5)

11. Consider a translation so that $P(x, y)$ is translated to $P'(x + 2, y - 3)$.

(a) On graph paper, draw a ray that corresponds to this translation.
(b) If $\triangle ABC$ has coordinates $A(2, ⁻1)$, $B(3, 4)$, and $C(⁻4, ⁻2)$, find the coordinates of the image of $\triangle ABC$ under this translation.

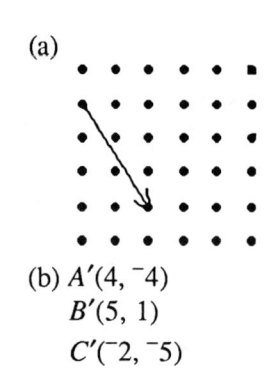

(a)

(b) $A'(4, ⁻4)$
$B'(5, 1)$
$C'(⁻2, ⁻5)$

12. Rectangle $\square RSTU$ has coordinates $R(a, b)$, $S(a, ⁻b)$, $T(⁻a, ⁻b)$, and $U(⁻a, b)$. Find the coordinates of the vertices of its image under the translation from Exercise 11.

$R'(a + 2, b - 3)$
$S'(a + 2, ⁻b - 3)$
$T'(⁻a + 2, ⁻b - 3)$
$U'(⁻a + 2, b - 3)$

Section 3 Slides or Translations 349

13. Consider the translation $T_{\overrightarrow{AB}}$ of $\triangle RST$ given below. Find two parallel lines so that a composite of two reflections will give the same image for $\triangle RST$.

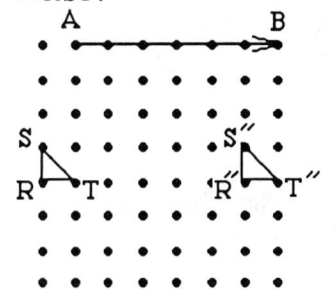

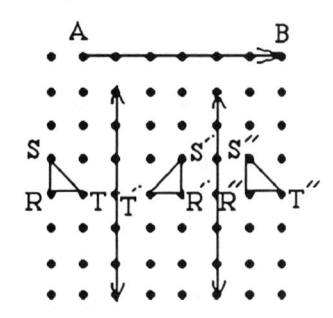

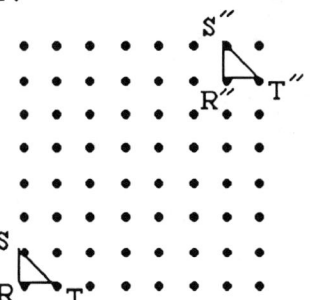

$m\,(\overline{RR''}) = 2 \cdot$ distance between parallel lines
$6 = 2 \cdot d$
$3 = d$

14. Consider the translation $T_{\overrightarrow{AB}}$ of $\triangle RST$ given below. Find two parallel lines so that a composite of two reflections will give the same image for $\triangle RST$.

350 CHAPTER 13 Coordinate Geometry and Transformations

Let $d =$ distance between parallel lines
$m\ (\overline{RR''}) = 2 \cdot d$
$6\sqrt{2} = 2 \cdot d$
$3\sqrt{2} = d$

15. In the figure below a translation maps A onto A'. Give the coordinates of the image of

   (a) C
   (b) X
   (c) F
   (d) R

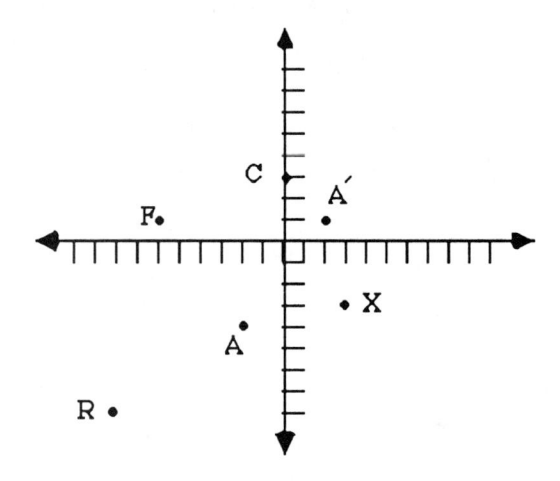

   (a) (4, 8)
   (b) (7, 2)
   (c) (⁻2, 6)
   (d) (⁻4, ⁻3)

Section 4 Rotations and Successive Motions

Cover the right side of the page and work on the left, then check your work

1. For each of the following, find a 90° counter-clockwise rotation of the given figure about the point P.

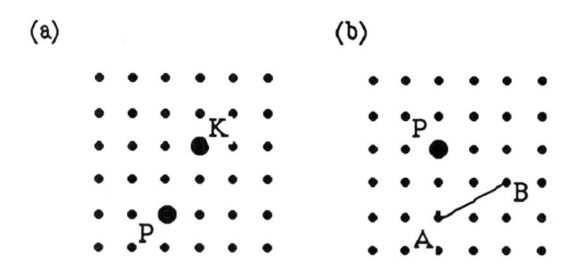

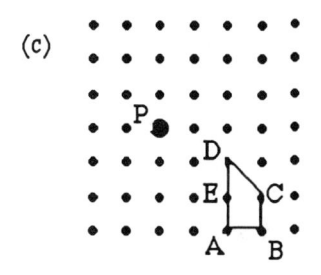

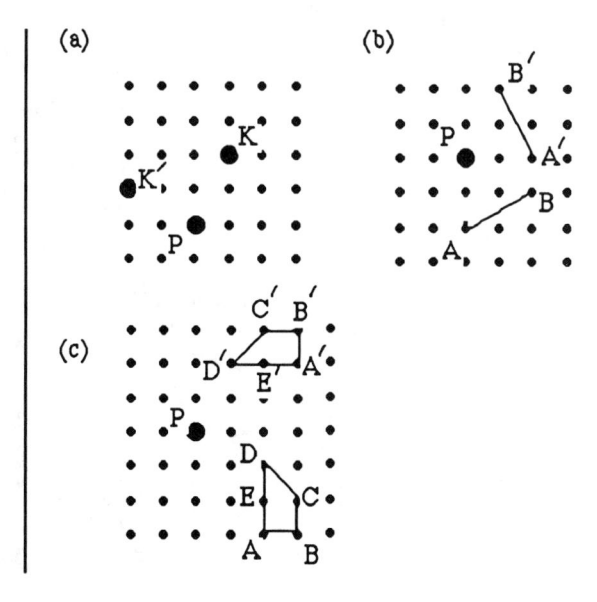

   (a)

   (b)

   (c)

2. Find a 90° clockwise rotation of the figures from Exercise 1 about point P.

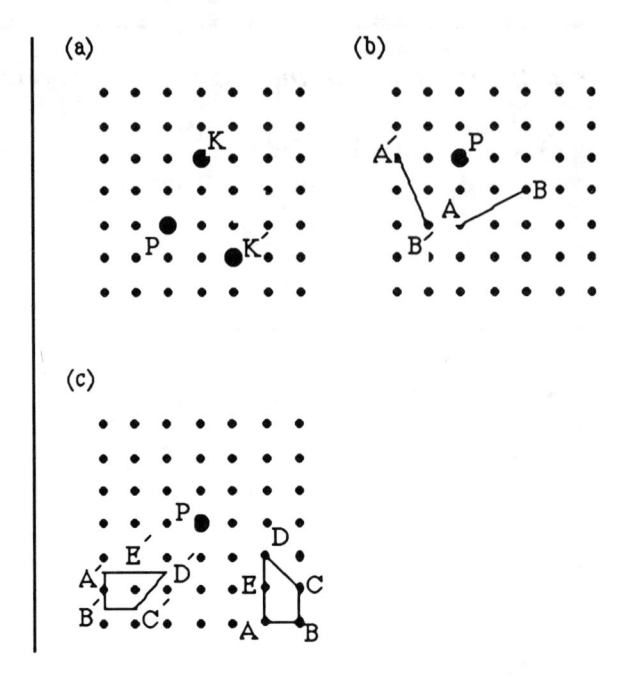

352 CHAPTER 13 Coordinate Geometry and Transformations

(a)

(b)

(c)

3. Find a 60° clockwise rotation of the figures from Exercise 1 about point P.

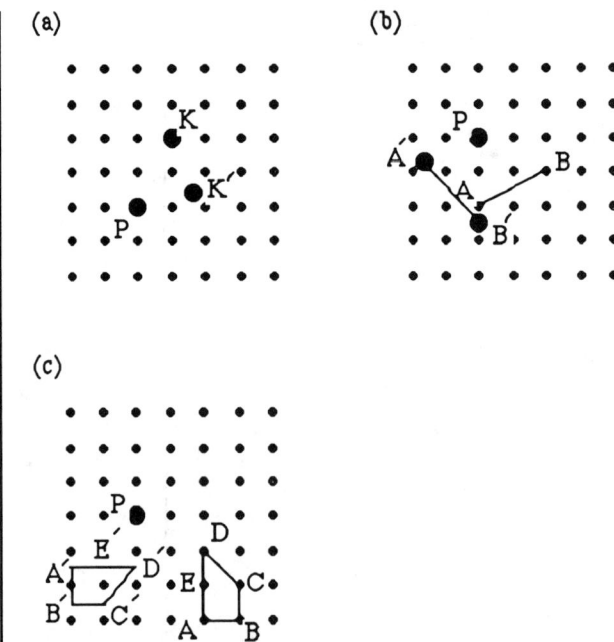

(a)

(b)

(c)

4. Draw the image of each figure under the given rotation about point P.

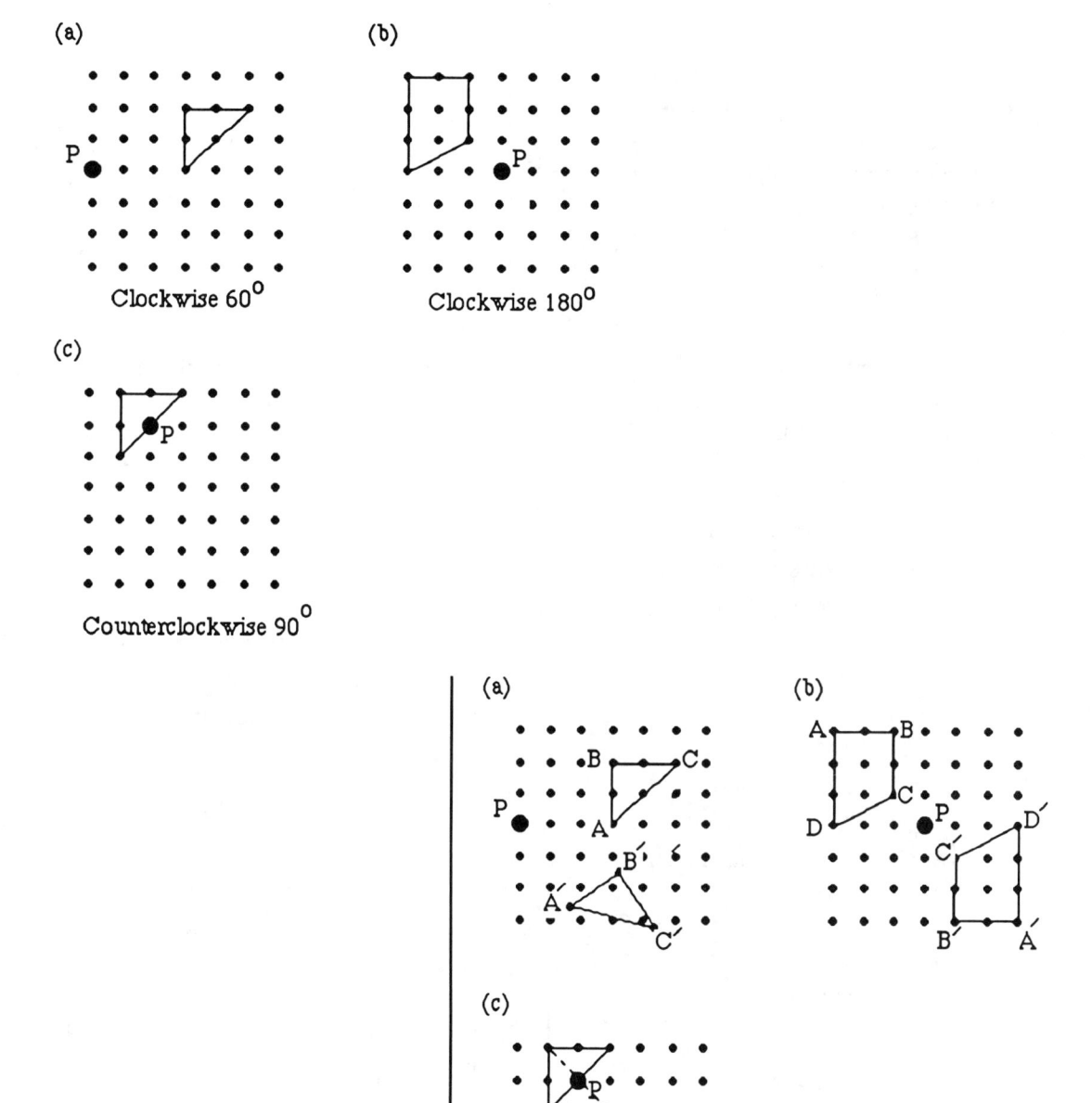

Section 4   Rotations and Successive Motions   353

(a) Clockwise 60°

(b) Clockwise 180°

(c) Counterclockwise 90°

(a)

(b)

(c)

5. The dashed figure is the image of the solid figure.
   Name the rotation for each.

(a)

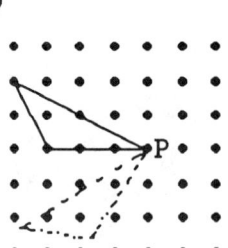

(b)

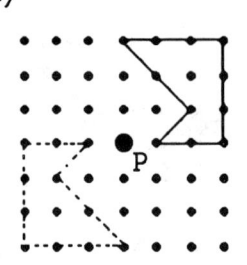

(a) counterclockwise 60^0
(b) clockwise or counterclockwise 180^0

6. Find an equivalent rotation for each of the following.

    (a) clockwise rotation of 120^0
    (b) counterclockwise rotation of 180^0
    (c) clockwise rotation of 250^0

(a) counterclockwise rotation of 240^0
(b) clockwise rotation of 180^0
(c) counterclockwise rotation of 110^0

7. Rectangle $\square RSTU$ has vertices $R(^-3, 3)$, $S(5, 3)$, $T(5, ^-3)$, and $U(^-3, ^-3)$. Find the coordinates of the vertices of the image of this rectangle under each rotation.

    (a) clockwise rotation of 90^0 about $P(1,0)$
    (b) counterclockwise rotation of 180^0

(a) $R'(4, 4)$, $S'(4, ^-4)$, $T'(^-2, ^-4)$, $U'(^-2, 4)$
(b) $R'(5, ^-3)$, $S'(^-3, ^-3)$, $T'(^-3, 3)$, $U'(5, 3)$

8. Rotate each of the following figures 135^0 clockwise about the point P.

    (a)

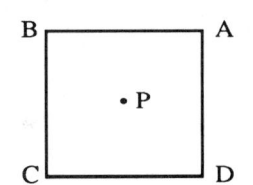

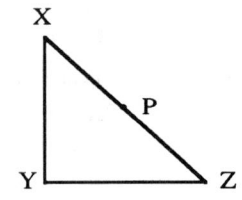

(c)

(a)

(b)

(c)

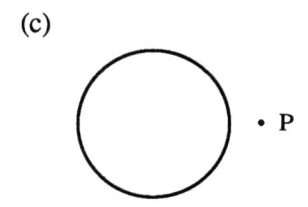

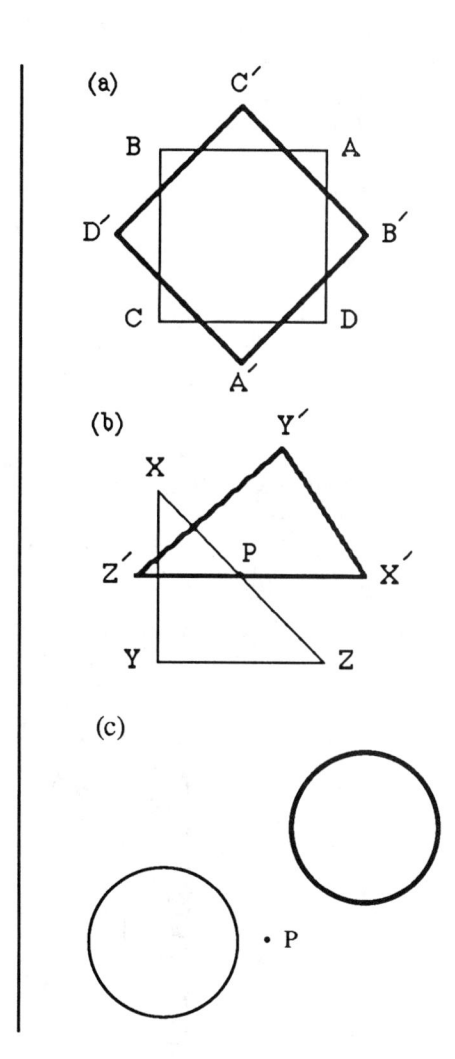

9. Perform a glide reflection on $\triangle XYZ$ for $T_{\overrightarrow{AB}}$ where the line of reflection is l.

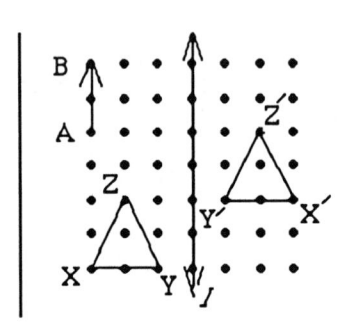

356   **CHAPTER 13**   Coordinate Geometry and Transformations

10. Find the coordinates of the image of each point.

    (a) Reflection of A about l followed by rotation of 90^0 clockwise about P
    (b) Rotation of 90^0 clockwise for B about P.

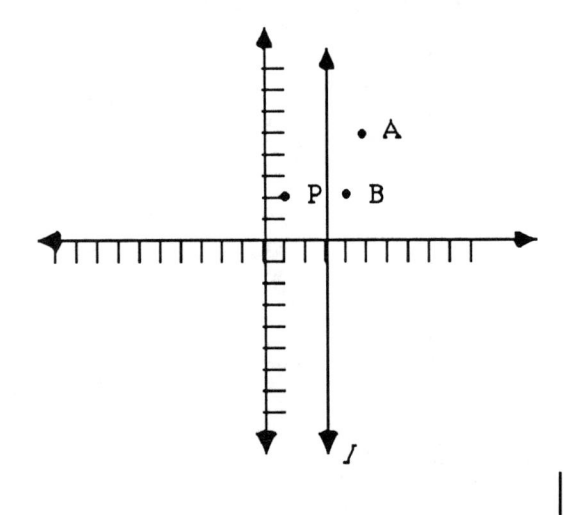

    (a) (4, 2)
    (b) ($^-1$, $^-1$)

11. Describe a transformation or composition of transformations that maps the given figure to each of the following.

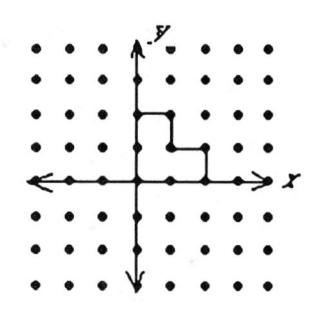

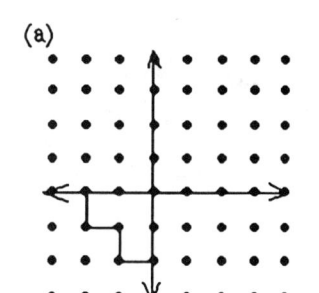

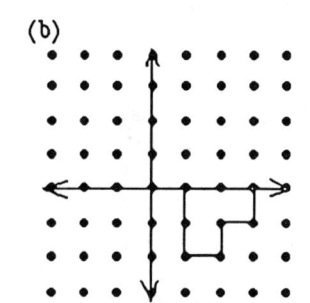

Section 4 Rotations and Successive Motions 357

(a) (b)

(a) Clockwise or counterclockwise rotation of 180^0 or reflection about line $y = {}^-x$
(b) Translation one unit to right and reflect about the x-axis

12. Draw the image of $\triangle ABC$ under the glide reflection $T_{\overrightarrow{XY}}$ where the line of reflection is $\overleftrightarrow{XY}$.

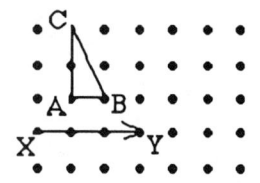

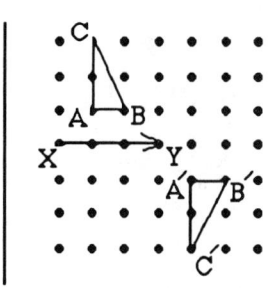

13. Draw the image of $\triangle ABC$ under the glide reflection $T_{\overrightarrow{XY}}$ followed by the glide reflection $T_{\overrightarrow{XY}}$ where the line of reflection is $\overleftrightarrow{XY}$ (from Exercise 12).

14. Find a single transformation that would give the same image as that of Exercise 13.

| Translation of 6 units to the right.

15. The rectangles in the figure are images under two reflections about lines m and n.

(a) Find a rotation that will accomplish the same result.
(b) What is the measure of ∠1?
(c) How do the answers to (a) and (b) compare?

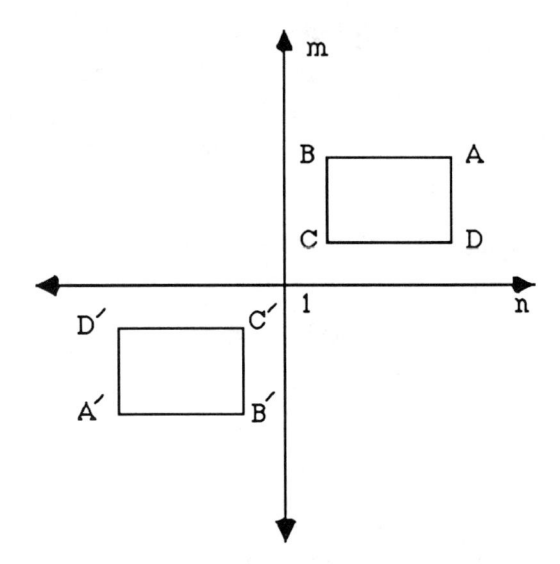

(a) Counterclockwise of 180°
(b) 90°
(c) The rotation is two times the measure of the angles formed by the lines of reflection.

Section 5 Transformations, Congruence, and Similarity

Cover the right side of the page and work on the left, then check your work

1. With *P* as the center, draw a magnification with scale factor 3 of segment $\overline{AB}$.

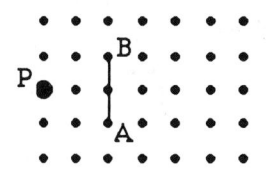

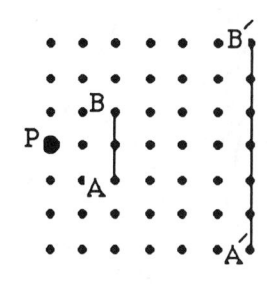

2. With *P* as the center, draw a magnification with scale factor 2 of $\triangle ABC$.

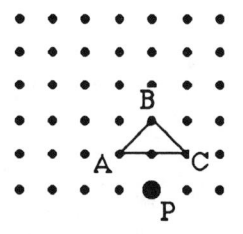

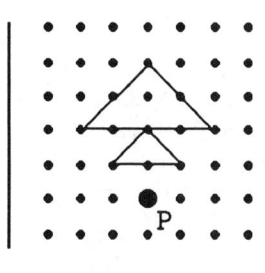

3. In Exercise 2, shrink with a scale factor of $\frac{1}{2}$.

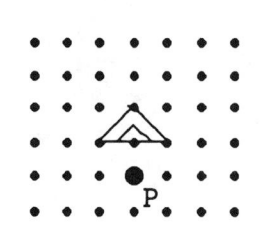

4. Describe the motion that will produce the image on top from the one on the bottom.

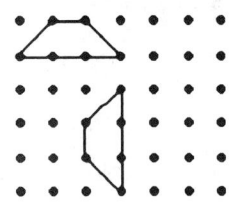

360 CHAPTER 13 Coordinate Geometry and Transformations

Rotation 90° clockwise folowed by a translation of 1 unit up.

5. For each pair of congruent triangles, name one composite of reflections, translations, or rotations that will map the figure on the left onto the figure on the right.

(a)

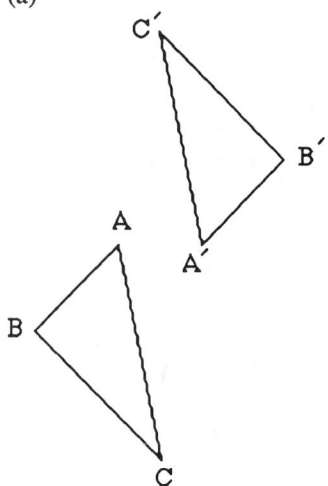

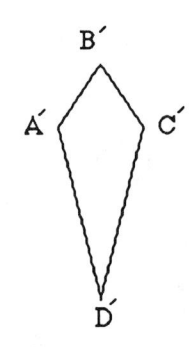

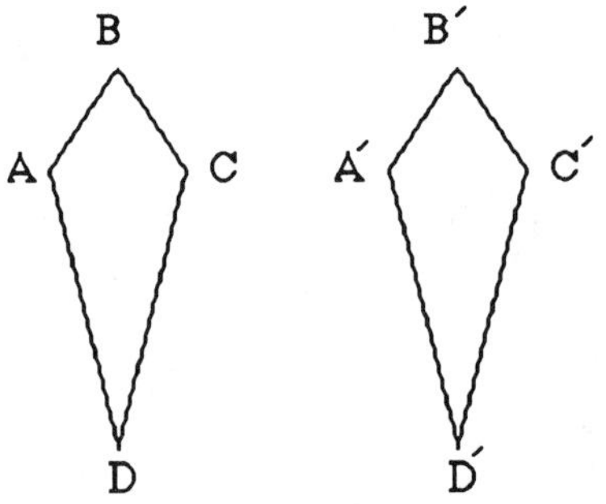

(a) rotation
(b) reflection

6. Draw two congruent rectangles □ABCD and □A′B′C′D′, such that □ABCD can be mapped on to □A′B′C′D′ by a

   (a) reflection
   (b) rotation
   (c) translation
   (d) glide reflection

(a)

(b)

D ▭ C
A   B

   C' ▯ B'
   D'   A'

(c)

D ▭ C   D' ▭ C'
A   B   A'   B'

(d)

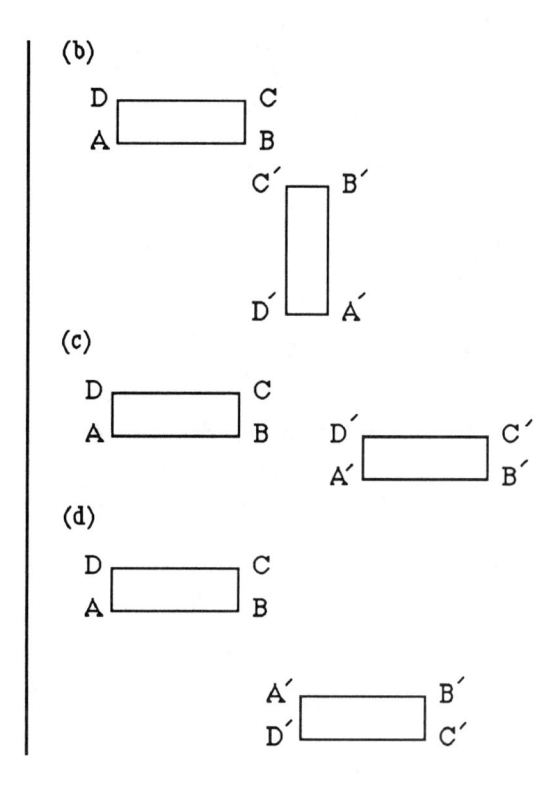

D ▭ C
A   B

A' ▭ B'
D'   C'

7. Find the image of △RST under a shrinking with a scale factor of $\frac{1}{2}$. The center is P.

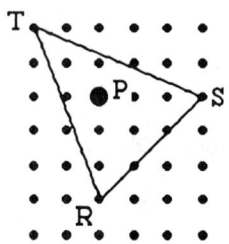

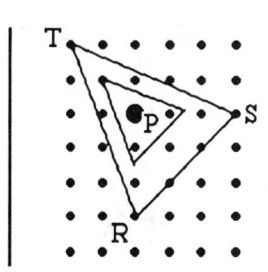

8. Find transformations that will map △ABC to △A'B'C'.

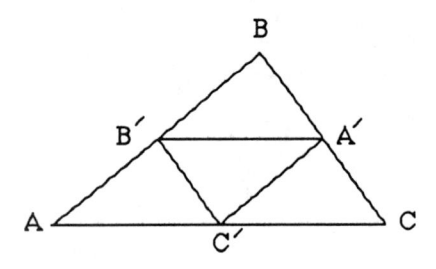

Rotation followed by a shrinking with a scale factor of $\frac{1}{2}$.

9. Given $\triangle R'S'T'$ as the unique of $\triangle RST$ under a magnification, locate points R, S, and T so that R' is the center of the transformation and $m\,\overline{ST} = \frac{1}{2}m\,(\overline{ST'})$.

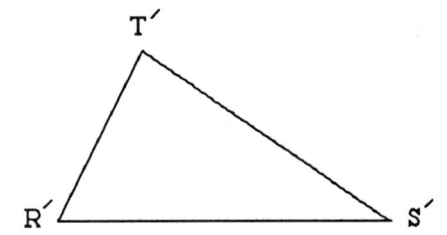

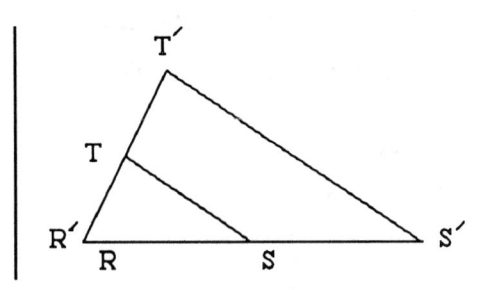

10. In the figure below, perform a magnification so that $(1, 3)$ moves to $(^-9, 7)$. Draw the image.

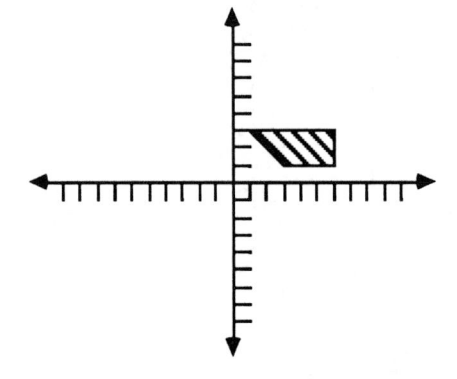

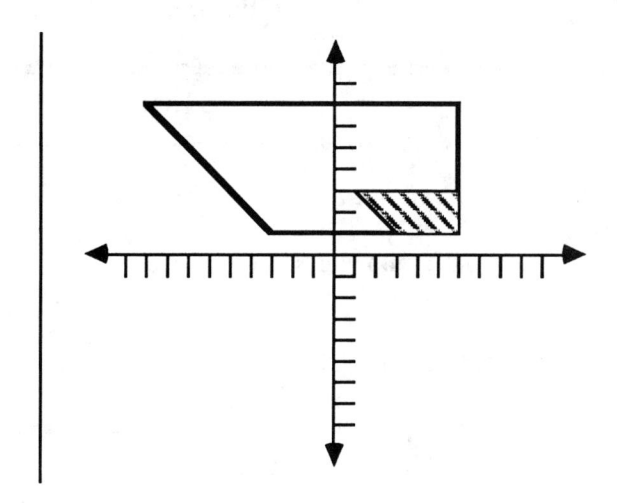

Section 5 Transformations, Congruence, and Similarity **363**

Chapter 14
Algebra and Geometry

Section 1 Functions

Cover the right side of the page and work on the left, then check your work

1. Which of the following sets of ordered pairs are functions?

   (a) {(1,6), (2, 7), (3, 8), (4, 9)}
   (b) {2, 5), (3, 6), (2, 6), (4, 7)}
   (c) {(2, 4), (3, 4), (4, 4), (5, 4)}
   (d) {(2, 3), (2, 4), (2, 5), (2, 6)}

   (a) function
   (b) not a function, (2, 5) and (2, 6) appear
   (c) function
   (d) not a function

2. Classify each of the following as a function or not a function.

   (a)

x	1 2 3 4
y	1 2 3 4

   (b) Domain   Range
       1   1
       2   1
       3   3
       4   3

   (c) {(5, 25), (6, 36), (7, 49), (8, 64)}

   (a) function
   (b) function
   (c) function

3. Which of the following graphs represent functions?

(a)

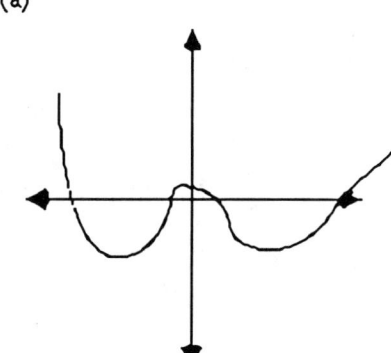

(b)

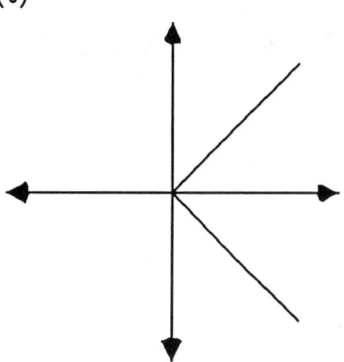

(c)

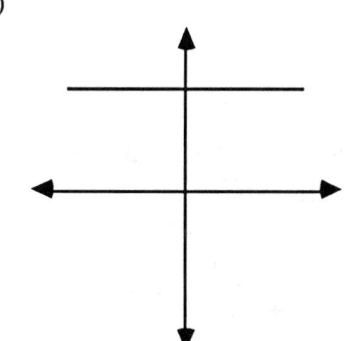

(a) function
(b) not a function

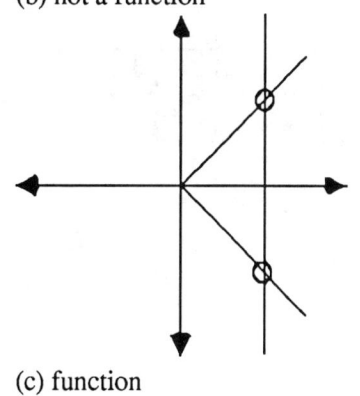

(c) function

4. Find the domain and range of each function.

(a) {(2, 1), (3, 1), (4, 5), (5, 6)}
(b) {(⁻1, 3), (0, 4), (1, 5), (2, 6)}

(a) $D = \{2, 3, 4, 5\}$
   $R = \{1, 5, 6\}$
(b) $D = \{^-1, 0, 1, 2\}$
   $R = \{3, 4, 5, 6\}$

366　CHAPTER 14　Algebra and Geometry

5. Given the function rules, compute $f(1), f(2), f(^-2)$.

   (a) $f(x) = 3x + 5$
   (b) $f(x) = \dfrac{x}{x+3}$
   (c) $f(x) = 3x^2 - 1$

   > (a) $f(1) = 8$
   > $f(2) = 11$
   > $f(^-2) = {}^-1$
   > $f(^-2) = 3(^-2) + 5 = {}^-6 + 5 = {}^-1$
   > (b) $f(1) = \dfrac{1}{4}$
   > $f(2) = \dfrac{2}{5}$
   > $f(^-2) = {}^-2$
   > (c) $f(1) = 2$
   > $f(1) = 3(1)^2 - 1 = 3 \cdot 1 - 1 = 2$
   > $f(2) = 11$
   > $f(^-2) = 11$
   > $f(^-2) = 3(^-2)^2 - 1$
   > $= 3 \cdot 4 - 1 = 11$

6. Complete the following table for the function $f(x) = x^2 + 3$.

x	-2	-1	0	1	a	$a+1$
$f(x)$						

   > | x | -2 | -1 | 0 | 1 | a | $a+1$ |
   > |---|---|---|---|---|---|---|
   > | $f(x)$ | 7 | 4 | 3 | 4 | $a+3$ | $a+2a+4$ |
   >
   > $f(a+1) = (a+1)^2 + 3$
   > $= (a+1)(a+1) + 3$
   > $= a^2 + 2a + 1 + 3$
   > $= a^2 + 2a + 4$

7. Given $f(x) = (2x^3 + 3x)(x^2 - 4)$, find the following.

   (a) $f(0.5)$
   (b) $f(^-1.2)$
   (c) $f(c+1)$

   > (a) $^-11$

Section 1   Functions   367

(b) 1.167
(c) $2c^5 + 10c^4 + 12c^3 - c^2 - 8c - 15$
$f(c + 1) = [2(c + 1)^3 + 3] [(c + 1)^2 - 4]$

8. Find a rule or equation that describes each function.

   (a) {(1, 6), (2, 7), (3, 8), (4, 9), (5, 10)}
   (b) {(1, 5), (2, 7), (3, 9), (4, 11)}
   (c) {(0, ⁻5), (2, ⁻4), (4, ⁻3), (6, ⁻2)}

(a) $f(x) = x + 5$
(b) $f(x) = 2x + 3$
(c) $f(x) = \frac{1}{2}x - 5$ or $\frac{x}{2} - 5$

9. A five-minute long-distance phone call costs $2.50. A 10-minute call costs $6.25. The first 3 minutes, however, only cost $1.00. Write an equation for the cost $C(n)$, of a call lasting n minutes longer than 3 minutes.

$C(n) = \$1.00 + \$0.75(n)$
Cost for 5 minute call = $1.00 + (0.75)2$
                         ↑   ↑
                      first 3   last 2
                      minutes   minutes

10. For what temperature will the Celsius and Fahrenheit values be equal?

⁻40°
$C(F) = \frac{5(F - 32)}{9} = F$
$5(F - 32) = 9F$
$5F - 160 = 9F$
$-160 = 4F$
$^-40 = F$

11. Given that $f(x) = x - 7$ and $g(x) = x + 11$ and using only these values, start with the number 50 and generate the number 65. As an example, we can start with 99 and generate 100 as follows.

$99 \rightarrow f(99) \rightarrow g(92) \rightarrow f(103) \rightarrow g(96) \rightarrow f(107)$
$99 \rightarrow 92 \rightarrow 103 \rightarrow 96 \rightarrow 107 \rightarrow 100$

368 CHAPTER 14   Algebra and Geometry

One solution is:

$$50 \xrightarrow{g} 61 \xrightarrow{f} 54 \xrightarrow{g} 65$$

12. The value of a computer in dollars is given by the function below, where x is measured in years since the purchase of the computer.

$$V(x) = 50{,}000\left(1 - \frac{x}{30}\right)$$

(a) What is the initial value of the computer?
(b) When will the computer be worth half its initial value?
(c) When is the computer worth nothing?

(a) $50,000
When $x = 0$
(b) 15 years
$$25000 = 50000\left(1 - \frac{x}{30}\right)$$
$$0.5 = 1 - \frac{x}{30}$$
$$^-0.5 = -\frac{x}{30}$$
$$15 = x$$
(c) 30 years

13. If an object is thrown upward with an initial velocity of 32 feet per second, then its height after t seconds is given by:

$$h(t) = 32t - 16t^2$$

(a) Find $h(0.5)$.
(b) Find $h(2)$.
(c) Find the maximum height attained.

(a) 12
$h(0.5) = 32(0.5) - 16(0.5)^2$
$= 16 - 16(0.25) = 16 - 4 = 12$
(b) 0
(c) 16 ft.
Use guess and check to find:
When $t = 1$, $h(1) = 32(1) - 16(1)^2$
$= 32 - 16 = 16$

14. The function $P(x) = 150x - x^2$ represents profit P in terms of the number of units made.
    (a) Find $P(30)$
    (b) Find $P(70)$
    (c) Find $P(82)$
    (d) What happens to the profit for $70 \le x \le 82$?

    (a) $3,600
    $P(30) = 150(30) - (30)^2$
    $= 4500 - 900 = 3,600$
    (b) $5,600
    (c) $5,576
    (d) It increases to a maximum and then begins to decrease.

15. Connie runs a donut shop. From past result, she has found that the cost of operating her shop is represented by the following function

    $$C(x) = 3x^2 - 42x + 160$$

    where x represents the number of hundreds of donuts sold.

    (a) Find the cost when 400 donuts are sold.
    (b) Find the cost when 900 donuts are sold.
    (c) Find the number of donuts that must be sold to produce the lowest cost.

    (a) $184
    (b) $25
    $C(9) = 3(9)^2 - 42(9) + 160$
    $= 243 - 378 + 160 = 25$
    (c) 700 donuts sold produces a cost of $13

Section 2 Expressing Ideas with Linear Functions

Cover the right side of the page and work on the left, then check your work

1. Find five solutions for each of the following equations and inequalities.

   (a) $2x - 2y = 8$
   (b) $2x + y = 5$
   (c) $3x + 4y \geq 12$
   (d) $2x - 3y < 0$
   (e) $4x - 6 = y$

(a)

x	y
0	-4
1	-3
2	-2
3	-1
4	0

(b)

x	y
0	5
1	3
2	1
3	-1
4	-3

(c)

x	y
0	4
1	3
2	2
3	1
4	0

(d)

x	y
0	0
1	0
2	2
3	2
4	2

(e)

x	y
0	-6
1	-2
2	2
3	6
4	10

2. Construct a graph for the points satisfying each of the following.

   (a) $x + y + 1 = 0$
   (b) $3x - 4y = {}^-12$
   (c) ${}^-x + 2y \geq 4$
   (d) $y > {}^-x$

Answers follow:

(a)

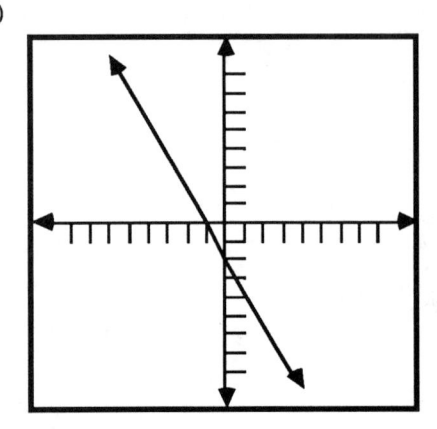

(b)

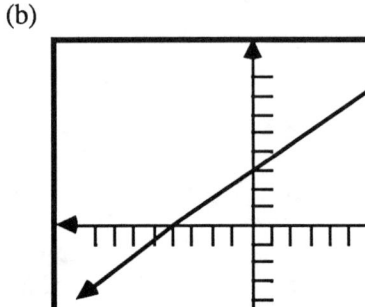

(c)

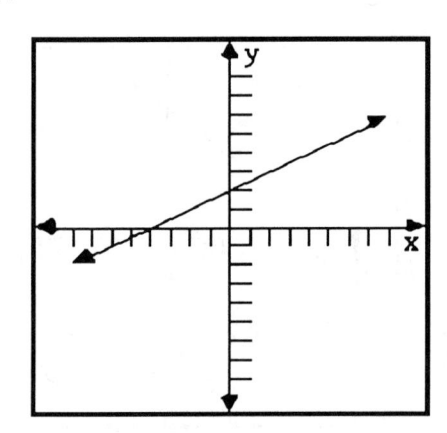

(d)

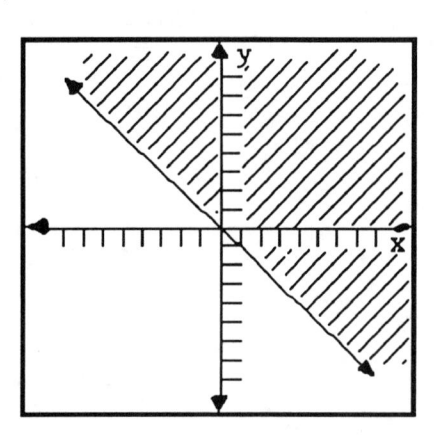

3. Find the x and y-intercepts of the following equations.

(a) $^-4x - 2y = 0$
(b) $^-3x + y = 8$
(c) $y = ^-4x - 3$
(d) $2x - 4y - 8 = 0$

(a) x-intercept = 0
 y-intercept = 0
(b) x-intercept = $-\frac{8}{3}$
 y-intercept = 8
(c) x-intercept = $-\frac{3}{4}$
 y-intercept = $^-3$
(d) x-intercept = 4
 y-intercept = $^-2$

4. Graph $y = x + 2$ for $^-2 \leq x \leq 3$.

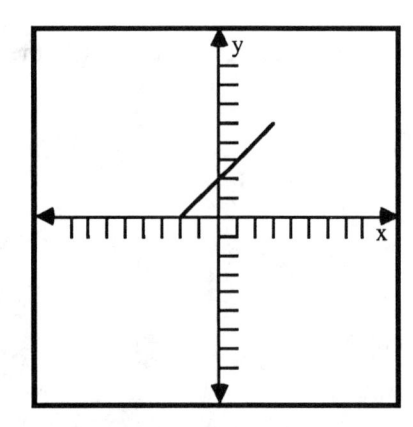

372 CHAPTER 14   Algebra and Geometry

5. Graph $y = 0.5x - 0.75$ for $^-1.5 \leq x \leq 1.5$.

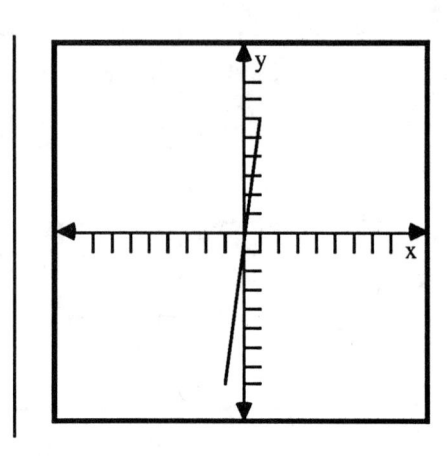

6. Find an equation of the line with x-intercept 7 and y-intercept $^-2$.

     | $2x - 7y = 14$

7. In 1897 a physics professor, A. E. Dolbear, suggested that temperature T, in degrees Fahrenheit, is given by

$$T = \frac{1}{4}x + 40$$

where x is the number of cricket chirps per minute.

(a) What is the temperature for 60 cricket chirps per minute?
(b) Find the T-intercept.

     | (a) 55^0
     | $T = \frac{1}{4}(60) + 40 = 15 + 40 = 55$
     | (b) T-intercept $= 40^0$, when $x = 0$

8. Graph each pair of equations on the same set of axes.

(a) $y = 3x + 2$
$y = \frac{-1}{3}x + 4$

(b) $y = {}^-2x - 3$
$y = \frac{1}{2}x + 1$

Answers follow:

(a)

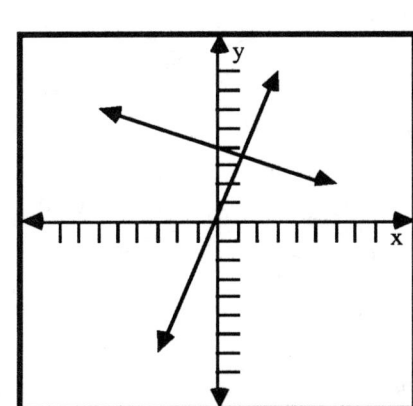

(b)

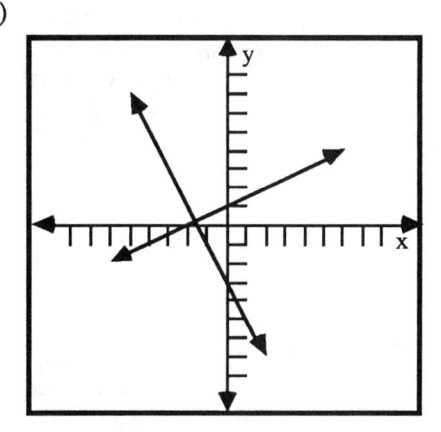

9. How do the graphs of the pairs of lines in Exercise 8 appear to be related?

They are perpendicular.

10. The number of days, D, in a year on which the ground is covered with snow is given by

$$D = 0.155H + 11$$

where H is the elevation in meters.

(a) How many days of snow cover are there at sea level?
(b) At what elevation will there bre snow all 365 days a year?

(a) 11
$H = 0$
(b) 2,284 m
$365 = 0.155H + 11$
$354 = 0.155H$
$2284 \approx H$

11. Maximum heart rate, in beats per minute, during exercise is given by

$$r_{max} = 0.981\ r_5 + 5.948$$

where r_5 is the heart rate within 5 seconds after stopping exercise. [Reported in <u>Science and Sport</u> by Thomas Vaughn (Boston: Little, Brown and Co., 1971)].

374   **CHAPTER 14**   Algebra and Geometry

(a) A jogger's heart rate is 130. Find her maximum heart rate during exercise.

(b) A swimmer's maximum heart rate is 120. What is his heart rate 5 seconds after swimming a race?

(a) 133
$r_{max} = 0.981(130) + 5.948$
$= 127.53 + 5.948 \approx 133$
(b) 116

12. Entering freshmen test scores are decreasing linearly. In 1975 the average score was 550 and in 1990 the average score was 500.

(a) Graph the linear equation that relates average test score to time.

(b) Estimate the average test score in the year 2000.

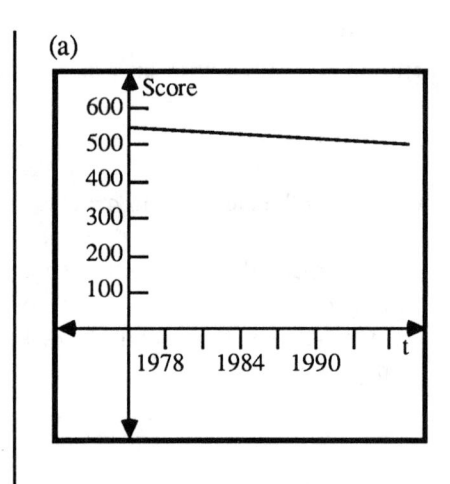

(a)

[Graph showing Score vs. t, with a line decreasing slightly from around 550, axes marked 100–600 on Score axis and 1978, 1984, 1990 on t axis]

(b) 467

13. Graph $2x + 2y \geq 10$ for $x \leq 0$ and $y \geq 0$.

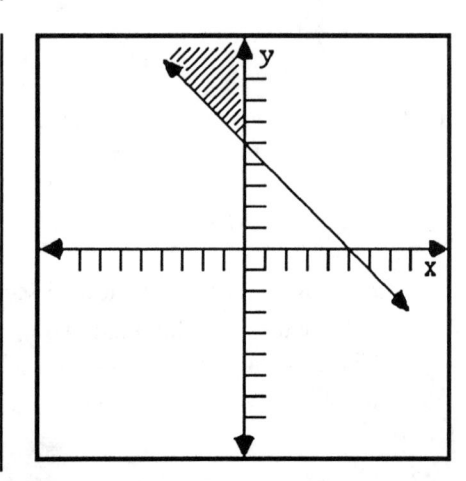

[Graph showing shaded region in second quadrant above the line, with line extending into fourth quadrant]

14. A vending machine has a value at the end of *y* years given by

    $V = 3000 - 200y$

    Complete the table showing the value of the machine at the end of a given year.

Year	0	1	2	3	6	8	14
Value	3000	2800					

Year	0	1	2	3	6	8	14
Value	3000	2800	2600	2400	1800	1400	200

15. From Exercise 14,

    (a) What is the value of the machine when it is purchased?
    (b) When will the machine have a value of 0?

    (a) $3,000
    (b) 15 years after purchase
    $0 = 3000 - 200y$
    $200y = 3000$
    $y = 15$

Section 3 Slopes and Linear Equations

Cover the right side of the page and work on the left, then check your work

1. Find the slope for the line through each pair of points.
   (a) (5, 2), (4, ¯3)
   (b) (2, ¯6), (1, 0)
   (c) (3, 7), (¯6, 7)
   (d) (2, 5), (¯3, 4)
   (e) (2, ¯3), (2, 0)

   (a) 5
   $$m = \frac{-3-2}{4-5} = \frac{-5}{-1} = 5$$
   (b) ¯6
   (c) 0
   (d) $\frac{1}{5}$
   $$m = \frac{4-5}{-3-2} = \frac{-1}{-5} = \frac{1}{5}$$
   (e) Undefined

2. Graph the line through the point (1, 3) with the indicated slope.

   (a) $\frac{1}{3}$
   (b) 2
   (c) ¯2
   (d) $\frac{3}{4}$

   Answers follow:
   (a) 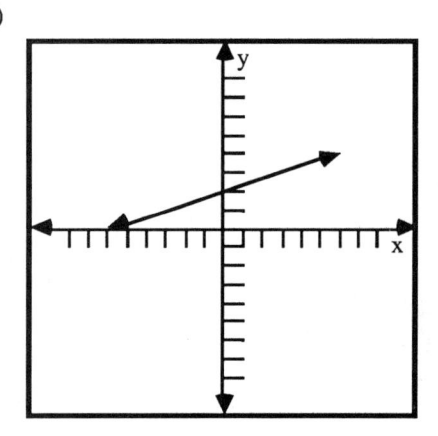
   (b)

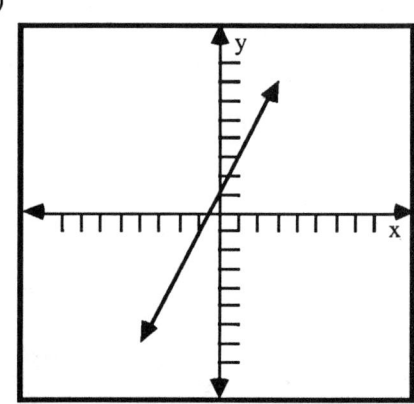

(c)

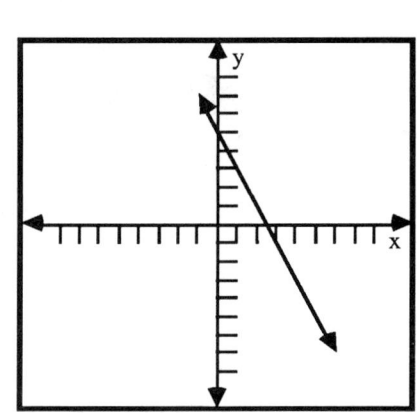

(d)

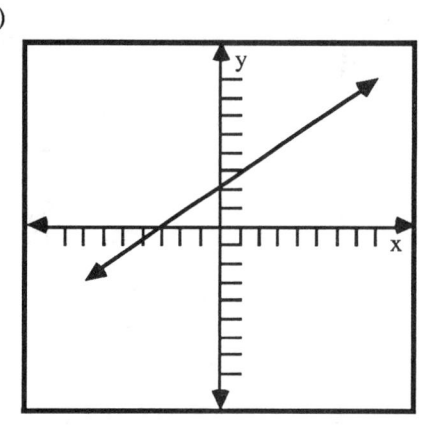

3. Find an equation for the line that has:

(a) a slope of 2 and goes through the point (4, 5).
(b) a slope of ¯1 and goes through the point (3, ¯3)
(c) a slope of $\frac{1}{4}$ and goes through the point (0, 3)

(a) $y = 2x - 3$
$y - 5 = 2(x - 4)$
$y - 5 = 2x - 8$
$y = 2x - 3$
(b) $y = ^-x$
(c) $y = \frac{x}{4} + 3$
$y - 3 = \frac{1}{4}(x - 0)$
$y - 3 = \frac{x}{4}$
$y = \frac{x}{4} + 3$

4. Graph each of the lines from Exercise 3.
(a)

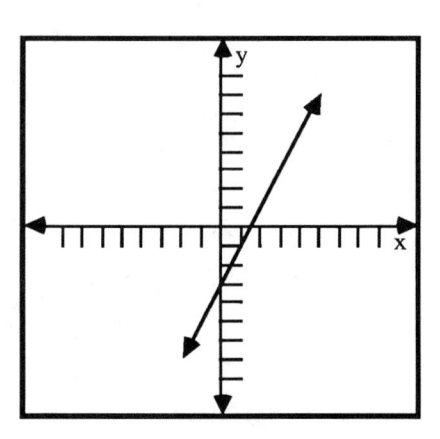

(b)

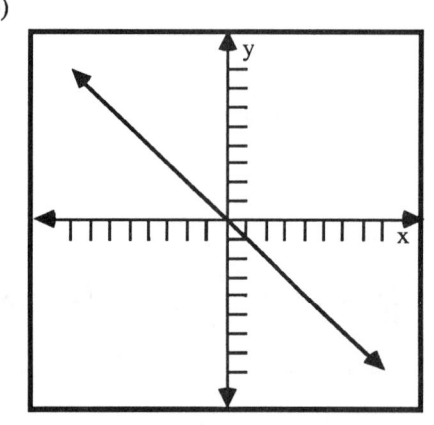

378　CHAPTER 14　　Algebra and Geometry

(c)

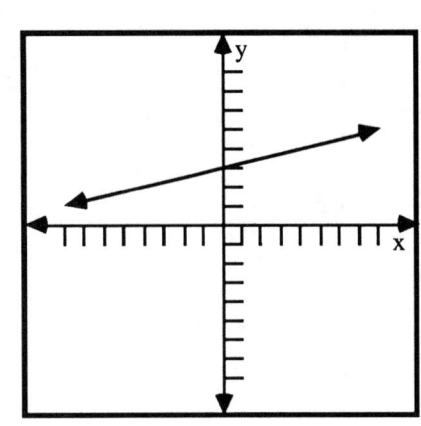

5. For each of the following, find the slope and y-intercept.

   (a) $y = 5x - 3$
   (b) $2x - 4y - 7 = 0$
   (c) $y = \dfrac{x + 3}{4}$
   (d) $^-4x - 2y = 6$

   (a) slope = 5
   　　y-intercept = $^-3$
   (b) slope = $\dfrac{1}{2}$
   　　$2x - 4y - 7 = 0$
   　　$^-4y = ^-2x + 7$
   　　$y = \dfrac{x}{2} - \dfrac{7}{4}$
   　　y-intercept = $-\dfrac{7}{4}$
   (c) slope = $\dfrac{1}{4}$
   　　y-intercept = $\dfrac{3}{4}$
   (d) slope = $^-2$
   　　$^-4x - 2y = 6$
   　　$^-2y = 4x + 6$
   　　$y = ^-2x - 3$
   　　y-intercept = $^-3$

6. Graph each of the lines from Exercise 5.

   Answers follow:

(a)

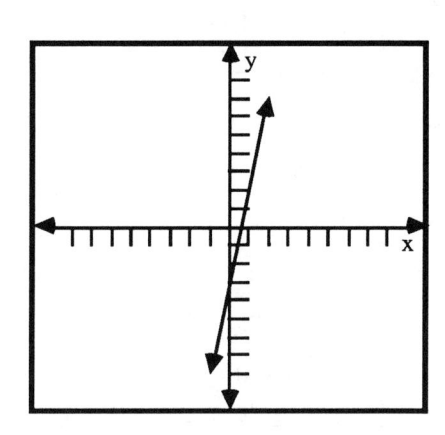

(b)

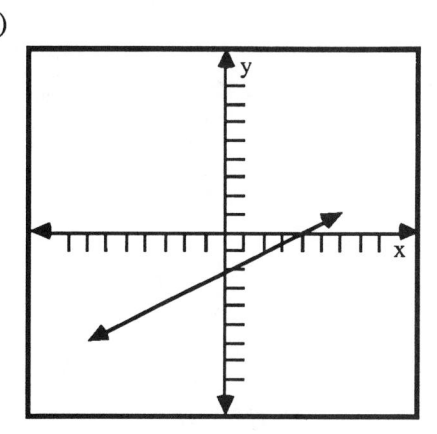

(c)

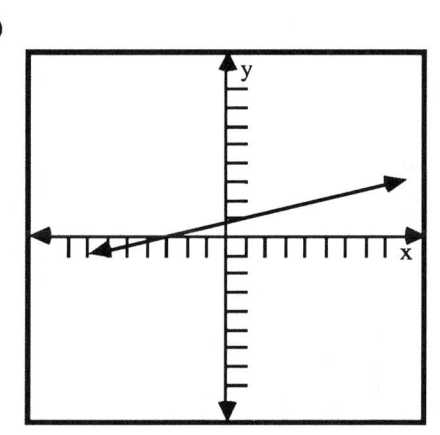

(d)

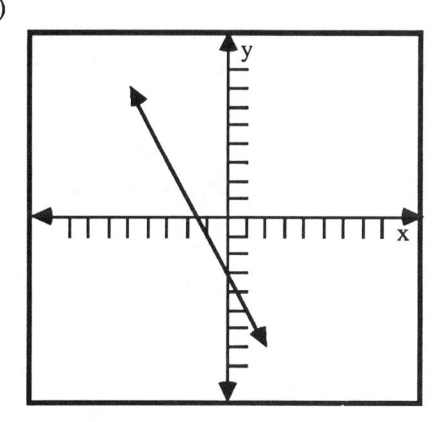

7. Answer each of the following.

(a) What does the graph of $y = mx + b$ look like when m is positive?
(b) When b is positive, what does the graph of $y = mx + b$ look like?
(c) What does the graph of $y = mx + b$ look like as b gets larger?

> (a) slopes upward to the right
> (b) crosses y-axis above the origin
> (c) crosses y-axis at higher and higher points

8. Find an equation of the line through each of the following pairs of points.

(a) (2, 3), (6, ⁻5)
(b) (3, 4), (⁻2, 4)
(c) (0, 0), (a, b)

CHAPTER 14 Algebra and Geometry

(a) $y = {}^-2x + 7$
$m = \dfrac{-5-3}{6-2} = \dfrac{-8}{4} = {}^-2$
$y - 3 = {}^-2(x - 2)$
$y - 3 = {}^-2x + 4$
$y = {}^-2x + 7$
(b) $y = 4$
(c) $y = \dfrac{b}{a}x$
$m = \dfrac{b-0}{a-0} = \dfrac{b}{a}$
$y - 0 = \dfrac{b}{a}(x - 0)$
$y = \dfrac{b}{a}x$

9. Find an equation for each of the following lines.

   (a) Line with x-intercept of $^-3$ and y-intercept of 5.
   (b) Line containing the point $(0, {}^-4)$ with a slope of $\dfrac{1}{2}$.

(a) $3y - 5x - 15 = 0$
$\dfrac{x}{-3} + \dfrac{y}{5} = 1$
$15\left(\dfrac{x}{-3} + \dfrac{y}{5} = 1\right)$
${}^-5x + 3y = 15$
$3y - 5x - 15 = 0$
(b) $y = \dfrac{1}{2}x - 4$

10. Use the graph to find the slope of the line.

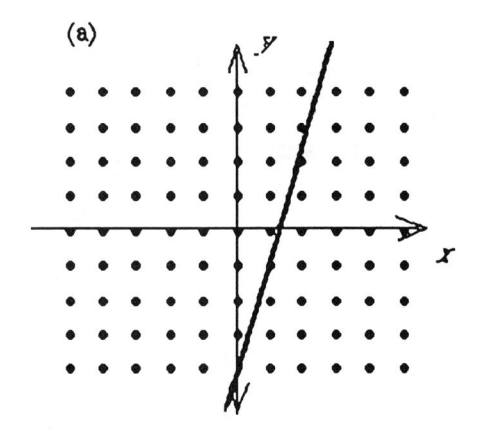

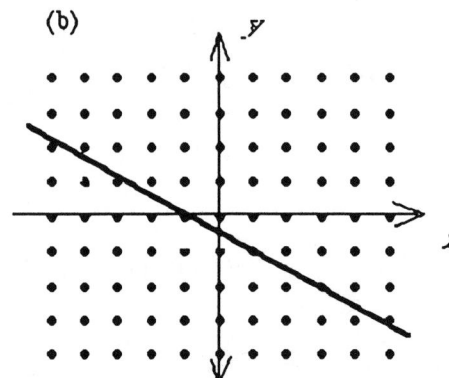

   (a)

   (b)

(a) slope = 3
(b) slope = $-\frac{1}{2}$

11. Which of the following lines are parallel to each other and which are perpendicular to each other?

   (a) $y - 3x - 2 = 0$
   (b) $2y - 6x + 14 = 0$
   (c) $y + 2x + 3 = 0$
   (d) $y - 2x - 4 = 0$
   (e) $y - 3x = 0$

(a), (b), and (e), all have slope = 3
(c), (d), both have slope = $^-2$

12. Find an equation of the line through (1, $^-2$) and parallel to $2y = 6x - 4$.

$2y = 6x - 4$
$y = 3x - 2$ slope = 3
$y - (^-2) = 3(x - 1)$
$y + 2 = 3x - 3$
$y = 3x - 5$

13. Find an equation of the line through (2, $^-1$) and perpendicular to $x - 2y + 3 = 0$.

$y = {}^-2x + 3$

$x - 2y + 3 = 0$
$^-2y = {}^-x - 3$
$y = \frac{x}{2} + \frac{3}{2}$
need a slope of $^-2$ since $\frac{1}{2} \cdot {}^-2 = {}^-1$
$y - (^-1) = {}^-2(x - 2)$
$y + 1 = {}^-2x + 4$
$y = {}^-2x + 3$

14. If P dollars is invested at the simple interest rate, r, for t years, the value of the investment is given by

   $A = P(1 + rt)$.

   Suppose you invest $5,000 at 9% simple interest per year.
   (a) Graph the amount you will have in t years.
   (b) What is the slope of this graph?

(c) What is the *A*-intercept of this graph?

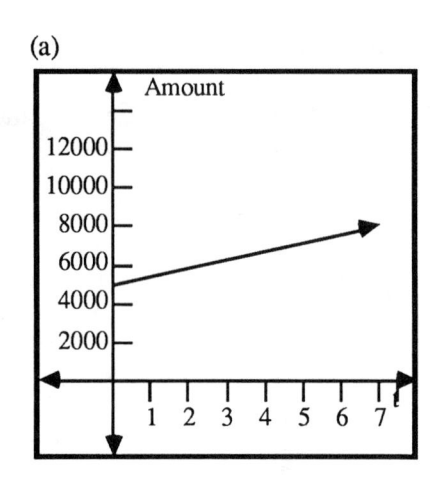

(a)

[Graph showing Amount vs. time with y-axis marked 2000, 4000, 6000, 8000, 10000, 12000 and x-axis marked 1 through 7, with a line starting at about 5000 and rising]

(b) $A = 5000(1 + 0.09t)$
$ = 450t + 5000$
slope $= 450$
(c) 5,000

15. It costs $100 for the maintenance and repairs to drive a five-year-old car for 2,000 miles and $700 to drive the car for 7,000 miles.

   (a) Assume the relationship between the number of miles driven (*m*) and cost for maintenance and repairs (*C*) is linear. Write an equation expressing this relationship.

   (b) What is the cost for driving 20,000 miles?

(a) (2000, 100), (7000, 700)
slope $= \dfrac{700 - 100}{7000 - 2000} = \dfrac{600}{5000} = \dfrac{3}{25}$
$C - 100 = \dfrac{3}{25}(m - 2000)$
$C - 100 = \dfrac{3}{25}m - 240$
$C = \dfrac{3}{25}m - 140$
(b) $2,260
$C = \dfrac{3}{25}(20,000) - 140$
$ = 2400 - 140 = 2,260$

Section 4 Systems of Linear Equations

Cover the right side of the page and work on the left, then check your work

1. Solve the following systems of equations graphically.

    (a) $3x + 4y = {}^-5$
      $2x - y = 4$
    (b) $2x + y = 2$
      $3x - 2y = {}^-4$

Answers follow:

(a) (1, -2)

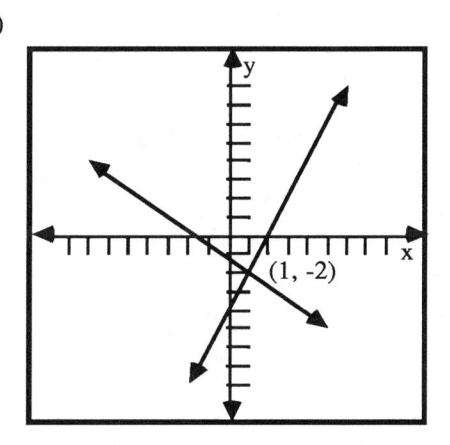

(b) (0, 2)

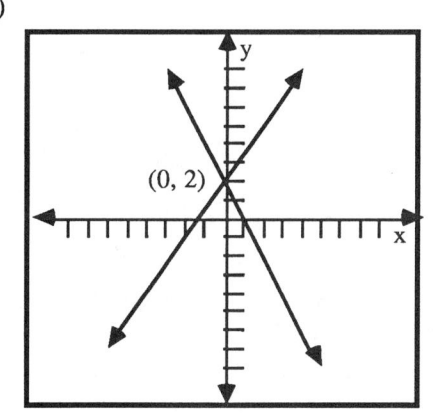

2. Solve the following systems by eliminating a variable.

    (a) $x - y = 9$
      $x + y = 3$
    (b) $2x - 2y = 1$
      $3x + 5y = 11$
    (c) $4x - y = {}^-1$
      $x + 3y = {}^-9$
    (d) $\frac{1}{3}x - \frac{1}{6}y = 3$
      $\frac{2}{3}x + \frac{5}{6}y = 1$

(a) $x = 6, y = {}^-3$

$x - y = 9$
$\underline{x + y = 3}$
$2x = 12$
$x = 6$

$6 - y = 9$
$^-y = 3$
$y = {}^-3$

(b) $x = \dfrac{27}{16}, y = \dfrac{19}{16}$

$2x - 2y = 1$
$\underline{3x + 5y = 11}$

$10x - 10y = 5$ [5 · equation 1 + 2 · equation 2]
$\underline{6x + 10y = 22}$
$16x \qquad = 27$
$\qquad x = \dfrac{27}{16}$

$3\left(\dfrac{27}{16}\right) + 5y = 11$

$\dfrac{81}{16} + 5y = 11$

$5y = \dfrac{176 - 81}{16}$

$5y = \dfrac{95}{16}$

$y = \dfrac{19}{16}$

(c) $x = -\dfrac{12}{13}, y = \dfrac{61}{13}$

(d) $x = \dfrac{48}{7}, y = -\dfrac{30}{7}$

$\dfrac{1}{3}x - \dfrac{1}{6}y = 3$
$\dfrac{2}{3}x + \dfrac{5}{6}y = 1$
$2x - y = 18$ [Multiply both equations by 6]
$4x + 5y = 6$

$10x - 5y = 90$ [5 · equation 1 + equation 2]
$\underline{4x + 5y = 6}$
$14x \qquad = 96$
$\qquad x = \dfrac{48}{7}$

$y = -\dfrac{30}{7}$

Section 4 Systems of Linear Equations

3. Solve the following system of equations for x and y.

$$\frac{1}{x} - \frac{1}{y} = \frac{1}{4}$$
$$\frac{4}{x} + \frac{3}{y} = 3$$

Hint: First solve for $\frac{1}{x}$ and $\frac{1}{y}$.

$$x = \frac{28}{15}, y = \frac{7}{2}$$

$$\frac{1}{x} - \frac{1}{y} = \frac{1}{4}$$

$$4\left(\frac{1}{x}\right) + 3\left(\frac{1}{y}\right) = 3$$

$$3\left(\frac{1}{x}\right) - 3\left(\frac{1}{y}\right) = \frac{3}{4} \quad [3 \cdot \text{equation 1} + \text{equation 2}]$$

$$4\left(\frac{1}{x}\right) + 3\left(\frac{1}{y}\right) = 3$$

$$\overline{7\left(\frac{1}{x}\right) \qquad = \frac{15}{4}}$$

$$\frac{1}{x} = \frac{15}{28}$$

$$x = \frac{28}{15}$$

$$\frac{15}{28} - \frac{1}{y} = \frac{1}{4}$$

$$-\frac{1}{y} = \frac{1}{4} - \frac{15}{28}$$

$$-\frac{1}{y} = -\frac{8}{28}$$

$$y = \frac{28}{8} = \frac{7}{2}$$

4. A person has 40 coins, consisting of dimes and quarters. The total value of the coins is $6.25. How many of each type of coin does she have?

Number of dimes = 25
Number of quarters = 15

Let d = number of dimes
q = number of quarters

$d + q = 40$
$10d + 25q = 625$

386 CHAPTER 14 Algebra and Geometry

$$\begin{aligned} \Downarrow \\ {}^-10d - 10q &= {}^-400 \\ \underline{10d + 25q = 625} \\ 15q &= 225 \end{aligned}$$

$$q = 15$$
$$d = 25$$

5. An airplane flies 3,000 miles from New York to Los Angeles in 5 hours against a wind. From Los Angeles to New York the trip takes 4 hours with a tail wind of the same velocity. Find the speed of the plane and the speed of the wind.

Plane's speed = 675, Wind speed = 75

Let p = speed of the plane
w = speed of the wind

$5(p - w) = 3000 \Rightarrow 5p - 5w = 3000$
$4(p + w) = 3000 \Rightarrow 4p + 4w = 3000$
$$\Downarrow$$
$$\begin{aligned} 20p - 20w &= 12000 \\ \underline{20p + 20w = 15000} \\ 40p &= 27000 \\ p &= 675 \end{aligned}$$

$4(675) + 4w = 3000$
$2700 + 4w = 3000$
$4w = 300$
$w = 75$

6. Two solutions of antifreeze and water are to be mixed to form a new solution. The first solution is made form 5 L of antifreeze mixed with 10 L of water. The second solution is made from 7.5 L of antifreeze mixed with 20 L of water. How many gallons of each solution should be combined to form a new solution of 9 L with $\frac{4}{9}$ L of antifreeze per liter of solution?

Liters of first solution = 5
Liters of second solution = 4

Let x = number of liters of first solution
y = number of liters of second solution

$x + y = 9 \quad \Rightarrow \quad x + y = 9$

Section 4 Systems of Linear Equations

$$\frac{1}{2}x + \frac{3}{8}y = \frac{4}{9}(9) \Rightarrow \underline{4x + 3y = 32}$$

$$\begin{array}{l} {}^-3x - 3y = {}^-27 \\ \underline{4x + 3y = 32} \\ x \qquad\quad = 5 \\ y = 4 \end{array}$$

7. The sum of two numbers is 23. Four times the first minus two times the second is 8. Find the numbers.

First number = 9, Second number = 14

Let x = first number
y = second number

$x + y = 23$
$4x - 2y = 8$
$\Downarrow$
$2x + 2y = 46$
$\underline{4x - 2y = 8}$
$6x \qquad\quad = 54$
$\qquad x = 9$
$\qquad y = 14$

8. In the city of Megopolis a taxi ride costs $1.00 for the first $\frac{1}{5}$ of a mile and 10¢ for every $\frac{1}{5}$ of a mile thereafter. If x represents the number of fifths of mile in a ride, then the total cost of a ride is given by
$$y = 10(x - 1) + 100$$

Suppose in another city a taxi ride costs 75¢ for the first $\frac{1}{5}$ of a mile and 15¢ for each $\frac{1}{5}$ of a mile thereafter.

(a) Write an equation for the cost of a taxi ride for the second city.
(b) Solve the two equations simultaneously and explain what the solution means (represents).

(a) $y = 15(x - 1) + 75$
(b) $x = 6$ fifths of a mile, $y = \$1.50$

$y = 10(x - 1) + 100$
$y = 15(x - 1) + 75$

388   CHAPTER 14   Algebra and Geometry

$$\Downarrow$$

$$^-10x + y = 90$$
$$^-15x + y = 60 \quad \text{[Subtract second equation from first]}$$
$$5x = 30$$
$$x = 6$$

$$y = 15(6 - 1) + 75$$
$$= 15(5) + 75$$
$$= 150$$

For a ride of $\frac{6}{5}$ of a mile the fare in each city will be $1.50.

9. A store manager wants to make 100 kg of coffee that is a blend of coffee that sells for $6/kg with coffee that sells for $3/kg. If he wants the blend to sell for $5/kg, how much of each coffee should go in the blend?

$66\frac{2}{3}$ kg of the $6 coffee

$33\frac{1}{3}$ kg of the $3 coffee

10. Jenny has $25,000 to invest. She deposits part of the money at 10% annual interest and the rest at 6% annual interest. Her total annual income from the two investments is $2,020. How much is invested at each rate?

$13,000 invested at 10%, $12,000 invested at 6%

Let x = amount at 10%
    y = amount at 6%

$$x + y = 25000$$
$$0.10x + 0.06y = 2020$$

$$\Downarrow$$

$$x + y = 25000$$
$$10x + 6y = 202000$$

$$^-6x - 6y = ^-150000$$
$$\underline{10x + 6y = 202000}$$
$$4x = 52000$$
$$x = 13{,}000$$
$$y = 12{,}000$$

Section 5 Quadratic Functions and the Parabola

Cover the right side of the page and work on the left, then check your work

1. Identify the vertex and the equation of the line of symmetry for each parabola shown below.

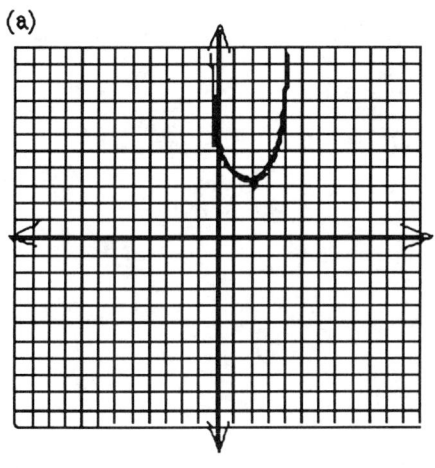

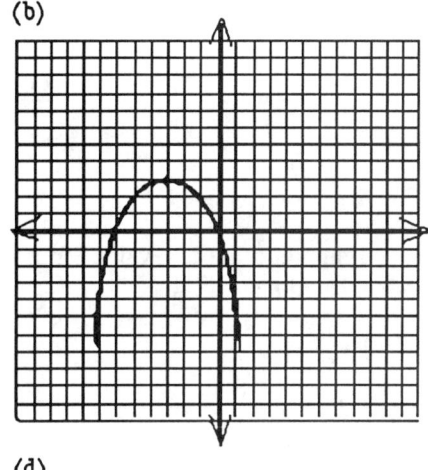

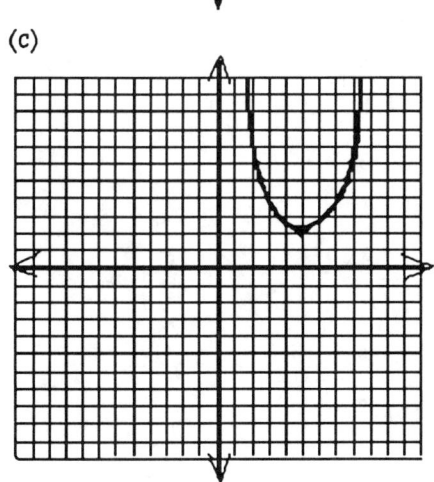

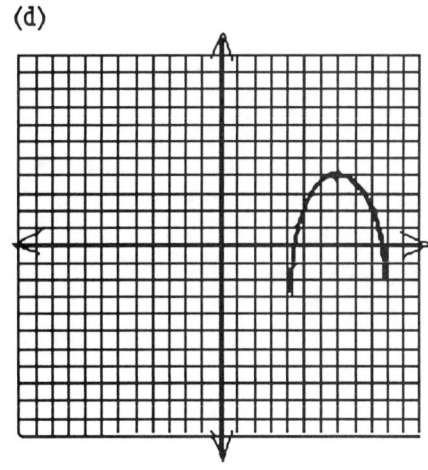

   (a)

   (b)

   (c)

   (d)

	(a) vertex: (2, 3)
	axis of symmetry: $x = 2$
	(b) vertex: (⁻3, 3)
	axis of symmetry: $x = {}^-3$
	(c) vertex: (5, 2)
	axis of symmetry: $x = 5$
	(d) vertex: (7, 4)
	axis of symmetry: $x = 7$

2. For the following functions find the equations of the lines of symmetry of their graphs.

390 CHAPTER 14 Algebra and Geometry

(a) $f(x) = x^2 + 2x - 3$
(b) $f(x) = {}^-x^2 + 2x + 3$
(c) $f(x) = 2x^2 + x - 2$
(d) $f(x) = 3x^2 + 6x + 2$

(a) $x = {}^-1$
$x = -\dfrac{b}{2a} = -\dfrac{2}{2 \cdot 1} = {}^-1$
(b) $x = 1$
(c) $x = -\dfrac{1}{4}$
(d) $x = {}^-1$

3. Find the maximum or minimum value for each of the following functions.

(a) $f(x) = x^2 + x - 6$
(b) $f(x) = 7x^2 + 14x - 30$
(c) $f(x) = x^2 - x - 1$
(d) $f(x) = {}^-x^2 + 4x$

(a) $^-6\dfrac{1}{4}$
Minimum at $x = -\dfrac{1}{2}$

$f\left(\dfrac{^-1}{2}\right) = \left(\dfrac{^-1}{2}\right)^2 + \left(\dfrac{^-1}{2}\right) - 6$
$= \dfrac{1}{4} - \dfrac{1}{2} - 6 = {}^-6\dfrac{1}{4}$

(b) $^-37$
(c) $^-1\dfrac{1}{4}$
Minimum at $x = \dfrac{1}{2}$

$f\left(\dfrac{1}{2}\right) = \left(\dfrac{1}{2}\right)^2 - \dfrac{1}{2} - 1$
$= \dfrac{1}{4} - \dfrac{1}{2} - 1 = {}^-1\dfrac{1}{4}$

(d) 4

4. State whether the graphs of the functions from Exercise 2 will turn upward or downward.

(a) upward
(b) downward

Section 5 Quadratic Functions and the Parabola **391**

(c) upward
(d) downward

5. Draw the graphs of each part of Exercise 3, indicating the line of symmetry and labeling the maximum or minimum point.

(a)

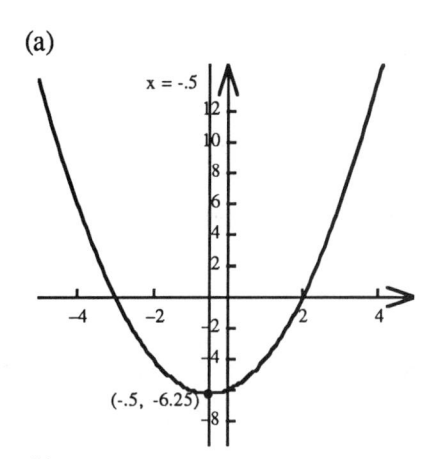

x = -.5

(-.5, -6.25)

(b)

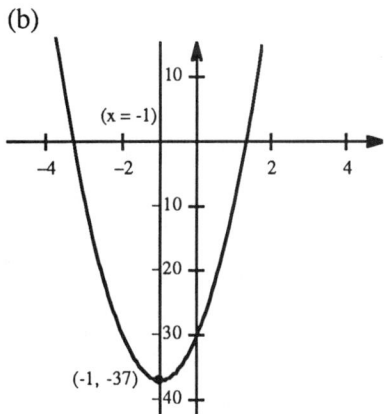

(x = -1)

(-1, -37)

(c)

x = .5

(.5, -1.25)

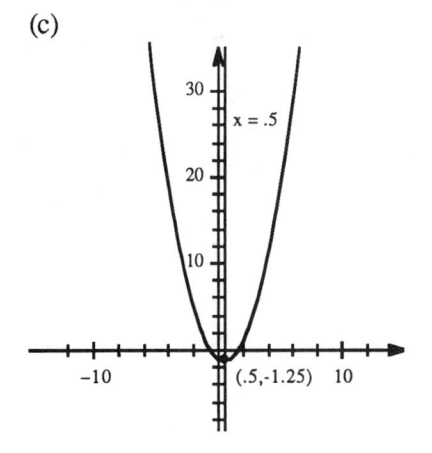

392 **CHAPTER 14** Algebra and Geometry

(d)

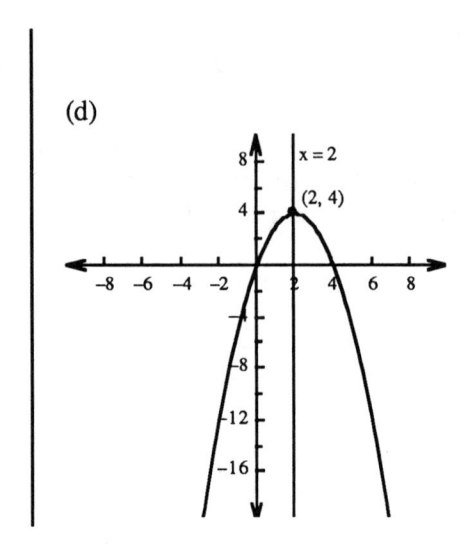

6. Draw the graphs of each part of Exercise 2, labeling the x and y-intercepts.

(a)

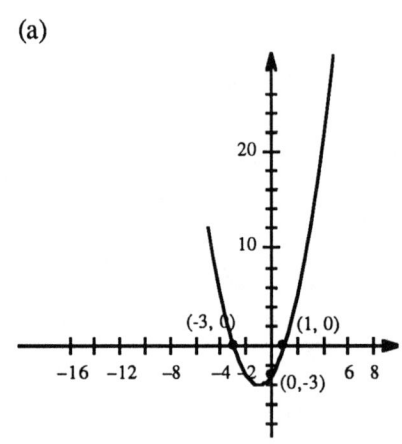

(b)

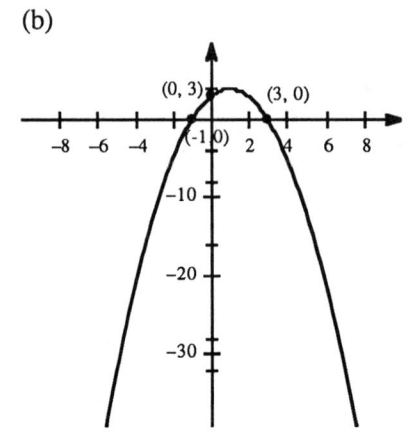

(c)

$$\left(\frac{-1-\sqrt{17}}{4}, 0\right) \qquad \left(\frac{-1+\sqrt{17}}{4}, 0\right)$$

$(0, -2)$

(d)

$$\left(\frac{-3-\sqrt{3}}{3}, 0\right) \qquad \left(\frac{-3+\sqrt{3}}{3}, 0\right)$$

$(0, 2)$

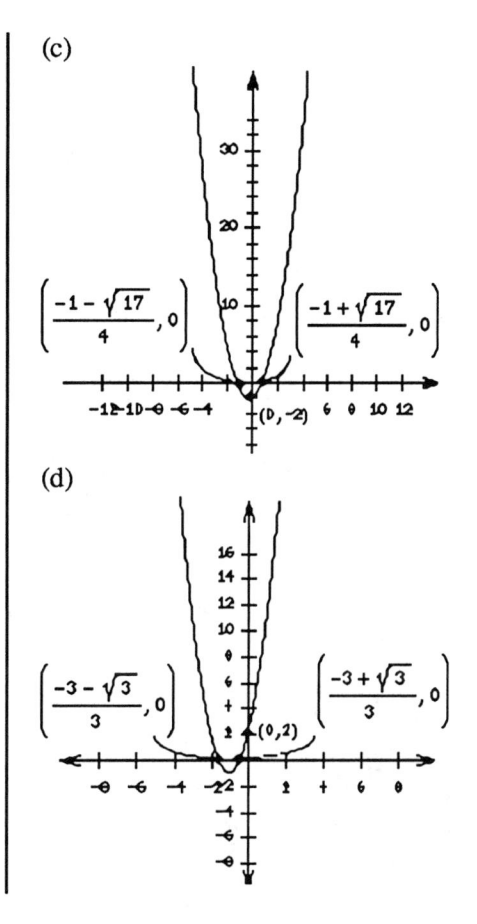

7. (a) Solve the equation $y = x^2 + 2x - 5$ using the quadratic formula. How many solutions did you get?
   (b) Draw the graph of $y = x^2 + 2x - 5$. How does this graph relate to part (a)?

(a) Two solutions:

$$x = \frac{-2 \pm \sqrt{4 - 4(1)(-5)}}{2}$$

$$= \frac{-2 \pm \sqrt{24}}{2} = -1 \pm \sqrt{6}$$

CHAPTER 14 — Algebra and Geometry

(b)

Two solutions show two points of intersection of the graph of the function and the x-axis.

8. Use the quadratic formula to solve the following quadratic equations.

   (a) $6x^2 + 7x - 3 = 0$
   (b) $x^2 + 5x = 6$
   (c) $x^2 = 4x - 1$
   (d) $2x^2 + 2x - 3 = 0$

$$\text{(a) } x = \frac{1}{3}, -\frac{3}{2}$$

$$x = \frac{-7 \pm \sqrt{49 - 4(6)(-3)}}{12}$$

$$= \frac{-7 \pm \sqrt{49 + 72}}{12}$$

$$= \frac{-7 \pm \sqrt{121}}{12}$$

$$= \frac{-7 \pm 11}{12} = \frac{1}{3} \text{ or } -\frac{3}{2}$$

(b) $x = 1, -6$
(c) $x = 2 + \sqrt{3}, 2 - \sqrt{3}$
(d) $x = \frac{-1 + \sqrt{7}}{2}, \frac{-1 - \sqrt{7}}{2}$

9. Solve each of the following equations.

   (a) $(x - 1)(x + 3) = 0$
   (b) $x^2 - 5x = 0$
   (c) $x^2 - 7x + 12 = 0$
   (d) $(x + 2)(2x - 1) = 0$

(a) $x = 1, ^-3$
(b) $x = 0, 5$
$x^2 - 5x = x(x - 5) = 0$
(c) $x = 3, 4$
$x^2 - 7x + 12 = (x - 3)(x - 4) = 0$
(d) $x = ^-2, \dfrac{1}{2}$

10. Solve each of the following equations.

(a) $x^2 - 9 = 0$
(b) $2x^2 = 20$
(c) $3x^2 - 75 = 0$

(a) $x = 3, ^-3$
$x^2 - 9 = 0$
$x^2 = 9$
$x = \sqrt{9} = 3, ^-3$
(b) $x = \sqrt{10}, ^-\sqrt{10}$
(c) $x = 5, ^-5$

11. The revenue R for a bus company is given by
$$R(x) = 5000 + 50x - x^2$$
where x is the number of unsold seats.

(a) Find the number of unsold seats that produces maximum revenue.
(b) Find the maximum revenue.

(a) $x = 25$
$x = \dfrac{-b}{2a} = \dfrac{-50}{-2} = 25$
(b) $R(25) = \$5,625$

12. Find a value k so that $x^2 - 10x + k = 0$ has exactly one solution.

$k = 25$
$x = \dfrac{10 \pm \sqrt{100 - 4(1)(k)}}{2}$
To have one solution, we must have
$100 - 4(1)(k) = 0$
$k = 25$

13. A rectangular lot 100 m long by 70 m wide is to be made into a park surrounded by a sidewalk of width x.
(a) Express the area of the park as a function of x.
(b) How wide should the sidewalk be if the park is to have an area of 6,499 m²?

(a) $A(x) = (100 - 2x)(70 - 2x)$

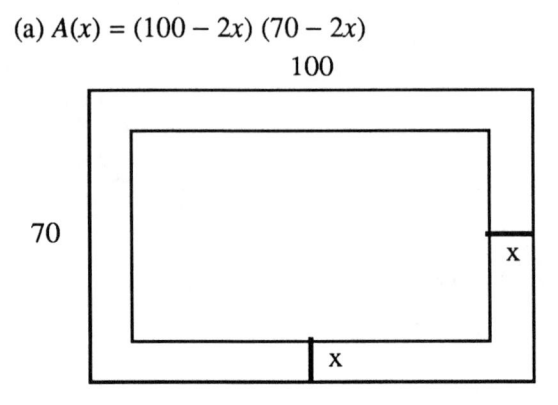

(b) $(100 - 2x)(70 - 2x) = 6499$
$7000 - 200x - 140x + 4x^2 = 6499$
$4x^2 - 340x + 501 = 0$

$$x = \frac{340 \pm \sqrt{(340)^2 - 4(4)(501)}}{8}$$
$$= \frac{340 \pm \sqrt{115600 - 8016}}{8}$$
$$= \frac{340 \pm 328}{8}$$
$= 83.5$ or $\boxed{1.5}$ m

14. A company that plans bus tours charges a fare of $8 per person and carries 300 people per day. The company estimates that it will lose 100 passengers for each $1 increase in the fare. What is the most profitable fare for the company to charge?

$19

Let x = fare $- 8$ and $R(x)$ = revenue
$R(x) = (8 + x)(300 - 10x)$
$= 2400 - 80x + 300x - 10x^2$
$= {}^-10x^2 + 220x + 2400$
$x = \frac{-220}{-20}$
x-intercept of the vertex of the parabola = 11
fare $- 8 = 11$
fare $= 19$

Chapter 15
An Introduction to Computers

Section 1 An Introduction to the BASIC Language

Cover the right side of the page and work on the left, then check your work

1. Evaluate each of the following expressions.

   (a) 3 + 7 * 5
   (b) 8 * (8 − 2)/4
   (c) 5 + 2 ^ 3 − 4
   (d) 9 ^ 2 − 4 ^ 3

   > (a) 38   Multiply, then add
   > (b) 12   Subtract in parentheses, then multiply
   > (c) 9
   > (d) 29

2. Write each of the following as a base-ten number in standard form.

   (a) 3.25E +04
   (b) −7.30 E+03
   (c) 2.91 E−04
   (d) −1.23 E−07

   > (a) 32,500
   > (b) −7,301
   > (c) 0.000291
   > (d) −0.000000123

3. Write each, using E notation.

   (a) 3746.2
   (b) −0.0346
   (c) 319,000
   (d) 0.0081

   > (a) 3.7462 E+03
   > (b) −3.46 E−02
   > (c) 3.19 E+05
   > (d) 8.1 E−03

4. Write each of the following BASIC expressions.

   (a) The product of 12 and 36.
   (b) The division of the cube of 9 and the square of 4.
   (c) The third power of the sum of 3 and 16.8.

CHAPTER 15 Introduction to Computers

(a) 12 * 36
(b) 9 ^ 3 / 4 ^ 2
(c) (3 + 16.8) ^ 3

5. Write a BASIC expression for each of the following.

(a) $X + Y - \dfrac{X^2}{Y^2}$

(b) $\dfrac{19}{X(3Y^2 - 4)}$

(c) $\dfrac{X^2 - 3X}{X^3 + X^2}$

(a) X + Y − (X ^ 2 / Y ^ 2)
(b) 19 / (X * (3 * Y ^ 2 − 4))
(c) (X ^ 2 − 3 * X) / (X ^ 3 + X ^ 2)

6. What will be the output be for each of the following?

(a) 10   A = 4
    20   B = 7
    30   PRINT A, B
    40   END

(b) 10   Y = 7
    20   Z = 9
    30   X = Y * Z − 3
    40   END

(c) 10   C = 5
    20   D = 8
    30   PRINT "THE VALUE OF C"
    40   PRINT "IS " ; C
    50   PRINT "THE VALUE OF D"
    60   PRINT "IS " ; D
    70   END

(a) 4   7
(b) 3600
(c) THE VALUE OF C IS 5
    THE VALUE OF D IS 8

7. Which of the following statements is incorrect?

(a) 10   PRINT "APPLES AND ORANGES
(b) 10   X + 7 = X
(c) 10   LET X = (3)(5)
(d) 10   57 = X

(a) Incorrect
(b) Incorrect
(c) Incorrect
(d) Incorrect

8. Correct each of the incorrect statements from Exercise 7.

(a) 10   PRINT "APPLES AND ORANGES"
(b) 10   X = X + 7
(c) 10   LET X = 3 * 5
(d) 10   X = 57

9. Determine the output of each of the following programs.

(a) 10   A = 7
    20   A = A + 1
    30   PRINT A
    40   END

(b) 10   A = 6
    20   B = 1
    30   B = B + A
    40   PRINT A,B
    50   END

(c) 10   X = 3
    20   Y = 2
    30   Y = X ^ Y
    40   X = Y ^ X
    50   PRINT X + Y
    60   END

(d) 10   S = 4
    20   S = S + S
    30   S = 2 * S + 1
    40   PRINT S
    50   END

(a) 8
(b) 6   16
(c) 17
(d) 17

10. Write a program for calculating and printing the value of Y, when

$$Y = 3A + \frac{A^2}{5} + 7 \qquad \text{where } A = 2.$$

10 LET A = 2
20 Y = (3 * A) + (A ^ 2 / 5) + 7
30 PRINT Y
40 END

Section 2 Some BASIC Statements

Cover the right side of the page and work on the left, then check your work

1. What is the output of the following programs?

   (a) 10   FOR K = 1 TO 5
       20   PRINT K ^ 3
       30   NEXT K
       40   END

   (b) 10   INPUT A, B, C
       20   S = A + B + C
       30   P = A * B * C
       40   PRINT S, P
       50   END
       ? 3, 4, 5

   (c) 10   FOR J = 2 TO 8
       20   X = J * J
       30   IF X ≤ 10 THEN 50
       40   PRINT J, X
       50   NEXT J
       60   END

   (d) 10   INPUT X, Y
       20   IF X < Y THEN 40
       30   PRINT X; "IS LESS THAN ";Y
       40   END
       ? 4, 9

(a)	1	
	8	
	27	
	65	
	125	
(b)	12	60
(c)	4	16
	5	25
	6	36
	7	49
	8	64
(d)	4 IS LESS THAN 9

2. Find a number or statement for ? to give the required output.

   (a) 10   FOR X = 1 TO 4
       20   ?
       30   NEXT X
       40   END

       The output is 3
                     6
                     9
                     12

   (b) 10   FOR X = ? TO ?
       20   PRINT X ^ 2 – 2
       30   NEXT X
       40   END

       The output is 7
                     14
                     23
                     34

   (a) PRINT 3 * X
   (b) FOR X = 3 TO 6

3. Write programs using FOR ... NEXT that will produce the following sequences of numbers.

(a) 5, 6, 6, ... , 20
(b) 75, 70, 65, ... , 10
(c) 3, 5, 7, ... , 21

(a) 10 FOR I = 5 TO 20
    20 PRINT I
    30 NEXT I
    40 END
(b) 10 FOR I = 15 TO 2
    20 PRINT I * 5
    30 NEXT I
    40 END
(c) 10 FOR I = 1 TO 10
    20 PRINT 2 * I + 1
    30 NEXT I
    40 END

Write a program to perform the tasks specified in each exercise.

4. The cost of renting a car is $75 plus 25¢ per mile for all miles driven above 200 miles. The program should allow the number of miles to be input. The mileage above 200 miles (if any) is printed along with the total cost.

    10 INPUT M
    20 MILES = M − 200
    30 COST = 75 + MILES * 0.25
    40 PRINT MILES, COST
    50 END

5. The program is to calculate and print the surface area and volume for a sphere. The radius, in centimeters, of the sphere should be input.

    10 INPUT R
    20 AREA = 4 * 3.14 * R ^ 2
    30 VOL = (4 * 3.14 * R ^ 3) / 3
    40 PRINT AREA, VOL
    50 END

6. Write a program to calculate and print the following sum, where *n* is input.

$$(1 \cdot 2)^2 + (2 \cdot 3)^2 + (3 \cdot 4)^2 + (4 \cdot 5)^2 + \ldots + [(n(n + 1)]^2$$

    10 INPUT N
    20 FOR I = 1 TO N
    30 SUM = SUM + (I * (I + 1))^2
    40 NEXT I
    50 PRINT SUM
    60 END

7. What is the output of the following program?

         10   FOR I = 2 TO 4
         20   FOR J = 1 TO 3
         30   X = I * J
         40   PRINT X
         50   NEXT J
         60   NEXT I
         70   END

    | 2
    | 4
    | 6
    | 3
    | 6
    | 9
    | 4
    | 8
    | 12

8. What is the output of the following program?
   Assume the input is 2.
         10   INPUT N, E
         20   FOR I = 3 TO N
         30   EXP = I ^ E
         40   PRINT EXP
         50   NEXT I
         60   END

    | 9
    | 16
    | 25
    | 36

9. All the items in a store are on sale for 30% off. Write a program that will take an input representing the selling price of an item. The program will print the selling price and the sale price (30% off the selling price).

    | 10   INPUT P
    | 20   SP = P − (0.3 * P)
    | 30   PRINT P, SP
    | 40   END

10. Write a program that will convert square feet to square yards.

    | 10   INPUT SF
    | 20   SM = SF / 9
    | 30   PRINT SM
    | 40   END

Section 3 Solving Problems using the Computer

Cover the right side of the page and work on the left, then check your work

1. Give the output of the following programs.

   (a) 10 I = 1
       20 WHILE I < 10
       30 PRINT I ^ 3
       40 I = I + 1
       50 WEND
       60 END

   (b) 10 INPUT X1, Y1, X2, Y2
       20 M = (Y2 – Y1) / (X2 – X1)
       30 WHILE M > 0
       40 PRINT "SLOPE = " ; M
       50 WEND
       60 PRINT "SLOPE NOT POSITIVE."
       70 END

   > (a) 1
   >   27
   >   125
   >   343
   >   729
   >
   > (b) If M > 0, say M = 4, the output will be
   >   SLOPE = 4
   >
   >   If M ≤ 0, the output will be
   >   SLOPE NOT POSITIVE

2. Write a program to calculate and print the compound amount for a given principal, interest rate, and period. The program should allow the input of P, r, n, and t, where
   $$A = P(1 + \frac{r}{n})^{nt}$$
   Recall that n is the number of compounding periods per year and t is the number of years.

   > 10 INPUT P, R, N, T
   > 20 AMT = P * ((1 + (R / N))^(N * T))
   > 30 PRINT AMT
   > 40 END

3. Write a program to determine the length of time it will take for an initial investment to double. The money is invested at 8% interest compounded daily. Input the amount of the initial investment and the interest rate.

   > 10 INPUT P, R
   > 20 WHILE A < 2 * P
   > 30 C = C + 1
   > 40 A = P * ((1 + (R / 365))^(365 * C))

```
                                        50  WEND
                                        60  PRINT "THE INVESTMENT
                                            WILL DOUBLE AFTER "C "YEARS."
                                        70  END
```

4. Write a program to compute the number of permutations of N objects.

```
        10  INPUT N          OR        10  INPUT N
        20  C = 1                      20  I = 1
        30  FOR I = 1 TO N             30  WHILE I < = N
        40  C = I * C                  40  C = C * I
        50  NEXT I                     50  I = I + 1
        60  PRINT C; "PERMUTATIONS"    60  WEND
        70  END                        70  PRINT C; "PERMUTATIONS"
                                       80  END
```

5. A pitcher's earned run average is found by dividing the number of earned runs by the number of innings pitched and multiplying the result by 9. Write a program that will calculate a pitcher's earned run average, given the number of earned runs allowed and number of innings pitched.

```
        10  INPUT NR, NI
        20  ERA = (NR / NI) * 9
        30  PRINT "EARNED RUN AVERAGE = "; ERA
        40  END
```

6. Rewrite any incorrect statement below.

   (a) IF X = 300 THEN 150
   (b) IF X = 50 PRINT "CORRECT"
   (c) IF Y = 30 THEN PRINT CORRECT
   (d) PRINT "OKAY" IF X > 30

```
        (a) Correct
        (b) IF X = 50 THEN PRINT "CORRECT"
        (c) IF Y = 30 THEN PRINT "CORRECT"
        (d) IF X > 30 THEN PRINT "OKAY"
```

7. The cost C of an item is to be input. A store owner wants to determine the selling price P of the item. If the cost of the item is greater than $100, there is a 20% markup. If the cost of the item is less than or equal to $100, there is a 30% markup. The program should print the cost and selling price of the item.

```
10  INPUT C
20  IF C > 100 THEN 50
30  P = C + 0.3 * C
40  IF E < 10 THEN 60
50  P = C + 0.2 * C
60  PRINT "THE COST IS " ; C
70  PRINT "THE SELLING PRICE IS " ; P
80  END
```

8. The cost of producing your company's product is given by the following function,

$$C(x) = 3x + 70 \text{ if } x \leq 20$$
$$= 2x + 80 \text{ if } x > 20$$

where x is the number of items produced. Write a program that will print the cost of producing x units.

```
10  INPUT X
20  IF X > 20 THEN 50
30  C = 2 * X + 70
40  IF E < 10 THEN 60
50  C = 2 * X + 80
60  PRINT "THE COST IS " ; C
70  END
```

9. Write a program that will calculate the number of combinations of N things taken R at a time. N and R will be input by the user.

```
10   INPUT N, R
20   A = 1
30   B = 1
40   C = 1
50   FOR I = 1 TO N
60   A = I * A
70   NEXT I
80   FOR J = 1 TO R
90   B = J * B
100  NEXT J
110  FOR K = 1 TO (N − R)
120  C = K * C
130  NEXT C
140  COMB = A/(B * C)
150  PRINT "NUMBER OF COMBINATIONS=";COMB
160  END
```

Section 4 More BASIC Tools for Problem Solving

Cover the right side of the page and work on the left, then check your work

1. Give the output of each of the following programs.

   (a) 10   LET A = 2
        20   PRINT A,
        30   LET A = A + 1
        40   IF A > 10 THEN 60
        50   GOTO 20
        60   END

   (b) 10   READ S, T
        20   DATA 9,12
        30   IF S < T THEN 60
        40   PRINT S + T
        50   GOTO 70
        60   PRINT S * T
        70   END

   (a) 2
       3
       4
       5
       6
       7
       8
       9
       10

   (b) 108

2. Correct any of the following statements that are incorrect.

   (a) 10   READ A, 10
   (b) 20   X = 3 (4 + 7)
   (c) 20   DATA A, 10

   (a) 10   READ A, B
   (b) 20   X = 3 * (4 + 7)
   (c) 30   DATA 5, 10

3. Rewrite the following program using a FOR-NEXT loop so that the output is

      3
      6
      9
     12
     15

   10   N = 1
   20   IF M > 10 THEN 60
   30   PRINT 3 * N
   40   LET N = N + 1
   50   GO TO 20

```
                                60    END
                                                                10    FOR N = 1 TO 5
                                                                20    PRINT 3 * N
                                                                30    NEXT N
                                                                40    END
```

For Exercises 4 - 7, give the output for the given program.

4.
```
        10    INPUT X, Y
        20    Z = (X + Y) / 2
        30    IF Z > 10 THEN 60
        40    PRINT Z
        50    GOTO 70
        60    PRINT Z ^ 2
        70    END
        ?  7, 9
```
           | 8

5.
```
        10    PRINT "WHAT YEAR WERE YOU BORN?"
        20    INPUT Y
        30    AGE = 1992 – Y
        40    PRINT "YOU ARE "; AGE; " YEARS OLD."
        50    END
        ?  1948
```
        | YOU ARE 44 YEARS OLD

6.
```
        10    READ X, Y
        20    Z = X * Y
        30    PRINT X " * " Y " = " Z
        40    GOTO 10
        50    DATA 3, 5, 7, 9, 8, 10
        60    END
```
        | 3 * 5 = 15
        | 7 * 9 = 63
        | 8 * 10 = 80

7.
```
        10    PRINT "INPUT TWO NUMBERS."
        20    INPUT A, B
        30    C = A / B
        40    R = INT (C)
        50    PRINT A "/" B "IS APPROXIMATELY " R
        60    END
        ?  37, 9
```
        | 37 / 9 IS APPROXIMATELY 4

8. Construct a program that will take any four-digit number as input and round the number to the nearest ten.

```
10  INPUT N
20  A = N + 5
30  B = INT (A / 10)
40  C = B * 10
50  PRINT N; "ROUNDED IS "; C
60  END
```

9. Below are two tables- one lists the price per person relative to the number of persons on the tour and the other lists the cost to the company per person relative to the number of persons on the tour. Write a program that will allow the user to input the number of passengers on a tour. The program will print the amount of profit for the company for that number of passengers.

Number of Passengers	Price per Person	Cost per Person
N ≤ 10	$100	$75
10 < N ≤ 50	$90	$60
50 < N ≤ 100	$75	$40
N > 100	$55	$15

```
10   INPUT N
20   IF N < = 10 THEN 70
30   IF N < = 50 THEN 90
40   IF N < = 100 THEN 110
50   PROF = 55 * N – 15 * N
60   GOTO 120
70   PROF = 100 * N – 75 * N
80   GOTO 120
90   PROF = 90 * N – 60 * N
100  GOTO 120
110  PROF = 75 * N – 40 * N
120  PRINT "THE PROFIT IS "; PROF
130  END
```

10. Write a program that will print out all the numbers between 1 and 100 that are divisible by a given number input by the user.

```
10  INPUT N
20  FOR I = 1 TO 100
30  A = I / N
40  IF INT (I / N) < > I / N THEN 60
50  PRINT I; " IS DIVISIBLE BY ";N
60  NEXT I
70  END
```

Section 5 An Introduction to Logo

Cover the right side of the page and work on the left, then check your work

1. Correct each of the following commands.

   (a) RT90
   (b) PD 50
   (c) FD

   > (a) RT 90
   > (b) FD 50 or PD
   > (c) FD 50   (any number)

   Determine what figure will be drawn by each of the following Logo programs.

2. CS
   RT 90
   FD 60
   RT 90
   FD 60
   RT 90
   FD 60
   RT 90
   FD 60
   HT

3. CS
   FD 70
   BK 35
   RT 90
   FD 30
   LT 90
   FD 35
   BK 70
   HT

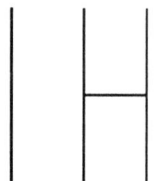

4. CS
   FD 40
   BK 40
   RT 90

FD 40
BK 20
LT 90
FD 25
BK 60
HT

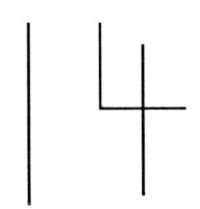

5. CS
   FD 50
   RT 120
   FD 50
   RT 120
   FD 50
   RT 120
   HT

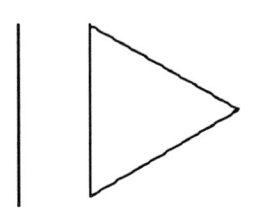

6. CS
   LT 90
   FD 100
   RT 90
   FD 30
   BK 60
   HT

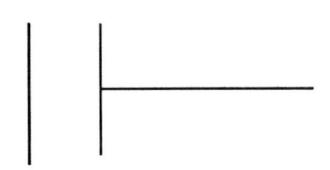

7. CS
   RT 30
   FD 50
   BK 100
   PU
   RT 60
   FD 20

```
LT 60
PD
FD 100
HT
```

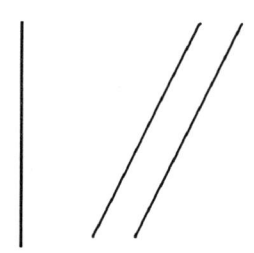

8. Modify the procedure in Exercise 6 to draw the following figure.

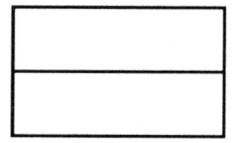

```
CS
LT 90
FD 100
RT 90
FD 30
BK 60
RT 90
FD 100
LT 90
FD 60
LT 90
FD 100
HT
```

9. Run the following Logo program. What does the HOME command do?

```
CS
FD 80
LT 90
FD 80
HOME
HT
```

Returns the turtle to its home position, drawing a line as it moves.

10. Write a procedure to draw the following figure.

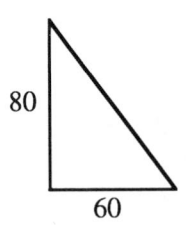

```
         80
              60
```

```
CS
LT 90
FD 60
RT 90
FD 80
HOME
HT
```

11. Write a procedure to draw the following figure.

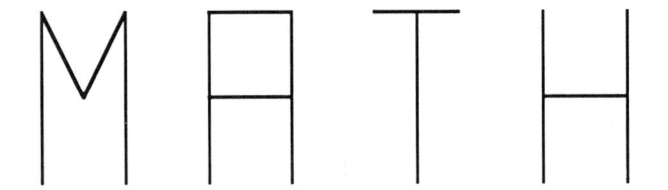

MATH

Answer follows:

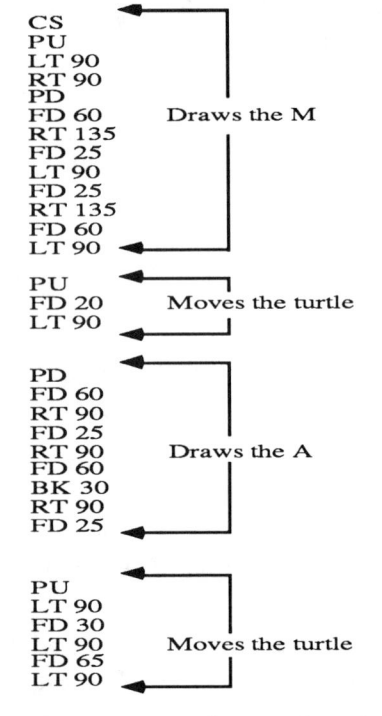

```
CS
PU
LT 90
RT 90
PD
FD 60        Draws the M
RT 135
FD 25
LT 90
FD 25
RT 135
FD 60
LT 90

PU
FD 20        Moves the turtle
LT 90

PD
FD 60
RT 90
FD 25
RT 90        Draws the A
FD 60
BK 30
RT 90
FD 25

PU
LT 90
FD 30
LT 90        Moves the turtle
FD 65
LT 90
```

(Move to top of next column)

```
PD
FD 60
LT 90
FD 15
BK 30
FD 15        Draws the T
BK 30
FD 15
LT 90
FD 60
LT 90

PU
FD 35        Moves the turtle
LT 90

PD
FD 60
BK 30
RT 90        Draws the H
FD 30
LT 90
FD 30
BK 60

HT
```

Section 6 Looping in Logo - The REPEAT Command

Cover the right side of the page and work on the left, then check your work

1. What will the turtle draw for each of the following?

   (a) REPEAT 6 [FD 30 BK 60 RT 90]
   (b) REPEAT 19 [FD 10 RT 10]

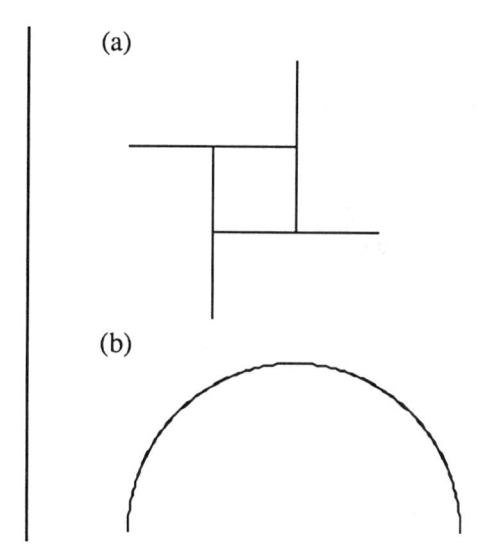

   (a)

   (b)

Use Logo to draw each of the following figures. Use the REPEAT command.

2.

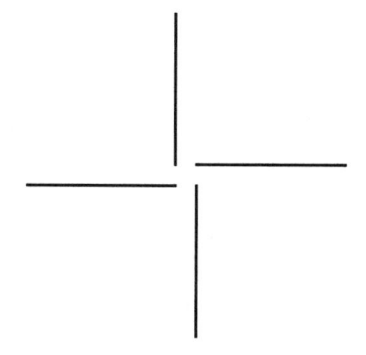

   REPEAT 4 [PU RT 90 FD 5 PD FD 50 BK 50]

3.

   RT 135 REPEAT 3 [FD 30 LT 90 FD 30 RT 90]

4. Rewrite the following program, using the REPEAT command.

    FD 30 PU RT 90 FD 10 PD
    FD 30 PU RT 90 FD 10 PD
    FD 30 PU RT 90 FD 10 PD

> REPEAT 3 [FD 30 PU RT 90 FD 10 PD]

5. Use a Logo procedure to draw each of the following.

(a) 12-sided polygon
(b) 20-sided polygon

> (a) TO TWELVE
>   REPEAT 12 [FD 20 RT 30]
>   END
> (b) TO TWENTY
>   REPEAT 20 [FD 15 RT 18]
>   END

6. Use REPEAT to produce each of the following.

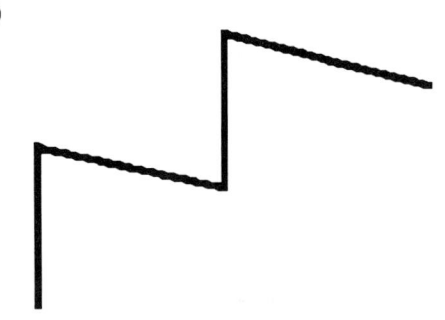

(a)   (b)

> (a) REPEAT 5 [FD 20 LT 72 PU FD 5 PD]
> (b) REPEAT 2 [FD 40 RT 110 PU FD 40 LT 110]

7. Use the TRIANGLE and the SQUARE procedures found on page 959 of your textbook to draw these figures.

(a)

(b)

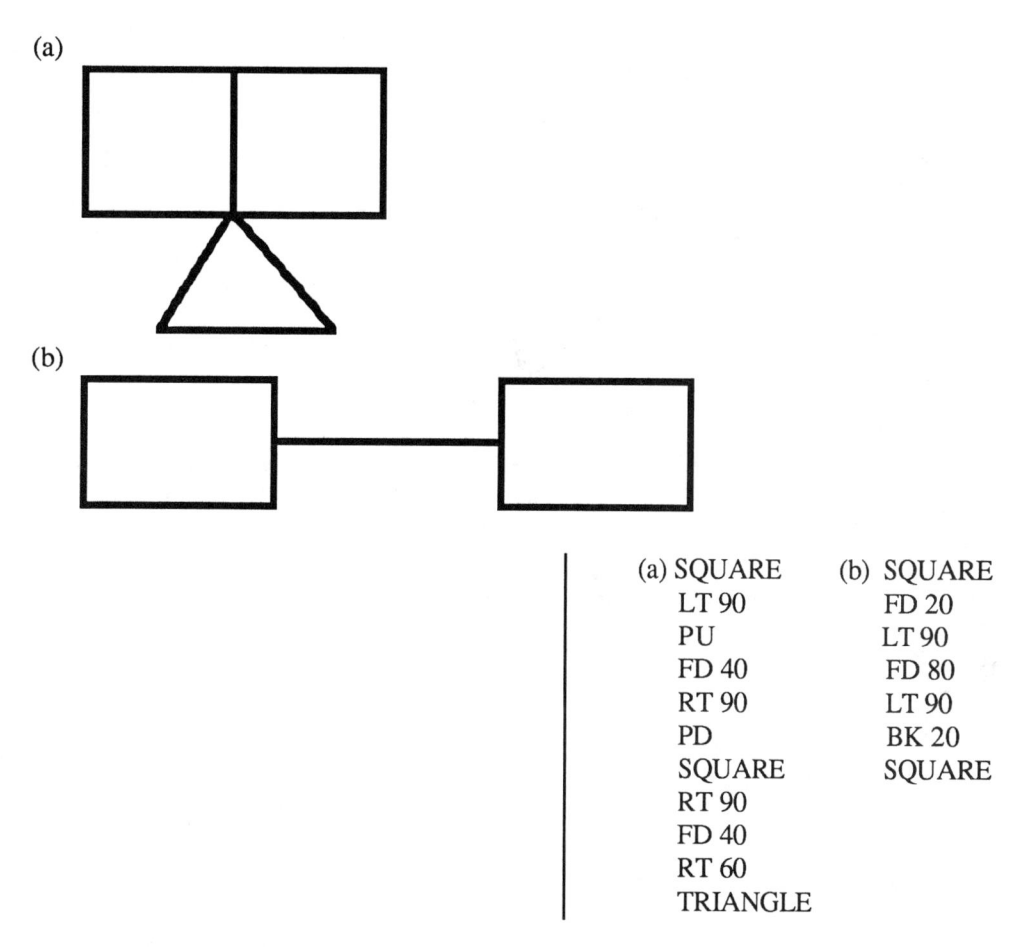

(a) SQUARE	(b) SQUARE
LT 90	FD 20
PU	LT 90
FD 40	FD 80
RT 90	LT 90
PD	BK 20
SQUARE	SQUARE
RT 90	
FD 40	
RT 60	
TRIANGLE	

8. Use nested REPEAT commands to draw each of the following.
   (a)

416 CHAPTER 15 An Introduction to Computers

(b)

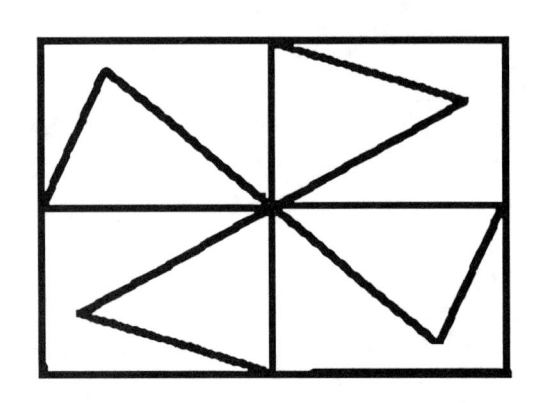

(a) REPEAT 10 [REPEAT 4 [FD 30 RT 90] LT 36]
(b) REPEAT 4 [REPEAT 3 [FD 30 RT 120] REPEAT 4 [FD 30 RT 90] LT 90]

9. Write a procedure to draw each of the following.

(a)

(b)

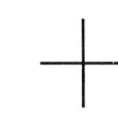

(c)

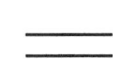

(d)

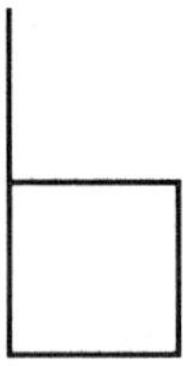

(a) TO THREE (b) TO PLUS
RT 90   FD 20
FD 20   BK 10
LT 90   RT 90
FD 20   BK 10
LT 90   FD 20
FD 15   END
BK 15
RT 90
FD 20
LT 90
FD 20

END

(c) TO EQUALS
RT 90
FD 20
BK 20
RT 90
PU
FD 10
LT 90
PD
FD 20
END

(d) TO SIX
RT 180
FD 60
REPEAT 3 [LT 90 FD 30]
END
END

10. Use the procedures from Exercise 9 to write a master procedure to draw the following.

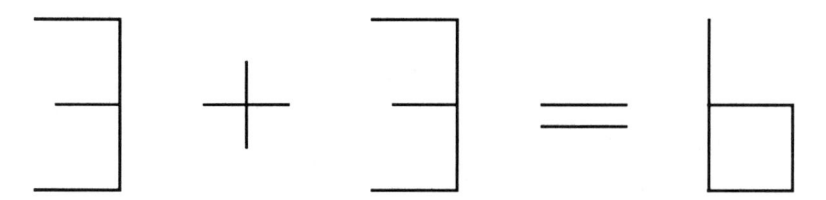

TO EQUATION
PU
LT 90
FD 100
RT 90
PD
THREE
PU
BK 40
RT 90
BK 30
PD
PLUS
PU
FD 10
LT 90

```
BK 20
PD
THREE
PU
BK 40
RT 90
BK 15
PD
EQUALS
PU
FD 20
LT 90
FD 25
PD
SIX
END
```